W0018101

About Island Press

Since 1984, the nonprofit Island Press has been stimulating, shaping, and communicating the ideas that are essential for solving environmental problems worldwide. With more than 800 titles in print and some 40 new releases each year, we are the nation's leading publisher on environmental issues. We identify innovative thinkers and emerging trends in the environmental field. We work with world-renowned experts and authors to develop cross-disciplinary solutions to environmental challenges.

Island Press designs and implements coordinated book publication campaigns in order to communicate our critical messages in print, in person, and online using the latest technologies, programs, and the media. Our goal: to reach targeted audiences—scientists, policymakers, environmental advocates, the media, and concerned citizens—who can and will take action to protect the plants and animals that enrich our world, the ecosystems we need to survive, the water we drink, and the air we breathe.

Island Press gratefully acknowledges the support of its work by the Agua Fund, Inc., The Margaret A. Cargill Foundation, Betsy and Jesse Fink Foundation, The William and Flora Hewlett Foundation, The Kresge Foundation, The Forrest and Frances Lattner Foundation, The Andrew W. Mellon Foundation, The Curtis and Edith Munson Foundation, The Overbrook Foundation, The David and Lucile Packard Foundation, The Summit Foundation, Trust for Architectural Easements, The Winslow Foundation, and other generous donors.

The opinions expressed in this book are those of the author(s) and do not necessarily reflect the views of our donors.

Forests for the People

Forests for the People

THE STORY OF AMERICA'S EASTERN NATIONAL FORESTS

CHRISTOPHER JOHNSON

DAVID GOVATSKI

ISLANDPRESS

Washington | Covelo | London

Library of Congress Cataloging-in-Publication Data

Johnson, Christopher, 1947 September 13–
    Forests for the people : the story of America's eastern national forests /
Christopher Johnson, David Govatski.
        pages   cm
    Includes bibliographical references and index.
    ISBN-13: 978-1-61091-009-5 (cloth : alk. paper)
    ISBN-10: 1-61091-009-5 (cloth : alk. paper)
    ISBN-13: 978-1-61091-010-1 (pbk. : alk. paper)
    ISBN-10: 1-61091-010-9 (pbk. : alk. paper)   1. Forest reserves—East (U.S.)—
History—20th century.   2. Forest conservation—East (U.S.)   I. Govatski,
David, 1949–   II. Title.
    SD428.A2E27   2012
    333.75′110974—dc23   2012030240

Printed on recycled, acid-free paper

Manufactured in the United States of America

10 9 8 7 6 5 4 3 2 1

Keywords: Island Press, Allegheny National Forest, Appalachian, Boundary
Waters Canoe Area Wilderness, Clarke-McNary Act, conservation, eastern
national forests, endangered species, forest, forest fire, Franklin D. Roosevelt,
Gifford Pinchot, Green Mountains, Green Mountain National Forest,
Hiawatha National Forest, Huron-Manistee National Forests, hydraulic
fracturing, Holly Springs National Forest, invasive species, John Weeks,
Lake States, Monongahela National Forest, New Deal, Ottawa National
Forest, Pisgah, prescribed burn, stream flow, timber, timber famine, U.S.
Forest Service, Weeks Act, White Mountains, wilderness, Wilderness Act
of 1964, wolf

To our wives, Kathi and Barbara;
to Chris's parents, Charles and Jacqueline;
and to the forest champions who made
the eastern national forests possible
and care for the forests today

CONTENTS

ACKNOWLEDGMENTS

In writing *Forests for the People*, we have benefited from the knowledge and wisdom of dozens of people who are active in the forest conservation movement. They answered questions and guided us to the information and resources that were so critical in writing about America's eastern national forests. We sincerely appreciate the advice and support that Char Miller provided. For part I, in which we relied primarily on library research, we would like to acknowledge the valuable assistance of two people at the Forest History Society in Durham, North Carolina: Jamie Lewis, the historian, and Cheryl Oakes, the librarian. Also providing valuable sources were the Weeks Memorial Library in Lancaster, New Hampshire; Sarah Jordan and Terry Fifield at the White Mountain National Forest; the Northwestern University Library; the University of Chicago Libraries; the Research Center at the Minnesota Discovery Center; the Tuck Library at the New Hampshire Historical Society; and Rauner Special Collections at the Dartmouth College Library. We also thank Marcia Schmidt Blaine of Plymouth State University for sharing information about some of the key players in the movement to save the White Mountains.

On-site research was invaluable to writing part II, and we benefited greatly from the knowledge of people in the U.S. Forest Service, state agencies, nongovernmental organizations, and citizen-activists. Larry Chambers, a media relations officer for the Forest Service, put us in contact with personnel in various national forests. At Holly Springs National Forest, Joel Gardner, Caren Briscoe, and Buddy Lowery of the U.S. Forest Service provided us with substantial information. In addition, Ann Philippi, Andy Mahler, Joe Glisson, and Ray Vaughan all took the time to recount for us the events that led to reform in timber-harvesting practices in Mississippi. In Florida, Steve Parrish, Mike Herrin, Chuck Hess, David Dorman, and Mike Drayton of the U.S. Forest Service shared their extensive information about prescribed burning. In West Virginia, Mary Wimmer, Beth Little,

ACKNOWLEDGMENTS

and Mike Costello readily shared their experiences in protecting wilderness in the Monongahela National Forest.

In Minnesota, Kate Surbaugh provided us with numerous contacts on the Boundary Waters Canoe Area Wilderness, including Bill Hansen and Bruce Kerfoot. Chel Anderson of the Minnesota Department of Natural Resources guided us through the ecology of the area, and Paul Dancic and Kevin Proescholdt afforded us with additional insight into the protection of the Boundary Waters. In our research on wolf recovery in Michigan, Tom Weise, Pat Hallfrisch, and Jess Edberg all shared their experiences. In researching oil shale drilling in the Allegheny National Forest, we benefited enormously from the assistance and knowledge of Cathy Pedler and William Belitskus of the Allegheny Defense Project, and Vincent Lunetta of Pennsylvania State University helped us focus on the most salient issues regarding hydraulic fracturing. Therese Poland of the U.S. Forest Service in Michigan shared her extensive knowledge of the emerald ash borer and pointed us toward critical sources. And, in bringing our story to completion in Vermont and North Carolina, we would like to thank Jamey Fidel of the Vermont Forest Roundtable, David Brynn of Vermont Family Forests, and Brent Martin of the Wilderness Society.

We also availed ourselves of the expertise of several professionals. Carmine Fantasia helped us locate several of the photographs and other images, designer Chris Clark created the graphs, and cartographer Chris Robinson drew the maps. For reviewing and commenting on parts of the manuscript, we want to thank Jamey Fidel, Joel Gardner, Mike Herrin, Jamie Lewis, Cathy Pedler, and Mary Wimmer for their insightful suggestions.

Finally, we owe a huge debt of gratitude to Barbara Dean, Erin Johnson, and the staff of Island Press for their unstinting support and numerous supportive suggestions that helped guide us through the writing of this book.

Introduction

On August 9, 1902, two camp counselors in their early twenties led eight young men from Camp Moosilauke, New Hampshire, on an ambitious backpacking expedition into the White Mountains, which lay to the east. The two counselors were Knowlton Durham, of Columbia University, and Benton MacKaye, a young forester from Massachusetts who, nearly twenty years later, would brainstorm the idea for the Appalachian Trail. Their journey would take them through the Lost River valley and into the Pemigewasset basin, through Crawford Notch, up Mount Washington and the other Presidentials, and then back to Camp Moosilauke.

They tramped east and, on August 12, entered the vast basin of the Pemigewasset River, cradled between Franconia Notch and Crawford Notch. A logging train carried them for four miles along the East Branch of the Pemigewasset. When the train reached its terminus, they hopped off and continued on foot, following an abandoned railroad track.

Slowly they climbed to a ridge that rewarded them with spectacular views. To the north, the mountains rose and fell toward Crawford Notch. To the south lay the Sandwich Range, which in 1902 was still a remote section of the White Mountains. The views were wondrous, but MacKaye and Durham also spotted ugly patches of land that lumber operators had cleared completely of trees. MacKaye was appalled at how thoroughly the sides of the mountains had been stripped. "The beauty of this region," he later wrote, "the wildest of the White Mountains, was in great part destroyed, the slashes of the lumbermen branding the mountains like unsightly scars on a beautiful face."[1] The sight left him heartsick.

Flash forward to the late 1980s and the pine forests of Louisiana, Texas, Florida, the Carolinas, and Mississippi, the habitat of the endangered red-cockaded woodpecker. More than 75 percent of existing populations of the bird inhabited these piney woods, and wildlife biologists had identified more than two thousand colonies, which consisted of a mating pair and one or two other birds.

The population of the woodpecker had been declining dramatically in recent years, and wildlife advocates criticized U.S. Forest Service management practices, claiming that the agency allowed timber harvesters to clear-cut, or cut every tree in a stand, reducing the birds' habitat and further endangering the existing populations. John W. Thompson, a former manager of Johns Manville's industrial forests who had become a dedicated bird-watcher after his retirement, argued for increased protection of the woodpecker's habitat.

The U.S. Forest Service listened and, by the early 1990s, modified its management of the forests in an effort to protect the woodpecker. In Homochitto National Forest in Mississippi, forest managers directed timber harvesters to thin out stands of forest and leave the most mature trees for the birds to build their nests. The U.S. Forest Service also modified timber-harvesting practices in other southern forests, managing some 250,000 acres of national forest to protect the habitat of the woodpecker.[2]

Between the time of Benton MacKaye and John Thompson, a revolution had transformed attitudes and policies toward America's forests. In 1900, most Americans regarded the forests as a resource that could not possibly be exhausted, and loggers were cutting massive amounts of timber in the East, South, and Great Lake states. Gifford Pinchot, Theodore Roosevelt, Benton MacKaye, and others in the forefront of America's forest conservation movement warned, however, that the rapid logging would ultimately destroy the nation's forests. In the first decade of the twentieth century, activists joined with conservation-minded legislators to press for legislation to protect the forests. They triumphed in 1911, when President

William Howard Taft signed the Weeks Act, which for the first time provided the federal government with the power and resources to purchase privately owned forestlands for the purpose of protecting them. That law made possible the creation of most of the national forests east of the 100th meridian, or 100 degrees west longitude. This imaginary line, which runs through the Dakotas, Nebraska, Kansas, Oklahoma, and Texas, is the traditional division between the amply watered lands of the eastern United States and the arid lands of the West.[3]

In this book, *Forests for the People*, the term *eastern* is used in its broadest sense to distinguish the national forests that lie east of the 100th meridian. The U.S. Forest Service administers these forests in two regions: the Eastern Region, or Region 9, which reaches from Maine as far west as Minnesota and as far south as Missouri; and the Southern Region, or Region 8, which stretches from Virginia south to Florida and west to Oklahoma and Texas. (Puerto Rico's El Junque National Forest is in the Southern Region.) The regions include fifty-two national forests, encompassing more than twenty-five million acres in twenty-six states. Of these national forests, forty-one have lands acquired under the auspices of the Weeks Act.[4] These forests carpet the ancient mountains of New England, ride the spine of the Appalachians south to Georgia, and reach into the swamplands of Florida. They stretch across the Piedmont of the Carolinas and Virginia, the rolling hills of southern Ohio, Indiana, and Illinois, and the Ozarks of Arkansas. They comprise the formidable north woods of Minnesota, Wisconsin, and Michigan.

Part I will tell the story of how America's eastern forests were saved in the early twentieth century and how the system of national forests was created in the East, South, and Lake states. Then, part II will examine eight current issues facing the eastern national forests, using a case-study approach. Each case study has been carefully selected to shed light on a larger challenge facing the eastern national forests:

Chapter 7, "Holly Springs National Forest: A Study in Forest Management Reform," examines the debates surrounding timber harvesting in Mississippi.

Chapter 8, "Florida's National Forests: A Revolution in Prescribed Burning," explores the development and use of prescribed burning on Florida's three national forests.

Chapter 9, "Monongahela National Forest: Wilderness at Heart," explains how the Wilderness Act of 1964 and other wilderness legislation have affected the Monongahela National Forest in West Virginia.

Chapter 10, "Boundary Waters Canoe Area Wilderness: Preservation versus Multiple Use," examines the debates surrounding the creation of the Boundary Waters Canoe Area Wilderness in the Superior National Forest in Minnesota.

Chapter 11, "Ottawa and Hiawatha National Forests: The Return of the Wolf," discusses the the recovery of the wolf population in the two national forests in Michigan's Upper Peninsula and examines the implications of changing attitudes about wildlife.

Chapter 12, "Allegheny National Forest: The Challenges of Shale Oil Drilling," explores the controversies surrounding drilling for oil and natural gas in the Allegheny National Forest in western Pennsylvania.

Chapter 13, "Michigan's National Forests: The Invasion of the Emerald Ash Borer," examines the growing problem of invasive species as reflected in the rapid spread and destruction caused by the emerald ash borer in Michigan's Huron-Manistee National Forests.

Chapter 14, "National Forests of Vermont and North Carolina: Loving the Forests to Death," discusses economic development near Vermont's Green Mountain National Forest and examines the problems of forest fragmentation and parcelization that are consequences of growth.

Although the eastern national forests represent only 13 percent of the entire national forest system, which has about 192 million acres, they are critical to the nation's natural resources. These forests are very different from their vast counterparts in the West. For one thing, they followed a different path to protection than did the western forests, which were still in the public domain in the late nineteenth and early twentieth centuries, allowing the federal government to create forest preserves directly from them. In the eastern half of the country, however, most of the forestlands lay in private hands in the early twentieth century, and the federal government had to purchase them from private landowners. The Weeks Act was critical to this process because it created the legal procedures and allocated federal revenues for making the purchases.

The second distinguishing characteristic of the eastern forestlands was their deteriorating ecological condition in the early twentieth century. Many of the lands had been cut over or burned by massive forest fires, and the U.S. Forest Service undertook a long process of restoring them. The process of restoration has proven to be enormously successful, adding immeasurably to our understanding of forest ecosystems.

The third distinguishing factor is the proximity of the eastern national forests to large populations. According to the U.S. Forest Service, the Eastern Region includes more than 40 percent of the U.S. population, and the Southern Region encompasses the fastest-growing region of the country, with booming cities from Atlanta to Birmingham. Millions of people live within a day's drive of an eastern national forest, which translates into heavy recreational use. Many of the forests lie near major cities, and this proximity creates pressures to exploit forest resources, from timber harvesting to oil and natural-gas extraction.

The fourth characteristic setting the eastern national forests apart is their size. Eastern forests are often smaller than their counterparts in the West. Of the top fifty national forests in size, the Superior in Minnesota ranks sixteenth, the Ouachita in Arkansas and Oklahoma ranks twenty-seventh, and the Mark Twain in Missouri ranks forty-seventh. The relatively small size of the eastern forests has an effect, intensifying conflicts over their uses. For example, the decision to set aside pine forests in the Homochitto National Forest as habitat for the endangered red-cockaded woodpecker affected the timber industry, an effect exacerbated by the national forest having only 189,000 acres.[5] These and other distinguishing features of the eastern national forests will be woven into our examination of their history and current issues.

At the heart of our account is a central question: What caused Americans to decide, in the early twentieth century, that the eastern forests were worth protecting and restoring? The answers to that question reveal a great deal about the development of the American conservation movement and, later, the environmental movement. At least three answers suggest themselves, all of which will be woven into our story. First, scientific knowledge about forests expanded greatly throughout the twentieth century, and scientists came increasingly to understand the connection among trees, other vegetation, soil, water quality, air quality, and wildlife. Supporters of forest conservation drew on this growing body of knowledge to persuade the public and legislators that forests were critical environments that had to be protected. The increasing understanding of the ecological role of forests and their relationship to other ecosystems will be a unifying theme of this book.

Second, the American public's attitudes about nature and the environment changed dramatically, beginning in the late nineteenth century, when conservationists first grew alarmed about America's rapidly diminishing forests. According to environmental philosopher Max Oelschlaeger in *The Idea of Wilderness*, through most of American history, the American public took an instrumental view of nature and viewed its attributes in strictly utilitarian terms. Oelschlaeger wrote that "the natural world was analogous to a factory to manufacture an unending stream of products for human consumption, and thus the landscape had only instrumental and not intrinsic value."[6] In the late 1800s, however, increasing numbers of Americans began to view nature as intrinsically valuable, and these attitudinal changes continued throughout the twentieth century as more people embraced outdoor recreation for its physical, social, psychological, and spiritual benefits. The changes in attitude were particularly dramatic regarding forests, which had had negative connotations from colonial days, as when the Puritans regarded the thick forests of New England as the playpen of the devil.

Third, the conservation movement represented a robust expression of grassroots democracy. For example, in the case of the Weeks Act, people in New England and the southern Appalachians voluntarily joined together to search for ways to save the eastern forests from being completely savaged. Their voluntary actions reflected the observations of Alexis de Tocqueville in *Democracy in America*: "Thus the most democratic country on the face of the earth is that in which men have, in our time, carried to the highest perfection the art of pursuing in common the object of their common desires and have applied this new science to the greatest number of purposes."[7]

Grassroots democracy was pivotal in protecting and restoring the eastern national forests, and public involvement has continued to play a critical role in influencing the management of the forests. Indeed, the United States has benefited enormously from having a vibrant blend of publicly owned and privately owned forestlands. As the story of the protection and restoration of these forests unfolds, it will become increasingly clear how a healthy network of eastern national forests—owned by and for the public—has benefited the country's economy, environment, and social health. Today, these forests provide eloquent testimony to the passion of thousands of citizens who committed themselves to their preservation.

HOW THE EASTERN NATIONAL FORESTS WERE SAVED

Part I tells the story of one of the most remarkable environmental reclamation projects in world history: the restoration of the eastern forests of the United States. It has been said that in 1500, a squirrel could have scrambled across treetops from Maine to Minnesota without ever touching the ground. As European Americans settled the continent, however, they cut down millions of trees for a variety of forest products, including lumber to build a growing nation's homes and businesses and to print increasing numbers of newspapers and books.

By 1900, the forests of New England's White Mountains, the southern Appalachians, and the Lake states of Michigan, Wisconsin, and Minnesota were heavily depleted. Corporate executives and political leaders alike feared that the country faced a timber famine that would inhibit economic growth. At the same time, thousands of hikers, campers, hunters, and anglers despaired over the disappearance of beautiful vistas of forest-covered mountains.

During this key period, outdoor lovers, progressive political leaders, and forward-looking business leaders joined together to form a movement dedicated to the rescue and the restoration of the forests of the East, South, and Lake states, culminating in the passage of the Weeks Act in 1911. Here is the story of how that pivotal conservation law was passed and how it created a robust network of eastern national forests.

CHAPTER 1

The Disappearing Forests of the White Mountains

It was 1890 in the Pemigewasset valley of New Hampshire's White Mountains, and the loggers attacked the stand of trees with grim determination. Two men downed the trees with five-foot-long crosscut saws—known affectionately as "misery whips"—and then laid the hardwoods on the ground. They rolled the precious spruce and pines over them and down the side of the mountain to waiting sleds, where a tender carefully loaded them. A teamster snapped the reins and drove the horse-powered sleds through the woods to a river or railroad siding (figure 1.1). There the logs were sent on their way to waiting sawmills, to be turned into lumber for a growing nation with a voracious appetite for wood.

The men were clear-cutting, or taking all the trees no matter how small or immature they were. New chemical processes allowed paper manufacturers to transform even the smallest spruce into paper, and as a result, the loggers cut every single tree and delivered the entire harvest to paper mills. The manufacturers ground the spruce into pulp and produced enormous rolls of paper that fed the needs of newspaper and magazine publishers. Meanwhile, back in the forest, miles of slash—debris formed from branches, leaves, twigs, and stumps—lay strewn over the ground. The slash dried into kindling, waiting for a lightning strike, a spark from a

Figure 1.1 *Logging road in Lincoln, New Hampshire. During the nineteenth and early twentieth centuries in the White Mountains, loggers used horses to haul logs along dirt roads, causing considerable erosion. USDA Forest Service, Eastern Region Photo Archives.*

passing steam locomotive, or a carelessly thrown match to ignite it. If the slash caught fire on a dry and windy day, the entire mountainside could blaze into a holocaust within minutes.

The forest in which the crew of loggers were performing their labors was nestled in the Pemigewasset valley, bordered on the west by the gorgeous Franconia Range with its Old Man of the Mountain and on the east by equally picturesque Crawford Notch. The sparkling waters of the East Branch of the Pemigewasset River meandered gracefully through the valley. An emerald forest of spruce, pine, and hardwoods had once carpeted the entire valley, but now the sea of green was interrupted by large tracts of land stripped of nearly all vegetation.

Over the next several years, the heavy logging in the White Mountains and in other parts of New England would result in massive deforestation and devastating fires. The timber harvesting was a direct outgrowth of the rapid industrialization of the United States during the second half of the nineteenth century, when the country had an unquenchable thirst for natural resources—for coal, iron ore, oil—and for wood. In many ways, wood was the oil of the nineteenth century. In addition to supplying lumber to build the feverishly expanding cities of the United States, wood powered locomotives, warmed factories and homes, and built a thousand necessities of daily life, from clothespins to shoe lasts.

The rapid disappearance of millions of trees, however, was inspiring something new, something important: the birth of a movement to conserve, protect, and restore forests. Conservationists challenged conventional wisdom about natural resources with provocative questions. Were America's forests valuable only as cornucopias of timber and other resources, and were those resources truly inexhaustible? Were the forests equally important for their beauty, their opportunities for recreation, and the habitat they provided to wildlife? Were public needs, especially in the East, being served by private ownership of the country's timberlands? These questions would soon stir passionate debates in southern Appalachia and the Lake states, as chapters 2 and 3 will examine, but it was in New England that these issues first surfaced.

New Hampshire's Forest Heritage

Timber harvesting had a long and honorable tradition in the White Mountains, which rise and fall like granite-laden waves over one million acres in northern New Hampshire and western Maine. The industry had its origins in the earliest days of European-American settlement of New England, and it reflected the attitudes that Europeans brought with them that nature was to be subjugated and used for the service of humanity. When the colonists arrived, they found boundless forests of majestic white pines, which they downed to build log cabins and then houses with beams made from the tree's strong, knot-free lumber. A single tree could produce an astonishing amount of wood. Old-timers recalled tabletops that were 33 inches wide and beams 7 inches wide, all carved from the wood of a single white pine.

After the American Revolution, the pace of settlement in New Hampshire accelerated, as did the amount of logging. Human economic activities and technology transformed the New England landscape into what environmental historian William Cronon has called "a patchwork quilt on the landscape."[1] Settlers cleared fields, divided land into parcels, and constructed roads, fences, houses, and barns. The parcelization of the land made it economically productive and made room for the infrastructure for population growth, reflecting assumptions about land use that European-Americans brought with them. Another assumption was the virtual inexhaustibility of the northern forests. Frederick Kilbourne, a historian who wrote a classic history of the White Mountains, *Chronicles of the White Mountains*, observed, "So vast were formerly the forests in the valleys and on the lower slopes of the Mountains themselves that the supply of timber seemed inexhaustible. . . . No thought of a possible future scarcity ever entered the minds of the early lumbermen."[2] Infinitude was a driving article of faith of the nineteenth century.

After the Civil War, two developments spurred New Hampshire's timber industry even more. In 1867, Governor Walter Harriman decided that the state would sell the last 172,000 acres of prime acreage in the northern forests—land that was estimated to be worth hundreds of thousands of dollars—for a mere $26,000.[3] Speculators snapped up the cheap land and immediately started harvesting the mature trees. It is true that Harriman greatly undervalued the land, but government policy at that time was to sell lands in the public domain to spur economic development by private enterprise.

The other stimulus for the logging industry was the coming of the railroads to northern New Hampshire. Before the 1860s, most loggers had to limit their operations to tracts near rivers and streams, on which they floated logs downstream to the burgeoning sawmills in Portsmouth and other cities. The forest interiors remained relatively untouched, as loggers found it difficult to haul long and cumbersome logs over rough mountain terrain. After the Civil War, though, entrepreneurs laid railroad tracks farther north into New Hampshire, primarily to bring tourists into the northern reaches of the mountains but also to ship freight, including logs.

J. E. Henry: "The Heartless Lumber King"

As a result, New Hampshire offered ample opportunities for resourceful entrepreneurs, and one who grabbed his chance with a vengeance was James Everell Henry, or J. E. Henry for short. Henry was destined to become the most famous logging baron of New England and would be widely disparaged as "the Heartless Lumber King," "the Wood Butcher," and "the Mutilator of Nature."[4] He burned with the same relentless drive for success as Andrew Carnegie, John D. Rockefeller, Jay Gould, and other tycoons of the Gilded Age who had made their fortunes by supplying a growing nation with raw materials and industrial products. The nation craved wood and other forest products, and J. E. Henry was more than happy to oblige.

He was a native of the northern forest, born on April 21, 1831, in Lyman, New Hampshire, where his father struggled to eke out a living as a farmer in New Hampshire's rocky soil and mountainous terrain. When the father died from tuberculosis in 1845, the son had to scramble to support his mother and six siblings, so at the age of 15, he tackled one of the toughest jobs in the northern woods: hauling freight over treacherous mountain roads from Portland, Maine, to Montpelier, Vermont, and other points in northern New England.

Along the rocky path to manhood, Henry experienced setbacks that would have defeated anyone less determined. At one point, he bought a supply of popcorn to sell at a local fair and thought he had made a handsome profit when he realized with a shock that he had been paid with counterfeit money. Such incidents steeled his cynicism and tough-mindedness. He tried his hand at a variety of other enterprises, including growing wheat in Minnesota, but that venture turned out badly because of a run of bad weather. Before long, Henry realized that his greatest opportunities for wealth were in his native New Hampshire, and he returned home in the mid-1870s. The nation's demand for forest products was exploding, and there seemed no end to the uses for high-quality wood, including rifle stocks, railroad ties, bridges, roads, houses, and public buildings. Wood for houses was in peak demand, as the great cities of the East absorbed immigrants spilling onto America's shores and rural young migrating to urban factories.

Henry recognized the huge opportunity that railroads were opening up in the northern woods, and he began to apply the methods of industry to timber harvesting. In 1876, he allied himself with Charles Joy and A. T. Baldwin to form the company Henry, Joy and Baldwin. They organized their operation vertically, controlling forestlands, harvesting the timber, owning rail stock, and milling the logs into lumber. Later they added paper manufacturing to their ever-growing empire. The company snapped up properties with virgin timber in Carroll and Bethlehem, townships northwest of the Presidential Range.[5]

Fires in the Zealand Valley

One of Henry's earliest targets was the virgin timber of the Zealand valley, a splendid bowl of forestland west of Crawford Notch. The firm purchased property in the valley, and then Henry bought out Joy and Baldwin, gaining control of the company. Now he was the sole decision maker, and he pushed hard. He took trees that were more than ten inches in diameter, yet he also left the younger trees to mature. According to forest historian Bill Gove, "At Zealand, he apparently wasn't applying the clearcutting that later was to become so obvious and so criticized when he logged the Pemigewasset Valley."[6] In 1884, Henry reached an agreement with the Boston and Lowell Railroad (later part of the Boston and Maine Railroad) to build a ten-mile logging line into the valley. Two years later, the Zealand Valley Railroad was successfully carrying logs from the heart of the valley, which featured up-and-down terrain that necessitated locomotives to pull their loads up a grade of 5.4 percent. The railroad operated for only thirteen years, 1884 to 1897, but in that short span of time it hauled millions of board feet.[7]

Perched on the northern apron of the valley, near the confluence of the Zealand and Ammonoosuc Rivers, was Zealand Village, which was itself a product of Henry's drive to control all phases of his operations. The town boasted a high-capacity sawmill powered by steam, but it also contained small houses for dozens of Henry's laborers, a boardinghouse, shops for repairing locomotives and their parts, and a post office. A school master even taught the children of Henry's workers in one of the houses in the village.[8]

Scattered throughout the Zealand valley were also numerous logging camps, which one writer described as little more than a "primitive log cabin for wood choppers and a log stable nearby for the horses."[9] Operations peaked during the winter, when the loggers, who numbered as many as 250, felled trees and transported them on sleds over the thick blanket of New Hampshire snow to waiting railroad cars. The men worked eleven hours a day, but they earned a fair wage of $6 a week plus room and board. And they were productive. In 1886 and 1887, they cut an astonishing thirteen million board feet of timber.[10] (One board foot is one foot long, one foot wide, and one inch thick.) Henry insisted that his crew make productive use of everything, even the manure. He ordered his men to scoop up the odiferous by-product and load it onto railroad cars, which carried it to the southern reaches of the state to be sold as fertilizer. The company built charcoal kilns and turned hardwoods into charcoal, which was in growing demand as a cooking fuel and for manufacturing steel. Workers cut hardwoods, placed them closely together into compartments in kilns, and lit a fire beneath, which they carefully controlled to char the wood rather than burn it, creating charcoal.

During the summer of 1886, loggers were working zealously in the Zealand valley when disaster struck. Nary a drop of rain had fallen throughout that spring, and the slash lay dry and brittle on the earth. One morning, as one of Henry's logging trains backed into the valley to pick up a load of logs, a spark flew from the locomotive and into the slash, immediately igniting it. By the time Henry's workers noticed the flames, it was too late, and fire raced through the dried-out debris that smothered the cutover lands, enflamed standing timber, and raced up surrounding mountains.

It is difficult to overstate how fearsome fires like this one were during the nineteenth century. Today, wildfires consume anywhere from two million to ten million acres in forests every year, but then, fires consumed twenty million to fifty million acres a year.[11] Locomotives were the most common cause, spewing hot coals from smokestacks and ash pans into surrounding woods and wreaking incendiary havoc. Near Bretton Woods, New Hampshire, one fast-moving train rounded a curve and flung coals eight feet into the dry grass lining the sides of the tracks. In a matter of minutes, the grass was blazing, and the fire was racing toward the trees standing only a few yards beyond.[12] Railroads, however, were not the only culprits. Careless smokers dropped matches, farmers burned brush in dry and windy conditions, and campers failed to extinguish campfires properly. And then there were the firebugs, who started fires to wreak revenge against people they just plain did not like.

Logging increased the severity of fires because of the slash that loggers typically left on the forest floor. During dry spells, branches and tree canopies dried into kindling, and when a spark flew into the midst of the slash, a conflagration quickly ensued. Flames rushed through the slash and spread to living trees, and soon an entire stand was aflame. There was no U.S. Forest Service to guide firefighting efforts, no bulldozers to create earthen barriers against the spread of fire, no helicopters to dump water onto parched trees, no planes to drop fire-retarding chemicals.

Eventually, the Zealand fire incinerated an area seven miles long and burned down three logging camps. One group of men survived only because they had the presence of mind to leap into a river. After a week, rains fell and doused the conflagration, but the damage had been done.

The fire had swept through twelve thousand acres of land, blackened standing hardwoods and spruce, and destroyed approximately two million board feet of sawlogs that were on skids waiting to be milled.[13] After this devastating fire, Henry began looking for greener forestland to conquer, and just to the south he found it. It was the Pemigewasset valley, an enormous wilderness in the heart of the White Mountains that contained mile upon mile of virgin spruce and fir. After the Zealand fire, Henry surreptitiously began to purchase lands in the Pemi, as locals called it, until by 1892 he controlled virtually the entire valley and was poised to start operations on a more massive scale than before.

Henry turned to clear-cutting as his *modus operandi*. He was convinced that the young timber he had left standing in the Zealand valley had fueled the spread of the fire, and he had lost the value of that timber. Consequently, he made the decision not to leave *any* trees standing in the Pemi, no matter how small or immature they were.[14] His loggers downed thousands of acres at a time, leaving enormous tracts completely denuded of trees. A reporter wrote, "Trees crashed down everywhere, were stripped of their wonderful plumes, were dragged away to the landings. . . . Everything was coming down before those axes, and the 'slash' . . . lay in great heaps, black against the snow."[15] Clear-cutting paid handsome profits in the short term, but it left land that would take years to regenerate into a productive forest again.

Henry's operations in Zealand and the Pemigewasset made him the most famous of New England's lumber barons in the late 1800s, but he was far from the only one. George Van Dyke controlled an empire of four hundred thousand acres of timberland in northern New England.[16] The Russell Paper Company purchased all the forestlands in Waterville, which provided access to the still-unlogged Sandwich Range in the southern part of the region. Other large companies included the Brown Lumber Company of Whitefield and the Kilkenny Lumber Company. The Saco Valley Lumber Company bought the rights to forestlands in the Mount Washington valley, which boasted some of the most arresting scenic vistas in the entire region. Small companies thrived as well, with between fifty and one hundred operators buying up land, according to one estimate.[17]

Table 1.1. Timber Cut in Northern New Hampshire, July 1, 1902, to June 30, 1903

Industry	Number of Establishments	Total (1,000 Board Feet)	Spruce (1,000 Board Feet)	Pine (1,000 Board Feet)	Hemlock (1,000 Board Feet)	Hardwoods (1,000 Board Feet)
Paper and pulp	6	105,552	105,552	—	—	—
Lumber	65	155,570	120,195	17,213	6,679	11,483
Bobbin	9	6,709	—	—	—	6,709
Shoe peg	3	3,081	—	—	—	3,081
Crutch	4	150	—	—	—	150
Miscellaneous	9	2,500	—	—	—	2,500
Total	96	273,562	225,747	17,213	6,679	23,923
Percent		100.0%	82.5%	6.3%	2.4%	78.8%

Source: Chittenden, *Forest Conditions of Northern New Hampshire*, 80.

Throughout the 1890s, the timber harvest in New Hampshire skyrocketed, driven partly by the discovery of chemical processes to manufacture paper from wood pulp. The best pulp came from spruce, any size spruce, no matter how small or immature. As a result, the value of spruce cut in New Hampshire surged from $1,282,022 in 1889 to $7,244,733 ten years later and to $13,994,251 in 1909.[18] Meanwhile, the uses for forest products kept multiplying. Maples, beeches, and birches were used to manufacture bobbins for the sewing machines of Lowell, Massachusetts, and other textile centers. Birch was transformed into crutches. Table 1.1 delineates the multiple uses of wood from northern New Hampshire for the period July 1, 1902, to June 30, 1903.

By 1899, companies were hauling off 23,468,000 board feet of lumber every year. Eight years later, the amount had catapulted to an annual harvest of 58,107,000 board feet. The logging was so intense by 1907 that 1,363,711 acres of land were cut over, 244,036 acres were turned into agricultural land, and 120,495 acres lay completely barren. Only 12 percent of the total forest area remained as virgin forest. Lumber and paper companies controlled 900,000 acres of land in New Hampshire.[19]

Understanding the Effects of Deforestation

Industrial logging might be paying handsome profits, but as early as 1880, the assault on the forests had also begun to stir substantial alarm among summer tourists, outdoor sports enthusiasts, hotel owners, and local citizens whose livelihoods depended on the beauty of the White Mountains. New Hampshire's legislature shared the concern because tourism was growing in importance to the state's economy, and in 1881, it voted to establish a Forestry Commission to examine the extent of the damage that had been done to the state's forests. The reports issued by this and succeeding commissions documented the extent of deforestation and examined scientifically the effects that deforestation was having on forest soils, water quality, erosion, river navigation, wildlife, and other aspects of these natural systems.

In 1889, after the devastation of the Zealand valley fire, the state legislature challenged the Forestry Commission with a more aggressive purpose: to explore the possibility of having the state purchase forestlands that could be turned into public parks or reserves. In its subsequent report, issued in 1891, the commission described the characteristics of a healthy forest and examined the role of deforestation in erosion, flooding, and the buildup of silt in rivers. It also investigated the economic effect of the loss of forestland, supplying data that conservationists later used to build the case for federal purchase and protection of forests. According to the report, "The most fatal agency in destroying the soil of a mountain forest country, and in wrecking the mountains themselves, is that of fire."[20]

The commission's goal of creating state forests did not gain momentum, however, because New Hampshire could not afford the costs necessary to make such purchases. Meanwhile, timber harvesting and fires continued to consume large chunks of forestland. The year 1903 was particularly devastating, with 554 conflagrations reported throughout the region.[21] Very little rain had fallen that spring, and the slash smothering the mountainsides was as dry as tinder. The fires started in June, probably from a carelessly thrown match or a campfire that burned out of control. Once

Figure 1.2 *Forest fire on Owl's Head. In 1907, a lightning strike ignited a devastating fire on Owl's Head, a mountain in the Pemigewasset valley in the White Mountains. The valley was the site of heavy logging in the early twentieth century. USDA Forest Service, Eastern Region Photo Archives.*

started, the fires ravaged the White Mountain region, burning more than 10 percent of the forestlands and causing damage estimated at $200,000. They scorched 10,000 acres in the Zealand valley that had escaped the 1886 fire and approximately 18,000 acres in the northern mountain ranges, for a total of 84,244 acres.[22] Another round of fires caused more damage, including the burning of the forests on Owl's Head in the Pemigewasset valley (figure 1.2).

The fires had created a crisis atmosphere, and in 1905, New Hampshire's Forestry Commission and the U.S. Bureau of Forestry issued a report that was even more detailed than the state's previous reports had been.

Authored by Alfred K. Chittenden, an assistant forest inspector for the U.S. Bureau of Forestry, *Forest Conditions of Northern New Hampshire* thoroughly examined the effect of logging and forest fires on New Hampshire. His report contained chilling detail about the fires, which had not only destroyed thousands of trees but also did severe damage to the forest soils. According to Chittenden, "The influence of fire on the soil is due almost wholly to the destruction of the humus and other organic matter in it."[23] Humus had built up over centuries from decaying leaves, fallen trees, and other vegetable matter, and it contained the nutrients that trees needed to grow. The intense heat generated by the fires also destroyed seeds, small trees, and root systems. When rains came, they washed away the soils with their rich nutrients, often leaving bare rock.

Even in areas that were spared fire, erosion from road construction was a major problem because of soil disturbances. According to the Chittenden report, humus acted as a mulch, soaking up melting snow and rainwater, holding it like a sponge, and slowly releasing the moisture into the region's streams and rivers at an even rate. After forests were clear-cut, though, the forest canopy disappeared, exposing the soil to direct sunlight, drying it out, and destroying the sponge-like quality. When heavy rains pounded the earth, they washed away the soil into the region's rivers and streams, leaving deep gullies. Rain carried tons of soil into northern New England's streams and rivers, silting them up. The silt threatened navigation on several of the region's rivers, including the Connecticut and the Merrimack. Indeed, the threat to interstate commerce in the region would buttress later arguments in favor of a federal forest reserve in the White Mountains.

Some conservationists also began to suspect that the loss of forestland might be one cause of increased flooding in Manchester and other cities in southern New Hampshire. In early March 1896, a torrential downpour lasted thirty-six hours, swelling the Merrimack and Piscataquog Rivers to the highest marks they had ever reached. The rampaging Merrimack broke through a dam upriver from Manchester, flooded an electric light plant, and caused several thousand dollars in damage. The waters inundated the Amoskeag textile mills, forcing the mills to close for two months, throwing ten thousand textile workers out of jobs, and devastating southern New Hampshire's economy.[24] Conservationists hypothesized that the logging in the northern part of the state might have allowed water to run off more heavily into streams and rivers, exacerbating the effects of storms. These claims emerged from the new science of hydrology, which would play a central role in the coming debates over what to do to protect the forests.

Effect of Deforestation on Scenic Beauty and Tourism

The loss of forestland threatened something else that lay at the very heart of the appeal of the White Mountains: their extraordinary beauty. As early as the 1830s, the mountains had begun to find a place in the American heart as a mecca for aesthetic appreciation, physical challenge, and psychological regeneration. Word of the visual splendors of the mountains carried to the populous regions of the East Coast, word that the region boasted seventy-four mountains that were more than 3,000 feet high; eleven that soared 5,000 feet into the atmosphere; and one, Mount Washington, that at 6,288 feet towered over the rest of the mountains in the Northeast. The growing appeal of such natural scenery—and the opportunities that natural beauty offered for physical, mental, and spiritual renewal—would play a central role in the movement to protect and restore the eastern national forests.

The pioneers of New Hampshire's early tourism industry were the Crawford family. In the early 1790s, Abel Crawford had moved his family from Vermont to Bretton Woods, just north of the Notch of the Mountains, which was later renamed Crawford Notch in memory of the family. The scenery and remoteness of the region captivated Abel. There he raised his family, including his son Ethan, who grew into a strapping young man. Ethan became the foremost guide in the region, and people from near and far sought him out to venture into the mountains. In 1819, Abel and Ethan cut a path through the woods to the top of Mount Washington, which was the destination for which their visitors most often aimed. The task was arduous, but they succeeded, and today the Crawford Path remains the oldest continuously used hiking trail in the United States. Ethan advertised the path and accommodations in newspapers, started to take in travelers, and continued guiding them into the mountains. From this humble start, the White Mountains became, with the Adirondacks, the premier destination in the United States for outdoor adventure recreation during the mid-1800s.

Artists also spread the reputation of the White Mountains by creating extraordinary paintings that captured the sublime beauty of the mountains, emphasizing the wild tempestuousness of nature. When Thomas Cole, one of the founders of the Hudson River School of landscape painting, visited the region for the first time in 1828, he sang the praises of the scenic splendors: "On every side, prospects mighty and sublime opened upon the vision: lakes, mountains, streams, woodlands, dwellings and farms wove themselves into a vast and varied landscape."[25]

Cole produced canvases that captured the singular power and mystery of the mountainscapes. Other prominent artists followed him there, including Asher B. Durand, Frederic E. Church, and Albert Bierstadt. By the 1870s, the White Mountain School of Art, a subcategory of the Hudson River School, was firmly entrenched. Artists like Benjamin Champney maintained studios in the region, and tourists could take home their very own reasonably priced canvases of the renowned sights of the region, like the Old Man of the Mountain or the Flume, with its huge boulder suspended precariously between two walls until it fell into the fast-rushing stream below in 1883.

As a result of all these factors, the White Mountains became a popular summer destination for a wide range of Americans, drawn by what guidebook writer Moses Sweetser described as the "almost infinite variety of scenery, inexhaustible in its resources and unlimited in its manifold combinations."[26] Resort hotels spread luxury throughout the region and enticed thousands of tourists, transporting them into the heart of the mountains as into a different world.

An Emerging Forest Conservation Movement

By the early twentieth century, the destruction of the scenic beauty by J. E. Henry and other loggers angered legions of conservationists, local residents, tourists, business owners, and others who had come to love the region. Charles Sprague Sargent, director of the Arnold Arboretum at Harvard University, published an influential horticultural journal, *Garden and Forest*, from 1888 to 1897, and in its pages he published cutting-edge scientific information that detailed the effect of deforestation on watersheds. Articles that he and others authored described soil erosion, declining water quality, the silting of rivers and streams, and flooding. Natural beauty was also close to his heart, and he depicted the effect of deforestation on scenic beauty and recreation.

Sargent was born in 1841 to a merchant in Boston who had made a fortune in railroads. At a young age, he became interested in botany and horticulture, with a particular fascination with trees and how they enhanced the landscape. In 1872, he became director of the arboretum and was also a professor of arboriculture at Harvard. Sargent operated the arboretum with a firm hand, handing out directions and making sure his disciples followed his orders precisely. A prickly sort, he sometimes clashed with others in the conservation movement. Gifford Pinchot called him a "natural aristocrat" but added that he was "the foremost advocate of forest preservation" in the United States.[27]

In the very first issue of *Garden and Forest*, dated February 29, 1888, the historian Francis Parkman warned, "If the mountains are robbed of their forests they will become like some parts of the Pyrenees, which, though much higher, are without interest, because they have been stripped bare."[28] An issue a year later included a quotation from Manchester's *Daily Mirror and American*, which made clear the threat:

> When the woods have been cut away, the White Mountains are about as bleak and barren a section of country as we know of, and when there are no timber lots left the charm that was formerly theirs will be wanting. Even the Profile, the Franconia and Crawford Notches and Mount Washington would lose much of their glory and beauty with the destruction of the forests.[29]

While Sargent emphasized the aesthetic and ecological impact of deforestation, other reports and articles detailed the economic effect. The New Hampshire Forestry Commission emphasized that the White Mountains had become a summer playground because they healed "the weariness and exhaustion of vitality which so often result from excessive activity in the crowded life of towns and cities."[30] Hotels and boardinghouses were bulwarks of the state's economy. In 1889, the state counted 1,113 hotels and boardinghouses, with revenues exceeding $5 million. Indeed, tourism was approaching agriculture in importance to New Hampshire's economy. If scenery were demolished, the report concluded, "they [the tourists] will not come."[31]

In 1893, Julius H. Ward, a minister who frequently sojourned in the mountains, published an article in the *Atlantic Monthly* titled "White Mountain Forests in Peril." The piece grabbed a national audience and trained its eyes on the devastation. Tourists who traveled to Mount Washington, Ward warned, "overlooked what was once a magnificent wilderness, but where now the axe and the fire have combined to leave what looks like a frightful desolation. . . . One sees the same frightful slaughter of forest, the trees cut off entirely, and the land growing up with birch and cherry bushes, which show that the soil has been ruined."[32]

On July 4, 1900, Reverend John Johnson, a minister in Littleton, New Hampshire, penned a vivid pamphlet, *The Constrictor of the White Mountains* (figure 1.3), which attacked the New Hampshire Land Company for its logging activities in the region. The magazine *New England Homestead* then picked up the pamphlet and distributed it to more than forty thousand households in New England. Johnson successfully identified a villain that New England residents affected by the timber industry could blame.[33]

As critics were attacking the timber industry, conservation forces were beginning to stir in the region and search for solutions. Hotel owners, for example, were taking steps to protect their industry. The holocausts that blazed through the region in 1903 had come agonizingly close to several hotels, including the Mount Pleasant House, north of Crawford Notch. Visitors there even witnessed burning embers in the yard, but fortunately the strong winds that day had died down before the fire reached the hotel.[34]

Because of this and other incidents, several hotel owners began to embrace the conservation cause. They purchased forestlands surrounding

Figure 1.3 *The Constrictor of the White Mountains. In 1900, Reverend John Johnson of Littleton, New Hampshire, wrote a fiery pamphlet that attacked the New Hampshire Land Company for devastating forests in the White Mountains. From* The New England Homestead, *December 8, 1900.*

their hotels to preserve the views for their guests. They established fire-prevention measures, adopted principles of sustained-yield forestry, and pressed for legislative protection of the forests. To ensure that the Mount Pleasant House never again faced a fire threat, the manager, John Anderson, asked assistance from the federal Bureau of Forestry. The bureau recommended that the hotel build fire lines and trails and hire men, paid for by the government but living at the hotel, to take measures to prevent fires. In 1904, the hotel also constructed a fire lookout on Mount Rosebrook, one of the first lookouts in New Hampshire.

Although pressures for forest conservation and fire prevention were building, the majority of the logging companies remained adamant that private-property rights allowed them to cut the forests they owned in any way that they desired. By 1909, J. E. Henry was in his seventies, and after suffering a series of strokes, he relinquished control of his logging empire. When a reporter interviewed him that year for *Collier's Magazine*, however, he remained unrepentant about the effect he had had on the forests of the White Mountains. "I never see the tree yit," he growled, "that didn't mean a damned sight more to me goin' under the saw than it did standin' on a mountain."[35]

Attitudes like those of "the wood-butchers" reflected an aging paradigm, however. The timber harvesters assumed that forest resources were inexhaustible and that they could be exploited for maximum benefit with no thought of tomorrow. The New Hampshire Forestry Commission's 1891 report placed the blame for such attitudes not only on the logging industry but also on assumptions in American culture about the inexhaustibility of the nation's wondrous natural resources. "In [the loggers'] wasteful and destructive methods of cutting timber," the commission charged, "they only illustrated the want of foresight, of self-restraint, and of regard for the interests of posterity which are still, unhappily, far too prominent features of American civilization."[36]

During the same period, many Americans were coming to view their nation's natural beauties, from Yosemite valley to Yellowstone National Park, with growing national pride. The scenic vistas that graced the landscape had taken on a new importance in the nation's identity, and the preservation of their beauties was becoming a cause that drew increasing public support. In Americans' changing attitudes toward forests, the paradigm was shifting slowly toward sustainability, scientific understanding, and recognition that forests had intrinsic value. The same paradigm shift was about to occur in the southern Appalachians and the Lake states, regions that faced the same furious assault on their forests that had nearly devastated the White Mountains.

CHAPTER 2

Trees to Build the Lake States

While logging and fires were consuming forestlands in New Hampshire, the forests of Michigan, Wisconsin, and Minnesota remained largely intact through the Civil War. Even as early as the 1850s, however, politicians and entrepreneurs in the Lake states had recognized the enormous economic potential of their woodlands. In 1856, Ben Eastman, a congressman from Wisconsin, rose on the floor of the U.S. House of Representatives and proclaimed, "Upon the rivers which are tributary to the Mississippi, and also upon those which empty themselves into Lake Michigan, there are interminable forests of pine, sufficient to supply all the wants of the citizens for all time to come."[1]

Eastman's words resonated in the East, where thousands of lumber entrepreneurs and loggers heard the sounds of opportunity beckoning from the pine forests to the west. One who heard the call was Daniel G. Shaw, an ambitious young man born in 1812 on a farm in the aptly named town of Industry, Maine. From an early age he had an entrepreneurial bent and operated a small sawmill business in his hometown. The supply of trees in Maine was beginning to dwindle, though, and Shaw looked westward. In 1851, he uprooted his wife and two sons and migrated to Allegany, New York, where he built a larger mill.

In only four years, the ambition bug bit him again, and he traveled farther west, to view for himself the "interminable forests of pine" that Ben Eastman had boasted about. He found that the good congressman had not exaggerated. In the valley of the Chippewa River in northwestern Wisconsin, a virgin forest of white pine—*Pinus strobus*—unrolled before him to the horizon. The sight made Shaw's heart beat just a little bit faster.[2]

The majestic white pine—which thrived in the Lake states' moist soil of clay, loam, and sand—could soar as high as two hundred feet above the earth, reach a diameter of eight feet, and live for up to five hundred years. It was tall and straight, and it floated well down rivers, which was why loggers called it the "cork pine."[3] Carpenters loved its strength, lightness, straight grain, and absence of knots. It was easy to saw, yet it built long-lasting houses that stood up to punishment. Millions of board feet of white pine built the burgeoning cities of Chicago, Detroit, St. Louis, Des Moines, Cleveland, Minneapolis, and St. Paul. The white pine traveled on Great Lakes ships to the Erie Canal and from there to the giant markets of the East. Locals referred to their bountiful pine forests as the pineries.

Thrilled by what he saw, Shaw immediately purchased a large tract of forestland in the Chippewa valley, and in 1858, he formed Daniel Shaw and Company with his brother-in-law, Charles A. Bullen. Immediately they began logging and processing lumber. In only two years, the Shaw Company was processing twenty-five thousand board feet of lumber a day, which is enough to build a typical two-thousand-foot single-residence home. For the next fifty-five years, Daniel Shaw and his sons, George B. and Eugene, would operate the company through a boom period that saw the cities of the Lake states grow and the forests shrink. In the late 1800s, the company reached its peak production of 28,440,000 board feet a year, a quantity that placed it only in the middle ranks of Wisconsin lumber companies. In twenty-five years, Shaw exhausted the original ten thousand acres he had purchased along the Chippewa River. The company purchased more lands and more stumpage, or the right to cut trees on others' property. In the Lake states, Shaw's experience was repeated hundreds of times by hundreds of other lumber operators who rode the land boom that was rapidly transforming this part of the United States.[4]

Rapid Development of the Logging Industry in the Lake States

In New England, farmers and loggers had required two centuries to attack the forests and deplete them. In Michigan, Wisconsin, and Minnesota, near-depletion would take only sixty years. The logging industry in the Lake states started to take off after the Civil War ended in 1865, a period that coincided with rapid industrialization and urbanization. These three states generated 2.75 billion board feet in 1869, 5 billion board feet in 1879, and 7 billion board feet in 1889.[5] By the early twentieth century, though, the forests of all three states faced near-depletion and damage from forest fires, severe erosion, and the silting up of rivers and streams, duplications of the damage that had been done to the once-magnificent forests of the White Mountains.

The logging and lumbering industries of the Lake states were different from, yet connected to, those of New England. Like Daniel Shaw, many of the lumber entrepreneurs of the Lake states were transplants from New England, where they had learned the basics of the industry. They knew how to identify high-quality timber, build efficient sawmills, hire experienced loggers, set up logging camps, drive logs downstream, and negotiate with wholesalers to distribute the logs.

Similarly, a forest conservation movement would, by the late 1800s, begin to develop in the Lake states. As in New England, the movement developed partly from the growing understanding of the science of forests as natural systems, and many of the leaders of the movement would be trained in the emerging science of forestry. There was also a key difference, however. In New England, early conservationists were motivated by the strong desire to protect the region's tourist industry, which was well established. The Lake states, in contrast, did not have such a well-developed tourism industry in the nineteenth century. Instead, many of the early conservation leaders were active in state forestry associations or worked for state or federal government. The motivation driving them was the conservation of a valuable resource, timber, which they viewed as central to the continued economic development of their states.

33

Geographic Advantages of the Lake States

In the rapid development of its logging industries, the Lake states benefited from enormous geographic advantages. All three states boasted rich
soils and plentiful rainfall, which nurtured the forests that spread almost
unbroken across their northern shoulders. Michigan contained the greatest forests of white pine in the world, thanks to the soil's high sand con-
tent, which produced a peerless grade of white pine. North of an imaginary
line running from Muskegon to Bay City, the white pines predominated,
but the forests also contained jack pine, aspen, red pine, and balsam fir.
The southern part of the state, meanwhile, featured hardwood forests of
oak, hickory, elm, maple, birch, basswood, and beech. In all, Michigan had
approximately 380 billion board feet of timber, most of it of high quality.[6]

Wisconsin was blessed with approximately thirty million acres of
forestland containing more than 100 billion board feet, much of it in the
northern and northeastern parts of the state. Forests covered 75 percent of
the state and nearly 100 percent of the northern counties.[7] Filibert Roth,
who was a special agent for the United States Department of Agriculture,
estimated that by 1898, only 17.4 billion feet of those virgin forests were
still standing.

In Minnesota, the logging and lumbering industries were critical
to early settlement and economic development. The state was home to
31,500,000 acres of forestland, of which 5,800,000 acres were pine.[8] Hardwood forests, known as the "Big Woods," dominated the lands west of the
Mississippi. These hardwood forests were mostly oak, but they also contained maple, black walnut, elm, ash, butternut, and basswood.[9] East of
the Mississippi were the conifer forests—the North Woods—that stretched
from the Mississippi River to Canada. These pineries had red pine and jack
pine in addition to white pine.

The region also benefited from an extensive network of navigable
rivers for floating logs to market and powering sawmills. In Michigan's
Lower Peninsula, the Saginaw River and its tributaries contained 864 navigable miles that drained 3,500,000 acres of forestland into Lake Huron.[10]
In western Michigan, the Manistee, Muskegon, and Grand Rivers flowed

into Lake Michigan, allowing cheap and easy transportation to Chicago and Milwaukee. Wisconsin featured four major navigable rivers—the St. Croix, the Black, the Wisconsin, and the Chippewa—which fed into the Mississippi, opening access to markets in St. Louis and beyond. Minnesota was also favored with a tremendous network of rivers, including the biggest of them all, the Mississippi. One of the tributaries of the Mississippi, the St. Croix River, was the birthplace of the logging and lumber industry in Minnesota.

Logging and Laissez-Faire Capitalism

Given the existence of huge reserves of white pine, geographic advantages, and voracious markets created by expanding cities, the story of the logging and lumber industry in the Lake states was one of laissez-faire capitalism and the rise of big business. Captains of industry such as Frederick Weyerhaeuser and John Pillsbury rose to prominence, and large lumber companies created efficiencies of volume by acquiring vast tracts of forestland, building huge mills that used the latest technologies, and integrating businesses vertically to control all aspects of production. Large operators also had capital and access to financiers.

Ironically, though, the logging industry in the Lake states also took on the patina of romance through the tales of Paul Bunyan and the exploits of the loggers themselves. The Bunyan tales first enlivened evenings in Maine's logging camps but then traveled west to the bustling camps of Michigan, Minnesota, and Wisconsin, where they appeared as early as the 1850s. The stories were paeans to Production, with a capital P. Everything about Paul Bunyan was outsized, reflecting the difficult challenges of logging America's northern empire. He stood seven feet tall and covered seven feet in one stride. To summon his fellow loggers, he picked up a hollow log and bellowed through it, but the powerful sound of his voice inadvertently blew down ten acres of pine. To cut as many trees as possible, he tied the handle of his axe, which had a double bit, to a strong rope, allowing him to swing it with frightful momentum. In one stroke, he felled several acres of trees.[11]

The romanticism of such tales, however, should not disguise that the big lumber companies acted in their own economic self-interest, aided and abetted by cultural attitudes that rationalized subjugation of the wilderness and the taming of the frontier. In 1893, Frederick Jackson Turner, who had grown up in Portage, Wisconsin, and was a young associate professor of history at the University of Wisconsin, delivered one of the seminal lectures in U.S. history, titled "The Significance of the Frontier in American History." As Turner elaborated on the centrality of the frontier, he shed light on deeply embedded cultural assumptions: (1) the forces of civilization, embodied by the trapper, the pioneer, and the farmer, triumphed over wilderness, which was a nasty obstacle to progress; (2) the guiding purpose of American progress was to harness the continent's vast natural resources for the development of the nation's economy; and (3) those natural resources were without limit.

Michigan's Logging Industry

Economic forces and the ideology examined by Turner were amply evident in the development of Michigan's logging industry. Forest historians Donald I. Dickmann and Larry A. Leefers wrote in *The Forests of Michigan*, "During the last half of the nineteenth century the woods of the new state came under a mass assault not unlike war in its ferocity. This plunder was largely unregulated by laws and unfounded by the conventions of civilized society."[12] Logging in the state spread from the south to the northeast, the northwest, and then the Upper Peninsula. Because of the reach of its tributaries, the Saginaw River valley became the epicenter of logging in the northeastern part of the state. The river flowed north before emptying its waters into Lake Huron at Bay City, where ships loaded lumber and carried it to cities on the Great Lakes as well as through the Erie Canal to the hungry markets of the East. The Saginaw valley and Bay City supported 112 sawmills, employing twenty-five thousand men.[13]

Michigan's forests and high-quality pines attracted a slew of lumber operators, and competition among them was intense. At first, some operators engaged in selective cutting, removing high-quality trees and leaving the younger trees, but if a company used such conservative logging methods, it lost out in production to companies that had no compunction about clear-cutting the forests. Lumber thieves only sped up the cutting by routinely felling trees on government land. Between 1844 and 1854, they cut half a billion board feet in northern Michigan and eastern Wisconsin.[14]

As in the White Mountains, the coming of railroads accelerated the pace at which forestlands were denuded. Winfield Scott Gerrish attended the Philadelphia Exposition in 1876 and saw a Baldwin locomotive, a powerful steam engine. The enterprising Gerrish quickly purchased one of the locomotives and laid tracks to carry logs out of the forests surrounding the Muskegon River. As it happened, the winter of 1877 had little snowfall, yet Gerrish was still able to transport his logs to market even as his competitors waited and prayed in vain for snow. His competitors followed his lead, and by 1889 railroads were transporting thousands of logs out of Michigan's forests. Railroads increased the rate of cutting by as much as ten times.

Many of Michigan's legislators were concerned about the depletion of the forests, however, and in 1887, the legislature established a State Forestry Commission. Like New Hampshire's Forestry Commission, Michigan's began to build a scientific foundation for forest conservation by gathering data about the effect of heavy timber harvesting and forest fires. At the commission's initial meeting in January 1888 in Grand Rapids, keynote speaker Norman A. Beecher warned that if the state continued along its path, the entire supply of timber would be gone in another fifteen years. According to the commission, one thousand mills operated in Michigan, and the value of their annual output was $60 million. One state senator estimated that the state's output was five billion board feet, which, in his words, would be sufficient "to girdle the earth with fifteen board fences, each five feet high at the equator."[15] The white pines in the Lower Peninsula were nearly depleted, but the demand for spruce, balsam, jack pine, and poplar was growing because of the need for pulp. By the early 1900s, 161 billion board feet of pine had been cut and processed into lumber and other wood products. The lumber industry had cut another 50 billion board feet in hardwood, cedar, and hemlock, approximately one billion logs in total.[16]

Wisconsin's Logging Industry

Logging in Wisconsin followed a different trajectory from Michigan's because of geography. In the 1850s, lumber companies started to work in central Wisconsin because they could float logs downstream on the Wisconsin River to the 107 sawmills that, by 1857, lined the banks of the river.[17] The northern reaches of the Wisconsin had numerous bends, however, making it impossible to float logs. By the 1860s, Daniel Shaw and other logging entrepreneurs began to open up the virgin forests in the Chippewa River valley in northwestern Wisconsin, where the river and its tributaries drained six million acres of forestland, most of it dominated by white pine.

In 1877, the Wisconsin Central Railroad opened up north-central Wisconsin to logging when it finished construction of a line from Milwaukee to Ashland, and other railroad lines quickly followed. As in Michigan, the railroads increased the pace of cutting by a quantum leap. By 1890, Wisconsin loggers produced enough wood to build 180,000 bungalows with two bedrooms, and by 1899, the output could have built 290,000 bungalows, enough to house one million people.[18]

As in Michigan, Wisconsin loggers often started by cutting selectively, but competitive pressures led them to clear-cut. The Daniel Shaw Company, for example, gradually reduced the minimum diameters of trees cut. At first, in 1857, the company was very selective, felling only trees greater than sixteen inches in diameter. By the 1870s, the minimum diameter had decreased to twelve inches because virtually everything larger had been harvested. By the late 1880s, the minimum requirement was down to ten inches diameter, and by the 1890s, the company was felling all pine trees, no matter what their diameter.[19]

Wisconsin saw also the emergence of Frederick Weyerhaeuser, who began to build a logging and paper empire after the Civil War. Weyerhaeuser did much to consolidate the logging industry, in the process transforming it from a collection of small, independent entrepreneurs into large corporations. Consolidation only made the industry more efficient, leading to more rapid devastation of the forests. This strong-willed entrepreneur was born in Germany on November 21, 1834. In 1852, he immigrated to Erie, Pennsylvania, and then followed a cousin to Rock Island, Illinois, which was blessed with fecund prairie soil and access to the Mississippi River. There he went to work for a sawmill and quickly showed an acumen for business, at one point selling a supply of lumber for $60 while his boss was away at dinner. The employer was so delighted with young Frederick's initiative that he gave him the responsibility of overseeing sales and operations at the lumberyard.[20]

In 1860, he bought a sawmill in Rock Island with Frederick Denkmann, a talented machinist. The business prospered, but one obstacle they faced was ensuring a steady supply of logs from the northern forests. Weyerhaeuser led a syndicate of Mississippi River mill owners, incorporating in 1870 as the Mississippi River Logging Company. The syndicate began buying forestlands in the Chippewa valley—Weyerhaeuser, for instance, purchased fifty thousand acres of pine in 1875[21]—but the lumber companies in the Chippewa valley deeply resented Weyerhaeuser and other Mississippi mill owners, whom they referred to as "invaders" because they bid up prices for forestland and logs. The Weyerhaeuser syndicate grew too powerful to resist, though, and the Chippewa companies had little choice but to cooperate with him.[22]

By the 1890s, Wisconsin ranked as one of the top timber-producing states in the country, and one year it produced the most timber. The heyday did not last long, however. By the early twentieth century, many of the pineries were completely cut over, and operators began to cut other species, including maple and hemlock. By 1912, when the Daniel Shaw Lumber Company ceased operations, Wisconsin's forests were largely depleted.

Logging in the Land of Ten Thousand Lakes

In the early nineteenth century, Minnesota was a land of lakes and pristine forests that stretched like green waves as far as the eye could see. Even while Michigan's forests were being assaulted, Minnesota's remained largely untouched, protected by the relative isolation of the Minnesota Territory. When logging came to the territory's north woods, however, it came rapidly. On March 13, 1837, three men—Henry H. Sibley, Lyman M.

Warren, and William A. Aitkin—reached an agreement with the Chippewa Indians to log trees and build sawmills on lands along the Snake River and the St. Croix River, which joins the Mississippi just south of Minneapolis–St. Paul. Months later, in July 1838, a group of men, including Orange Walker and George B. Judd, rode the steamship *Palmyra* into St. Croix Falls, where they formed the St. Croix Lumber Company, the first logging and lumber company in Minnesota. There they built Minnesota's first commercial sawmill, which opened on August 24, 1839.[23]

In 1932, the last major mill in the St. Croix valley would close. During the preceding halcyon century, loggers had cut and shipped an astounding 67.5 billion board feet of pine. According to Clifford and Isabel Ahlgren, who provide an excellent history of Minnesota's forests in the book *Lob Trees in the Wilderness*, "As a result, one-third of the area . . . was cut over and interlaced with logging roads, trails, camps, and occasional railroad spur lines"[24] (figure 2.1).

Minnesota's logging and lumber industry benefited from Americans moving west and establishing farms on the extraordinarily rich lands of Iowa, Illinois, Missouri, the Dakotas, Nebraska, and Kansas. Operators were able to ship logs and lumber down the Mississippi to St. Louis, which became the distribution point for midwestern markets. In 1851, the federal government opened up new areas of central and western Minnesota to logging when it completed negotiations and signed the Treaty of Traverse des Sioux, purchasing twenty-four million acres of land from the Sisseton and Wahpeton bands of Lakota Indians. Nineteen million of those acres were in Minnesota, and the state started offering land for sale. Many lumber operators paid cash, but another way in which land passed into private hands was through military bounty land warrants. Through the warrants, the government granted land to soldiers who had served in the Mexican War. In 1849, for example, warrants were used to grant ten thousand acres of land. Minnesotans bristled about the warrants because they were controlled by the federal government rather than by local interests, yet the warrants provided a way to acquire forestlands without investing cash. Lumber companies purchased warrants from soldiers and then used them instead of cash to purchase forestlands. Thus, in 1850, warrants were used

Figure 2.1 *Logging in Minnesota. In Minnesota's north woods, loggers used horse log jammers to transport logs from the forest to rivers, where they were floated downstream to sawmills. Minnesota Historical Society.*

to purchase thirty-two thousand acres, whereas cash was used to buy only thirty-eight hundred acres.[25]

By 1857, the state had 150,037 people, an influx of population that created local markets for wood products wrought from the majestic white pine. The treaty with the Lakota and resulting population growth also spurred the growth of St. Anthony Falls, the first settlement in what eventually became Minneapolis. Because of its location on the Mississippi, St. Anthony became a magnet for the lumber industry. Not only did the town abound with sawmills, but it gave rise to dozens of factories that churned out a variety of wood products.[26]

Minnesota companies engaged in both clear-cutting and high-grading, or the removal of mature trees of high quality. According to Clifford and

Isabel Ahlgren, "Old timers recall that high-grading was usually limited to holdings adjoining clear-cut operations."[27] A premium was placed on speed and efficiency, a process aided by the skill of loggers who migrated from New England. One foreman boasted that his crew, which had been working near a tributary of the St. Croix, had logged eight thousand logs in 117 days, between two million and three million board feet of lumber.[28] These men used axes, but when the crosscut saw came in the 1870s, loggers harvested trees with even greater efficiency.

Minnesota's lumber industry achieved rates of growth that were nothing short of astonishing. At the St. Croix Boom, where logs were sorted and distributed, five million board feet were processed in 1840. By 1860, the facility was readying one hundred fifty million feet annually for market. In 1870, however, after only thirty years of insatiable logging, operators abandoned the area because the supply of trees was exhausted.[29] In 1860, the industry in Minnesota produced lumber worth $1,234,203. By 1870, revenues exploded to $4,299,162.[30]

The logging in all three states caused damage that was just as severe and extensive as in New England. Increase Lapham was a New Yorker who moved to Wisconsin in 1836 and was the first person in the state to conduct scientific studies of the forests and other natural systems. In 1855, he warned, "It is much to be regretted that the very superabundance of trees in our state should destroy, in some degree, our veneration for them. They are looked upon as cumberers of the land; and the question is not how they shall be preserved, but how they shall be destroyed."[31]

"Cut-and-Run" and Forest Fires

The majority of companies engaged in "cut-and-run" tactics and clear-cutting, leaving the landscape strewn with stumps and cluttered with slash, as much as two hundred tons per acre that were just waiting for the spark from a passing locomotive to ignite. Logging operators made little attempt to replant land that they had cut over. In fact, in many cases, they stopped paying taxes on lands after they were finished cutting, and the land reverted to the states. Without efforts at replanting, the magnificent forests of white pines and red pines had difficulty regenerating them-

Figure 2.2 *The Great Peshtigo Fire of 1871. This woodcut, published by* Harpers Weekly *on November 25, 1871, captured the devastation of the forest fire that destroyed Peshtigo, Wisconsin, and killed fifteen hundred people. Wisconsin Historical Society. (WHI–1784)*

selves, and sun-seeking species like aspens and birches dominated where the pines had once spread their emerald mantles.[32]

In 1871, immense fires blazed across the region and focused national attention on the destruction of the Lake states' forests. That summer had been plagued with drought, and in early October, a high-pressure system settled over the region while hot, strong winds swept up from the south. On October 7, the infamous Chicago Fire ignited and burned for four days, consuming 17,450 buildings ranging over 2,240 acres. During the same week, fires in northern Wisconsin obliterated the town of Peshtigo, causing the tragic deaths of fifteen hundred people and destroying 1,280,000 acres of forest[33] (figure 2.2).

The fires spread to Michigan, and on October 6, the *Detroit Free Press* reported, "The lurid sun and warm winds which have prevailed here for several days continue [and] fires in the woods keep up the smoky atmosphere which renders everything obscure."[34] Smoke from the fires severely reduced visibility for ships plying the waters of Lake Huron. Towns throughout the Lower Peninsula were completely vaporized, including Manistee, Glen Haven, and Holland on Lake Michigan. In Holland, fires burned the entire downtown area, destroying three hotels, five churches, and 210 homes. Forests also burned throughout the eastern part of the Lower Peninsula for several days, attacking the Saginaw valley. Arthur Hill, who later served on the Michigan Forestry Commission, recalled:

> Among the most vivid recollections of my early boyhood are those of certain days when smoke from the burning forests about Saginaw was so dense that children living in the outskirts lost their way in coming to and going from school. We boys played hide-and-seek during school recess and could stand in the open not more than 60 feet apart and yet not be recognizable.[35]

As long as clear-cutting and the buildup of slash continued, fires kept attacking the Lake states, blackening forests in the Upper Peninsula in 1877, 1883, 1885, 1891, 1893, and 1894. In 1896, flames threatened Onto-nagon, which sits on the edge of Lake Superior and was the home of the Diamond Match Company. Swamps south of town caught fire, and strong southerly winds blew the flames toward town. Even though the workers fought desperately to keep the fire from reaching the match factory, it finally did, igniting the matches and creating a blaze so powerful that winds blew burning boards over the town's river and against distant dwellings, pinning the boards against houses until the buildings caught fire. The town was destroyed, and in the surrounding forests, 228,000 acres had been blackened. Miraculously, only two people were killed.[36]

As had happened in the White Mountains, the situation showed the effect that heavy logging and fires had on every aspect of the forests' natural systems. For example, the forests had had a variety of large fauna—moose, bear, gray wolves, caribou, bison, elk, lynx, and martens—and the rivers and streams abounded with trout and numerous other species, but loss of habitat from logging, fires, and farming dramatically reduced populations by the end of the nineteenth century. By then, moose, bison, elk, caribou, gray wolves, and wolverines were virtually extinct from the Lake states.

Fish were equally affected. On cutover lands, heavy rains carried silt into rivers and streams, and sawmills dumped sawdust into waterways. In addition, as forest canopies disappeared, sunlight directly hit rivers and streams and warmed the water, sometimes by as much as 10 degrees Celsius. Fish that were native to a stream could no longer spawn, leading to a decline in population that was exacerbated by overfishing and industrial pollution spewed from the growing cities that lined the Great Lakes. The heavy logging also reduced the populations of bird species that thrived in heavy forests and encouraged those that preferred forest edges. In the late nineteenth century, the Audubon Society observed only one chestnut-sided warbler in this part of the country.[37]

Conversion of Cutover Lands to Agriculture

The depletion of the forests created an economic crisis for the Lake states, and as a result, state governments and agricultural colleges strongly pushed for the conversion of cutover lands to agriculture. William Henry of the College of Agriculture at the University of Wisconsin argued, "With farms supplanting the forest, northern Wisconsin will not revert to a wilderness with the passing of the lumber industry, but will be occupied by a thrifty class of farmers whose well directed, intelligent efforts bring substantial, satisfactory returns from field, flocks and herds."[38]

The University of Wisconsin's College of Agriculture issued *Northern Wisconsin: A Handbook for the Homeseeker*, which spelled out the advantages of farming. From the 1880s to 1900, hopeful farmers established twenty thousand new farms in the northern part of the state, and the population expanded from 120,000 to 400,000,[39] yet agriculture proved to be a difficult proposition that far north. Winters were long and cold, and rocks, boulders, and tree stumps littered the soil. Benton MacKaye, who would later conceive the idea of the Appalachian Trail, spent his early career in the U.S. Forest Service and traveled throughout northern Michigan, Wisconsin, and Minnesota in the early 1900s. He wrote, "The game was to sell these lands to the prospective settler, omitting to advertise the stumps. I reported to one of the outfits the large number I counted on one of his acres. His response was swift and pungent, 'For God's sake, don't tell anybody!'"[40]

The growing season was short, and the sandy soils of northern Wisconsin and Michigan proved to be not much good at growing anything except the splendid white pines that had once dominated the landscape. Consequently, in many of Wisconsin's northern counties, the transition to agriculture was a disaster. By 1928, 25 percent of the land in the seventeen northernmost counties was tax delinquent, and state governments could not give it away.[41] Even where the states attempted to regenerate the white pine forests, disaster ensued. In the early twentieth century, foresters started to replant white pines, but the demand exceeded the supply, forcing nurseries to import white pine seedlings from Europe. Those imported seedlings carried white pine blister rust, a disease that proved deadly to the white pines in the United States in an early example of the problems caused by invasive species. White pine blister rust spread throughout the Lake states over the next twenty years.[42]

Scientific Studies of the Forests

Thus, by the early 1900s, Michigan and Wisconsin faced depletion of their wondrous northern forests, and Minnesota was well on its way toward the same fate. At the same time, the new science of forestry was beginning to have an effect on the governments of all three states by encouraging the collection of data and urging changes in timber-harvesting practices to provide timber on a sustainable basis. All three states created forestry commissions, which performed a critical function by disseminating information. Michigan's Forestry Commission, for example, estimated that the state had originally contained 380 billion board feet of timber, of which 244 billion board feet had been logged, 35 billion feet had been cut to clear land for settlement, and 73 billion feet had been burned or wasted. All that remained in 1926 was 27.5 billion board feet. Of that total, 75 percent was in the Upper Peninsula.[43]

The forestry commissions also provided a platform for conservationists to educate political leaders and the public about forestry. After Wisconsin established its Forestry Commission in 1897, Filibert Roth emerged as the most knowledgeable and persuasive conservation figure, documenting the assault on Wisconsin's forests in *Forestry Condition and Interests of*

Wisconsin, published in 1898. The study also included an introduction by Bernhard E. Fernow, the chief of the U.S. Bureau of Forestry. Fernow was a German-trained forester who had migrated to the United States in 1876 and introduced the principles of forestry to his adopted country.

Roth's report emphasized the interconnectedness of the different parts of the forest, linking logging and fires to soil erosion and declines in water quality. His survey covered Wisconsin's northern twenty-seven counties, which had nineteen million acres of land. Logging interests had worked 50 percent of the forests in these counties, or 25 percent of the entire state.[44] Before logging started, the state had contained one hundred thirty billion board feet of pine timber, but over forty years, all but seventeen billion board feet had been cut. The state was still cutting two billion board feet every year, but forests were returning at the rate of only two hundred million board feet a year.[45]

Roth also depicted the destruction caused by forest fires. "Nearly half this territory," he wrote, "has been burned over at least once. Forest is completely gone from 3 million acres. Several million more acres are but partly covered by the dead and dying remnants of the former forest." Stands of hemlock and hardwood that were still standing were badly damaged, and in the pineries, "repeated fires have largely cleared the lands of all the slashings." There were huge parcels of "stump prairies" that were littered with decaying stumps and overrun with weeds, grasses, and bushes.[46]

In addition, Roth detailed the erosion, which was causing silt to build up in the region's rivers and streams, just as had happened in New England. He noted "striking changes in the drainage conditions" of rivers and streams, changes that were "all too intimately connected with the changes in the surface cover to leave in doubt the influence of this latter on the former."[47] The water flow on all rivers was reduced, making it much more difficult to float logs downstream, and rivers like the Fox in southern Wisconsin provided far less power to factories than they once had.

Minnesota formed its State Forestry Association in 1876 for the purposes of encouraging the planting of trees on the prairie and to protect at least some of the forestlands as a source of lumber for the future. The organization's first action was to urge the state of Minnesota to observe

Arbor Day, an idea that the state of Nebraska had pioneered in 1872. The association wanted to encourage Minnesota's farmers to plant trees to provide wood for the future.[48] The association, however, did not emphasize the regulation of timber harvesting or the creation of publicly owned forests. The men who were involved early in the forestry association—men like John Pillsbury, a cofounder of the Pillsbury Company—reflected the traditional attitude that the country was best served by allowing private individuals and companies to develop the nation's resources. They considered government regulation or ownership of the land to be a violation of private property rights.

The facts on the ground, however, were beginning to weaken these ideological assumptions about private property. In Minnesota, very little pine was still standing, even in the northernmost lake country of the state, by the 1920s, and sawmills were going out of business by the dozen. In the mid-1800s, before logging started in earnest, there were 334,080 acres of pine forests in the lake country bordering Minnesota and Canada. By the end of the logging era in the 1930s, only 26,560 acres of pine forest were left in that region. Moreover, surveys in Minnesota found that the forests did not have younger trees, indicating that they were not regenerating themselves.[49]

Also beginning to undermine the traditional assumptions about the development of natural resources was the human effect of deforestation, which was apparent in all three Lake states. Towns in the upper Midwest had grown on the shoulders of the lumber industry, but when areas were logged out, the effects were economically devastating. Muskegon, Michigan, was typical. When the timber boom ended in the early twentieth century, the owners of timber companies simply up and left, going on to northern Minnesota, the southern Appalachians, or the Northwest. The loggers themselves were left without work, however, and they often had to migrate to where their skills would be valued.

From the ashes and waste of devastated forests, though, the forest conservation movement of the Lake states region was born. The Michigan Forestry Commission's report on its initial meeting in 1888 thoroughly documented the devastation of the state's pineries but then added a hopeful note:

> It is an admitted fact that where one-fourth to one-third of the total area of timber is allowed to grow, we can raise more products of all kinds than where the country is denuded. . . . A ton of coal dug from the earth and consumed is a lost treasure, but not so in forest culture, for we can produce the same.[50]

In 1876, Gifford Pinchot, the first chief forester of the U.S. Forest Service and one of the country's most influential early conservationists, attended the Centennial Exhibition in Philadelphia with his father. In his autobiography, *Breaking New Ground*, Pinchot recalled, "It contained no forest exhibit of any sort or kind except from the single State of Michigan."[51] By 1900, that would change, as conservationists and foresters alike would awaken the American public to the devastation of the nation's forests and to the idea that those forests could be saved and restored. Before that could happen, however, the assault on the forests that had begun in New England and continued in the Lake states would now bring its full fury on the southern Appalachians.

CHAPTER 3

A Forest Crisis in the Southern Appalachians

The Great Smoky Mountains of eastern Tennessee and western North Carolina nurtured the Woodruff family with their undulating mountains, unbroken forests, and sparkling streams. Dorie Woodruff vividly recalled her childhood in those mountains in her beautiful memoir, *Dorie: Woman of the Mountains*, which her daughter Florence Cope Bush published in 1992. Dorie recalled that her parents' first home, in the 1890s, was a log cabin near the Oconaluftee River, in the Smokies. For their livelihoods, they drew on the bountiful resources of the surrounding forests. From maple trees, her father fashioned tables and chairs, and her mother wove the seats of the chairs from strips of white oak. They used gourds to dip and carry water, grew vegetables in their garden, and transformed herbs from the surrounding forests into homemade medicines. They lived the self-sufficient lives that mountaineers had followed since the Scots-Irish and other ethnic groups had settled in the southern Appalachians during the colonial era.

During the 1890s, however, wrenching changes began to transform the southern Appalachians and the lives of thousands of mountaineer families like Dorie's. As the immense forests of New England and the Lake states dwindled from heavy logging, lumber and paper companies looked to the untapped forest resources of the South and began to build railroads into the southern forest interiors. At the same time, textile companies, drawn by the promise of low wages and a nonunionized workforce, began relocating from New England to the South.

Gradually, the Woodruff family found itself caught up in enormous economic and social changes. By the late 1890s, Dorie's father was working in a local sawmill even as he and her mother maintained their small farm, but life was difficult. He did not make much money from the sawmill, their food supply dwindled, and they waited anxiously throughout the spring for the garden fruits and vegetables to come in. Frustrated by their poverty, he took a job at a cotton mill in Spartanburg, South Carolina, in 1906, and the entire family moved, embarking on an odyssey that would take them to many different locations over the next several years. In Spartanburg, Dorie's father hated the mill. According to Dorie, "Many of the mountaineers who thought the cotton mills were the pot of gold at the end of the rainbow soon found they didn't like what was in the pot. The air was bad, the water tasted funny and they didn't like being subjected to the whims of the overseers. The things they valued most, freedom and independence, were gone."[1]

The family returned to farming for a time but found it difficult to forge a living, so Dorie's father decided to move the family once again and take logging work with the Little River Lumber Company, which had set up operations in the Great Smokies at a site southwest of Gatlinburg. She recalled, "Most of the valuable hardwood was gone from the lower regions of our mountains. Cherry, ash, and walnut were being cut at higher elevations. It was necessary to build the camps [in] . . . the heart of the Smokies."[2] Colonel W. B. Townsend, the owner of the company, decided to build a railroad into the mountains, a railroad that needed trestles to cross the deep valleys that dotted the region. It was tricky, difficult work, but Dorie's father excelled at it because of his skill in working with wood. Townsend also set up semipermanent lumber camps in which he wanted families because camps populated only by young men could get pretty rowdy. Dorie's parents agreed that the opportunity to make money was too good to pass up, and in 1912, they uprooted the family again and moved to Fish Camp, on the edge of the wilderness. They lived in a boardinghouse, with the family quarters in one end and boarders at the other end. Dorie's mother was the caretaker of the boardinghouse, and Dorie helped her cook, clean, and do laundry for the lumbermen.

That same year, the Cope family had also taken work in the area, with another lumber company. Eventually their son, Fred, and Dorie began to court and were married. They remained in the Smokies, though, and Fred worked on mechanical skidders that pulled logs to waiting railroad cars. Soon after, Dorie gave birth to their first child. They lived a peripatetic existence as Fred moved from one logging site to another in the Smokies. In 1917, they reunited with Dorie's family at Higdon's Camp, just south of Elkmont, Tennessee.

That year, one of the most traumatic events of their lives occurred as a forest fire threatened the camp. Dorie believed that the fire started when a spark flew off from a skidder's tinder box, igniting the debris of dry leaves and limbs that were strewn on the ground after logging. Fred was caught behind the fire as it traveled fast up the mountain. He said a short prayer, looked up, and saw that a small hole had opened up as if by miracle. He raced through the opening and escaped the swirling flames. According to Dorie, "Trees were exploding, sending fiery debris sailing through the air into the virgin timber. Smoke filled the sky with a gray-black ceiling. It rolled into camp in waves."[3]

Fred raced back to the camp and helped other families evacuate, but in the face of the onrushing flames, Dorie's parents stayed to save as many belongings as possible. Quickly they formed a bucket brigade, passed buckets along, and poured water on the roof's tar paper and the sides of the cabin. "Smoke covered the sun," Dorie remembered. "The only light was an eerie, orange glow from the flames."[4] Through the night, they continued to pour water on the roof and sides of the boardinghouse, while their little baby, Wilma, slept intermittently. Finally, at dawn, the wind changed direction, blowing the fire back toward the mountain and away from the camp. A thunderstorm finally doused the fire, but the forest was completely devastated:

Everywhere we looked was desolate. Charred snags stood where tall, green trees grew yesterday. The forest floor was bare and ash-covered. Ferns, wildflowers, and velvety moss were no more. There wasn't a bird to be seen or heard. We'd saved our home, but now it stood in a barren wasteland. I'd never seen such ugliness where beauty had reigned.[5]

Terrified by the fire, Dorie's parents decided to leave the logging camp and return once again to farming, and her father used what remained of the family's savings to buy a farm north of Gatlinburg. Dorie and Fred, though, decided to stay in the camp and persevere with the logging life.

Dorie's story vividly captures the transformation of the southern Appalachians from the 1880s to the 1920s as logging companies harvested vast stands of virgin timber, with effects on the environment that echoed those of the White Mountains and Lake states: widespread fires, erosion, and flooding. In addition, though, logging and other industries like textile manufacturing radically changed the lives of people who had lived in the agrarian society of the southern Appalachians for generations. Dorie's narrative captures the enormous human dislocation that occurred but also the deep love that the people of Appalachia had for the natural beauty of the magnificent mountains and forests that they called home.

The Southern Timber Frenzy

The South was indeed blessed with extraordinary forests, predominantly hardwood forests, that carpeted the Appalachian ranges bumping like a rugged spine from West Virginia south through Georgia. Prominent ranges within the southern Appalachians include the Allegheny Mountains in Virginia and West Virginia; the Blue Ridge Mountains, which ripple across Virginia, western North Carolina, northwestern South Carolina, and northeastern Georgia; the Cumberland Mountains, in eastern Kentucky and parts of West Virginia, Virginia, Tennessee, and Alabama; and the Unaka Mountains, found in eastern Tennessee, western North Carolina, and southwestern Virginia. The Great Smoky Mountains are part of the Unaka Range and feature the greatest mass of mountains in the southern Appalachians, and in 1900, they boasted the most extensive forests.

Horace Kephart, a librarian in St. Louis who followed his love of the outdoors to the Great Smokies, penned one of the most revered books about the mountains, *Our Southern Highlanders* (1922). In it he captured the singular beauty of the southern Appalachians:

> Pinnacles or serrated ridges are rare. There are few commanding peaks. From almost any summit in Carolina one looks out upon a sea of flowing curves and dome-shaped eminences undulating, with no great disparity of height, unto the horizon. Almost everywhere the contours are similar: steep sides gradually rounding to the tops, smooth-surfaced to the eye because of the endless verdure. Every ridge is separated from its sisters by deep and narrow ravines.[6]

In 1890, those forests included yellow poplar, American chestnut, cherry, ash, white oak, and red oak, although in some areas, pine and spruce were abundant. Because of the nutrient-rich soils, warm climates, and ample rainfall, trees grew to enormous proportions. Poplars could be twelve feet in diameter, and American chestnuts could reach diameters of more than eight feet (figure 3.1). In the Smokies, the Baxter Cabin, built in 1889 and located today about fifty feet off the Maddron Bald Trail, is reputed to have been built from the wood of a single enormous American chestnut tree.[7]

Until the 1870s, logging in the southern Appalachians was essentially carried out by the local population. Any self-respecting farmer could wield an axe with skill and use it to harvest timber for houses, fences, ladders, chairs, tables, anything of wood that the household demanded. The logging was strictly small scale, and the locals engaged in selective cutting so that the forests regenerated themselves. Given the relatively small scale of the logging operations, the cutting had little effect on the forests of the southern Appalachians in the first two decades after the Civil War. The entire region had a lower output of lumber and fewer sawmills than the single state of New York. The timber resources were not developed for a number of reasons, including the agricultural basis of the southern economy, the lack of capital in the South, the remoteness of many of the forests, and the mountainous terrain, which made it difficult to bring trees to market (figure 3.2).

Figure 3.1 *American chestnut, ca. 1920. The Jim and Caroline Walker Shelton family stood before an enormous American chestnut in the Great Smoky Mountains. Chestnuts were virtually wiped out by the chestnut blight. Archives, Great Smoky Mountains National Park.*

Starting around 1880, however, the southern Appalachians experienced what historian Ronald C. Eller has referred to as "one of the most frenzied timber booms in American history," a boom that lasted until around 1920 and left the southern forests as denuded and ecologically damaged as the forests of New England and the Lake states. As in the other two regions, the lumber barons cut everything and left behind thousands of square miles of stripped land. In 1880, for instance, 67 percent of West Virginia was covered with old-growth forests, and the Cumberland Plateau, the Blue Ridge Mountains, and the Smokies boasted seemingly endless tracts of virgin timber. Forty years later, those forests were completely cut over.[8]

Three factors differentiated the logging industry in the South from that in New England and the Lake states, however: (1) the preponderance of hardwood forests, (2) the dependence on investment dollars from outside the region, and (3) an agrarian society of small farmers who were not well integrated into the U.S. market economy. These differences explain how deforestation occurred in the South, but they also help explain the unique way in which a conservation constituency developed in the region.

The predominance of hardwoods delayed interest in logging in the region. Until the late 1800s, softwoods, particularly pine and spruce, were much preferred to hardwoods because they were easier to fell, transport, and cut into lumber, and they were ideal for construction, which drove much of the demand for forest products. Hardwoods were more difficult to cut down and process into lumber, so the costs were higher to bring them to market. The mountain terrain only made the process of bringing logs to market more difficult, and until the late nineteenth century, the return on hardwoods did not justify the expense. Around 1900, though, hardwoods came into greater demand for furniture and started to fetch higher prices, justifying the extra costs of cutting them down and transporting them.[9]

The second factor, investment in the South by outside capital, became evident in the early 1880s, when, for two reasons, investors from outside the region began to pour money into logging operations in the South. First, southern political leaders realized after the Civil War that the region had to industrialize if it were to compete economically with the rest of the

Figure 3.2 *Logging operation in southwestern Virginia. Loggers used primitive carts drawn by horses, cattle, or oxen to transport logs through the hilly terrain of the southern Appalachians. Photograph courtesy of Museum of Appalachia, Norris, Tennessee.*

nation, yet the region had a shortage of investment capital. Consequently, political leaders aggressively pursued investment from northern entrepreneurs and Europe. Second, as the forests of the Northeast and Great Lakes were clear-cut, fears of a "timber famine" sent northern lumber operators packing to the South in search of virgin timberlands. Outside capital also seeded the development of other industries in the South, including furniture (because of the extensive stands of hardwoods in the region), coal mining, and textile manufacturing. Gradually, the region shifted from agriculture to industry, but the transition was painfully slow.

As logging companies first came into the region, they negotiated with farmer-landowners to purchase the rights to trees on the farmers' property. The Southern Homestead Act of 1866 had made the process of land acquisition difficult for companies because it gave preference for buying property to freedmen and other locals without land, but lumber companies found ways around the law. Dorie Woodruff Cope recalled that in West Virginia, the Little River Lumber Company had employees claim forestland and live on it to establish possession, allowing the company legally to cut logs on the land.[10]

After the end of Reconstruction in 1877, southern representatives to Congress pushed for private ownership of the forestlands. Such ownership, they theorized, would lead to better protection against fire and decrease tree theft, which had become a significant problem. Congress therefore repealed the Southern Homestead Act, opening the way for northern businesses to purchase timberlands. The repeal of the law, though, led to widespread land speculation in the South as five-and-a-half million acres of publicly held lands were put up for sale. Land speculators could make sizable profits. One group of investors in West Virginia bought thirty-three thousand acres of forestland in 1906 for $245,000. Three years later, they sold the same land for $630,000.[11]

In 1889, southern Congressmen, alarmed by the speculation, pressed once again for ownership restrictions that would favor farmers and other small landholders. By that time, however, most of the land had been transferred to corporate ownership. The change in ownership patterns in the South—and throughout the nation—was striking. In 1870, the public owned 75 percent of the timber in the United States. By 1911, though, 80 percent of the timber had gone into private hands.[12]

The third factor differentiating the southern Appalachians was the agrarian economy, which was underserved by communication and trade networks. Farmers in the southern Appalachians had difficulty bringing their crops to market because of a lack of roads, bridges, and other infrastructure. The timber industry spurred the building of roads and railroads and was a catalyst in the long transition to a market-based, industrial economy in the South.[13]

The growth of the logging industry in the southern Appalachians started slowly, around 1880, and then accelerated throughout that decade and the 1890s. Around 1885, Alexander A. Arthur created the Scottish Carolina Land and Lumber Company and, using investment dollars from Cape Town, South Africa, and Glasgow, Scotland, purchased 60,000 acres of hardwoods in eastern Tennessee and another 120,000 acres of forestland in the Blue Ridge Mountains. The Unaka Timber Company also purchased forestlands in the Blue Ridge Mountains. From the 1880s through 1920, six different companies conducted logging operations in the Great Smokies, starting with the Scottish Carolina Timber Company.[14]

Northern lumber companies sent representatives through the southern forests to identify high-quality hardwoods, including ash, black walnut, and yellow poplar, and hired local farmers to do the cutting, for which a farmer typically received $2 for a tree that was four feet in diameter. Sawmills sprung up throughout the forests, many of them small. West Virginia, for instance, had fifty thousand mills in 1909, but 75 percent of them processed fewer than one million board feet a year, and only 25 percent cut more than one hundred million board feet. Often, the sawmills sat idle, waiting for logs to arrive.[15]

As a consequence, logging companies sought ways to bring logs to mills more regularly. One innovation was the creation of barriers called splash dams made of earth, which caught the water running off during the spring and created small lakes. Loggers collected the logs in these temporary lakes and then blasted the dams open—often with dynamite—unleashing a wave of water to wash the logs downstream to larger rivers.[16]

When the logs reached the larger river, workers collected them into one huge group and drove them downstream. The splash dams were one way to drive logs to mills before the coming of the railroads, but the rudimentary technology caused extensive damage to the forest environment. Logs were unruly when propelled forward by a burst of water, and they tore at the banks of creeks while hurtling downstream.[17]

The pace of cutting in the South accelerated in the late 1890s, fueled not only by the growing demand for hardwoods and the depletion of northern forests but by the demand for pulp. The entire print run of the Sunday edition of a major-city newspaper, for example, consumed twelve acres of forestlands containing spruce trees. Companies purchased their own tracts of forestland rather than buy timber from local farmers and then built sawmills, hired logging crews, and established logging camps, creating a vertically integrated operation. Timber production soared, reaching its apex in 1909 when four billion board feet were felled and processed. In 1900, for example, 30 percent of the hardwoods in the United States came from the southern Appalachians. Ten years later, the region produced 40 percent of the nation's hardwood lumber.[18]

The phenomenal growth would never have been possible without the railroads, for in a region as mountainous as the southern Appalachians, shipping the logs efficiently to market was essential. In the mid-1880s, companies began to build railroads that were capable of negotiating the mountainous terrain. The railroads and lumber operations were highly interdependent, with stockholders often holding stock in both types of companies. The railroads needed large amounts of logs to justify the huge expense of building tracks and purchasing rolling stock.[19]

Asheville, North Carolina, reflected many of the changes wrought by the logging railroads. Before 1880, Asheville had been a sleepy town with no industry to speak of, but after twenty-five difficult years of construction, the Western North Carolina Railroad was finally completed and began serving the town on October 3, 1880. Local resident Dudley W. Crawford recalled that the railroad allowed Asheville to "get into step with other sections of the country that were going forward in the march of progress."[20] The railroad later expanded northwest into eastern Tennessee and southwest into northern Georgia. With those lines, logging companies could now start serious cutting in western North Carolina and northern Georgia. Soon, another line connected Asheville to Spartanburg, South Carolina. The construction of the railroad also brought another consequence, however: the rapid growth of tourism in Asheville as people searching for cures to their health afflictions sought out the pristine air of the mountains.[21]

An Accelerated Pace of Logging

In return for the enormous investments that logging and railroad companies made, they demanded a high rate of return. Most operators turned to clear-cutting, following the "cut-and-run" approach that lumber operators had used to such devastating effect in New England and the Lake states. In many cases, the companies that had applied this approach in the North had moved on to the South, where thousands of square miles of virgin timber awaited the lumbermen's axe. In his book *American Forestry*, forest historian William G. Robbins explained, "Easy access to capital and the continued opening of cheap sources of virgin timber well into the twentieth century brought new mills into production, encouraged waste, and acted as a deterrent to forest conservation."[22]

The consequences of this approach were just as destructive in the South as they had been in the North. In 1910, Reverend A. E. Brown, who was director of the Southern Baptist Convention's mountain school department and became one of the South's leading conservationists, criticized the companies in an interview:

> Unfortunately, the men who owned timberlands did not seem to realize they had any other value beyond what they could get for them from the lumbermen. . . . No thought was given to the effect which the cutting of the timber may have on the mountain regions or looking to reforesting the area.[23]

After 1900, companies expanded their operations in the mountain regions of Georgia, North Carolina, Tennessee, and the Great Smokies. In West Virginia, the most intense logging occurred north of the New River, but there were also operations in the southern part of the state. In south-western Virginia, logging followed the same arc, increasing throughout the 1890s and reaching its pinnacle in the first decade of the twentieth century. In eastern Kentucky, lumber interests often hooked up with coal interests. For example, the Stearns Coal and Lumber Company, which was based in Ludington, Michigan, snapped up 113,000 acres in eastern Kentucky and established logging and mining operations there. In Virginia, lumber operators supplied mining companies with wood for support beams, railroad ties, and buildings in company towns.[24]

Mechanization speeded up the logging process. Logging operators used steam skidders, in which they wrapped cables around the logs and then pulled them quickly along the ground to waiting railroad cars. The skidders were powerful, yanking trees that were twenty-five to thirty feet in circumference. The skidding, though, dug deep ruts into the ground and tore out any seedlings that might have survived after cutting.

Another piece of machinery was the Clyde overhead skidder, which could pull logs up to three thousand feet through the air, improving the difficult process of transporting logs from mountainsides and allowing operators to harvest timber at ever higher elevations on the mountains. After 1900, steam log loaders became common. A crane fixed to a railroad car extended cables that were wrapped around the log. The crane then lifted the log into place on a flatcar, greatly speeding up the loading process. Records showed that by using a steam loader, loggers could load as many as twenty-one cars in one day[25] (figure 3.3).

If one man came to embody logging in the southern Appalachians, it was William McLellan Ritter, "the dean of the Hardwood Lumbermen of America."[26] In many ways, he was the J. E. Henry of the South. Ritter had grown up on a farm in Hughesville, Pennsylvania, and said about his childhood that he "had to *work*—there is no doubt about *that*."[27] The farm he grew up on had a sawmill, and at an early age, he decided to go into that business.

Ritter and a partner raised money in New York and founded the Denman and Ritter Lumber Company, of which Ritter was one-third partner. In 1890, Ritter moved to West Virginia and lived with a local family near

Figure 3.3 *Champion Fibre skidder in North Carolina, ca. 1918. Lumber companies like Champion used steam-powered cables to pull logs though the forest to waiting trains. Mechanization increased logging efficiency but damaged the forest floor. Archives, Great Smoky Mountains National Park.*

a one-thousand-acre tract, where he rolled up his sleeves and helped cut down trees and load the logs. The venture was a success, and he aggressively grew the size of operations, building sawmills throughout the state. Next he moved into western North Carolina, purchasing two hundred thousand acres during the period when the state's timber industry was beginning to take off. Seeking total control, he founded the William M. Ritter Lumber Company, which eventually became the largest lumber company in the southern Appalachians, controlling forestlands in Tennessee, Kentucky, and Virginia. By 1913, Ritter's various companies owned lands that contained two billion board of hardwood timber.[28]

Ritter was the classic paternalistic boss. In an interview, one employee, Clarence O. Vance, said that the employees of the company felt as if they were part of a family.[29] In the company town of Proctor, on Hazel Creek,

West Virginia, Ritter built churches, schools, and community centers and provided a company doctor for whose services the company charged $1 a month to single men and $1.50 to families. The train brought mail, and employees purchased their food at the company store, paying higher prices, of course, for the convenience. The company showed silent movies, and at Christmas, executives delivered candy and toys to the children. Vance admitted, "We lived pretty good."[30] Ritter also published a monthly magazine, *The Hardwood Bark*, which included information about the births, marriages, and deaths in the families of employees. The magazine created a sense of community and aimed to build morale and pride. At the same time, however, workers earned only twelve cents an hour, a not-very-princely sum even for those preinflation days.

All these perquisites should not disguise that the work was hard and dangerous. One employee, Charles F. Patton, was loading ties and suffered an injury because the sawmill spewed the ties out faster than he could pile them up. N. A. Green lost a hand when a planing machine did not work properly. E. G. Tolley's clothes became enmeshed in the gears of a moving locomotive. They all sued Ritter's company, and they all collected. Other companies faced similar judgments because of negligence. Sometimes revenge went beyond the courtroom. In 1915, an unknown worker planted dynamite under a clubhouse and a salesman's house at the Norwood Lumber Company, blowing up a good portion of the facility.[31]

In 1906, Ritter's company even faced charges of debt peonage when prosecutors accused the company of forcing employees to continue working against their will to pay off debts. On July 12, 1907, trials started against Ritter, who pleaded guilty. Judge Alston G. Dayton was extraordinarily lenient, however, giving the company a slap on the wrist with the minimal fine of $1,000.[32]

The largest company in the Smokies and the Blue Ridge Mountains was the Champion Fibre Company. The founder, Peter G. Thompson, wanted pulp for his paper mill in Hamilton, Ohio. He bought three hundred thousand acres of chestnut, spruce, and balsam trees in North Carolina. In 1911, he founded the Champion Lumber Company and promptly bought one hundred thousand more acres, including lands in eastern Tennessee.

The company, which employed more than seven thousand people, consumed 300 to 350 cords of wood every day to manufacture two hundred tons of wood pulp, which it then shipped to the paper mill in Ohio. Because Champion Lumber had invested so much in the southern Appalachians, it actually began to practice sustainable forestry, planting seedlings on lands that had been clear-cut. Thompson knew—as others would learn—that he had to plan for the future to ensure the long-term viability of his company.

Champion's sustainable practice was decidedly in the minority, however, for "cut and run" was by far the prevailing approach, with devastating environmental consequences for the southern Appalachians. One of the pioneering conservationists in the South, Emory Wriston of West Virginia, noted, "If trees could talk and cuss, West Virginia would be a poor place for a preacher to go on a picnic."[33] As in the White Mountains and the Lake states, the devastation that followed in the wake of heavy logging was incredible. Tons of slash smothered the ground where forests had once proudly stood, tree stumps dotted the land, and silt-laden streams could not support fish populations. Game was largely depleted, victimized by loss of habitat and overhunting.

The Wilson Report

By the end of the 1800s, it was evident that data about the harm being done to the southern forests needed to be gathered and disseminated, but southern states lacked the resources to develop forestry agencies that could take the lead in fact finding. Consequently, southern conservationists began to turn to the federal government. In 1900, the Appalachian National Park Association, which a group of Asheville-based conservationists had formed in 1899, joined with New England's Appalachian Mountain Club to present to the United States Congress a memorial recommending the creation of forest reserves in the southern Appalachians. Congress responded by appropriating $5,000 for the federal government to "investigate the forest conditions in the Southern Appalachian Mountain Region of western North Carolina and adjacent States."[34] The report was issued on January 3, 1901, with the title *A Report of the Secretary of Agriculture in Relation to the Forests, Rivers, and Mountains of the Southern Appalachian Region*, although it was more popularly known as the Wilson Report after James Wilson, secretary of the U.S. Department of Agriculture, who guided it to completion and wrote sections himself. Few documents in U.S. conservation history have had as great effect as this one, for it marshaled facts and arguments that led ultimately to the passage of the Weeks Act in 1911.

The report was extraordinary in its thorough examination of the ecological damage caused by deforestation in the South. It described the forests as integrated natural systems, emphasizing the interdependence of vegetation, wildlife, soils, and waterways. In addition to adding to knowledge of forest ecosystems, the report helped build a constituency for forest conservation in the South by documenting the damage to the forests and building the case for the creation of forest reserves. In an appendix, federal foresters Horace B. Ayres and William W. Ashe added greater detail about the damage being done to the southern forests.

Early in the report, Wilson wrote, "I have myself twice visited this region, and have seen at first hand the destruction of the forests and the consequent enormous damage by floods."[35] Wilson, Ayres, and Ashe examined 5.4 million acres land, of which a little more than 4 million were still covered by forest, but only 303,000 acres, or 7.4 percent of the forestlands, still contained virgin timber.[36] Wilson warned that "within less than a decade every mountain cove will have been invaded and robbed of its finest timber, and the last of the remnants of these grand primeval Appalachian forests will have been destroyed."[37] One of the arguments threading through the report was that by clear-cutting, logging companies endangered the future of their own industry. Ayres and Ashe wrote, "The damages [the lumbermen] cause come not so much from the trees [the logger] cuts in culling the forest as from the additional trees and seedlings of valuable species which he destroys in his lumbering operations."[38]

The Wilson Report's authors found evidence that fires had burned more 4,500,000 acres. As Horace Ayres and William Ashe traveled through the region, they saw over and over the detritus of "dead trees, scorched butts, hollow trees, dead saplings and seedlings."[39] According to one estimate, fires started by locomotives caused $50 million in damages every year.[40] Another study claimed that locomotives caused 71 percent of the fires, and campers and sawmill employees were responsible for another 20 percent.[41]

Clear-cutting and forest fires created ugly gullies of erosion, reinforcing the theme of the interdependence of different natural systems in a forest. As Wilson explained, "The soil, once denuded of its forests and swept by torrential rains, rapidly loses first its humus [and] is washed away in enormous volume into the streams, to bury such of the fertile lowlands as are not eroded by the floods, to obstruct the rivers, and to fill up the harbors on the coast."[42] According to one estimate, erosion had carried off six million cubic miles of topsoil in the Southeast, soil that could have covered the entire country of Belgium.[43]

Exacerbating the problem of erosion in the southern Appalachians was that farmers cultivated land and grazed stock on mountainsides, resulting in the loss of grasses and other vegetation that kept soil in place. Farmers had originally cultivated the alluvial bottoms of valleys, but because of population pressures and soil exhaustion, they gradually moved up the sides of mountains to find arable land. They even cultivated fields on 30 percent to 40 percent grades, land that they had to plant by hand because animals could not pull plows on such steep inclines. The mountainsides had thick layers of humus and rich soil. "But," Ayres and Ashe wrote, "on cultivation and exposure to the sun and washing rains this organic matter is rapidly dissipated. In this process most of the soil is washed away."[44]

Mechanized technology only worsened the erosion. When loggers used steam-powered skidders to haul logs to waiting railroad cars, the logs dug deep trenches in the ground and destroyed seedlings and saplings. The resulting erosion had serious consequences for the region's waterways. The flow of water into rivers and streams became irregular, with reduced water flow during droughts and severe flooding during rainy seasons. The drying up of streambeds could have serious economic effects. In West Virginia, for example, eight sawmills on one creek had to cease operations because the streams were nearly dried up.

Floods had even more calamitous effects, which were national in scope because several major waterways had their sources in the southern Appalachians. The Ohio, the Monongahela, the Big Sandy, the Great Kanawha, and the Little Kanawha Rivers collectively carried more than 20 percent of the commerce on waterways in the United states in 1911, and the Monongahela was the most heavily trafficked river in the entire Western Hemisphere. West Virginia alone had 748.5 miles of waterways that were used for navigation.[45] In May 1900, flooding caused by storms in the Blue Ridge Mountains swept away two hundred miles of farmland and caused damages of $1.5 million. The same series of storms caused extensive damage along the Yadkin River in North Carolina, the New River in Virginia and West Virginia, and the upper Tennessee River. All told, flooding in the region caused $10 million in damages.[46]

The damage was human as well. As the logging, coal, textile, furniture, leather, and other industries pulled the South into the industrialized, market-based economy, far-reaching cultural changes affected the families of thousands of mountaineers. The odyssey of Dorie Woodruff Cope's family reflected that of thousands of other mountaineer families. Her husband, Fred, continued logging, but one day, he was severely injured as he unhooked logs from a skidder. Several logs broke loose, slid, and rolled onto his legs, causing extensive muscle damage. During his recovery, he decided that he had had enough of logging and began studying to become an electrician. When he recovered, he landed a job building electric lines to Gatlinburg and abandoned his efforts to make a living from the land.[47]

Dorie's family had been inextricably linked to the forests, but slowly, these links were dissolving as the forests disappeared. Southerners who were similarly affected by the devastation of the region's forest would begin to form a constituency to protect the forests of the region from further destruction. An important foundation of this movement was the scientific understanding of forests that was conveyed in the Wilson Report, which detailed the extensive effect that deforestation had on entire watersheds by carrying away soils, causing floods, and reducing wildlife habitat. Another motivating factor for this emerging conservation constituency in the South was the desire to preserve the agrarian traditions that had defined the region for generations. Southern conservationists sought not only to preserve their forests but to sustain their rural culture, which was so intertwined with the forest. Cultural preservation remains an important theme of southern conservation today, as chapter 14 will explain.

As the southern forests faced the same crisis that had spread across New England and the Lake states in the early years of the twentieth century, the status of those forests was about to enter a new stage. The effects of rapid deforestation were clear, and the benefits of forest restoration and sustainable forestry were becoming more evident. The fight to stop the devastating deforestation and restore the eastern forests was just beginning, however. It would prove to be a long, difficult, and dramatic struggle.

CHAPTER 4

Building a Forest Conservation Movement

On September 4, 1902, Reverend Edward Everett Hale strode to the front of an assembly of citizens gathered in Intervale, a picturesque village located in the White Mountains just north of the teeming tourist town of North Conway, New Hampshire. Hale was a national figure of liberal conscience: chaplain of the U.S. Senate, long-time minister of the South Congregational Church in Boston, and author of the short story "The Man without a Country," which he had written in 1863 as an allegory to promote patriotism in the United States.

For years, Hale had summered in the White Mountains, and, like thousands of other mountain lovers, he had been appalled by the heavy logging and massive fires that had annihilated thousands of acres of once-magnificent forestland. The only redeeming aspect was that, so far, the Presidential Range, which included Mounts Washington, Adams, Jefferson, Madison, Monroe, and Clay, had been spared the axe.

That was about to change. The Berlin Mills Company of Berlin, New Hampshire, had announced the purchase of a large stand of high-quality virgin timber in the Presidential Range, and local residents reacted in horror as they visualized the lumber operators scalping the sides of these extraordinary mountains. No one was in a better position than Hale to lead a protest against the heavy logging. Not only was he a renowned minister and author, but he had been the president of the Appalachian Mountain

Club, which a group of mountaineers and conservationists had founded in Boston in 1876. In preparation for the meeting at Intervale, he invited John Hay, secretary of the U.S. Department of State. "We have a public meeting here next Thursday," he wrote, "for the preservation of the forests on the Presidential range." The purpose was clear: to create "a National park of all the higher mountains," making them forever off-limits to the timber harvesters and paper manufacturers.[1]

The meeting was impassioned, its participants impatient for action, and from it emerged two strategies. First, the citizens called upon President Theodore Roosevelt, who advocated forest conservation, to urge Congress to pass legislation allowing the federal government to purchase the White Mountains and turn them into a national park. Second, the group supported ongoing efforts by conservationists in the South to create a national park in the southern Appalachians.

Science, Progressivism, and Recreation

Hale's meeting at Intervale was a reflection that a forest conservation movement was beginning to develop in the United States by the early twentieth century. This movement's leaders consisted of a blend of ordinary citizens and professionals: foresters, outdoor recreationists, amateur botanists and horticulturalists, wildlife enthusiasts, professors, museum managers, and journalists. At first, the movement was somewhat inchoate, but gradually it began to coalesce around three well-defined goals:

1. The creation of forest reserves owned or controlled by government at the federal, state, and local levels

2. The application of new methods of scientific forestry to manage forests for multiple uses, including the long-term production of timber

3. The passage of legislation to regulate the management of public forests

Many of the practical questions surrounding the creation of national forests remained unresolved, however. For example, was the emphasis to be more on resource extraction or on recreation and the preservation of wildlife habitat? Were public forests to be economically self-sustaining? How was the mission of the national forests to be different from that of the national parks, which Congress was creating during the same era? Finally, how were the eastern forests to be managed differently from the western forests, given their different sizes, climates, and proximity to population centers?

It is no accident that the forest conservation movement developed during the Progressive Era, which lasted roughly from 1890 to 1920. After all, the conservation movement mirrored the principles and values of progressivism. The Progressives believed in scientific progress, the rational organization of society to solve social problems, and the prudent and efficient use of natural resources to sustain the country's economic growth over the long term. Moreover, the Progressives had a deep suspicion of corporations like the Standard Oil Company and an accompanying belief that government must serve as a countervailing force to the immense power of the corporations. The suspicion of the growing power of corporations helped steer conservationists toward a strategy of creating a network of publicly owned forests that professionally trained foresters would manage for the benefit of the American people. In his first State of the Union message in 1901, Roosevelt alerted Congress and the world that forest conservation would be of the utmost importance in his administration. "The fundamental idea of forestry," the president exhorted, "is the perpetuation of forests by use. Forest protection is not an end in itself; it is a means to increase and sustain the resources of our country and the industries which depend upon them."[2]

George Perkins Marsh and *Man and Nature*

A major intellectual foundation of forest conservation in the United States—perhaps *the* major foundation—was *Man and Nature: Or, Physical Geography as Modified by Human Action*, by an American diplomat named George Perkins Marsh. In remarkably prescient ways, Marsh blended scientific observation of forests with the radical insight that humankind must be a partner with—not an exploiter of—the natural world.

Nature seeped into Marsh's bones at an early age. Born in 1801 in Woodstock, Vermont, George accompanied his father, a successful lawyer, on explorations of the surrounding woods and fields. Marsh remembered one day when he and his father were out riding and his father "stopped his horse on the top of a steep hill, bade me notice how the water there flowed in different directions, and told me that such a point was called a *watershed*."[3] The young Marsh proved to be an exceptional student. He attended Dartmouth College and soaked up languages, eventually mastering twenty of them. Disabled by eye problems, he studied law by listening to others read precedents aloud. Blessed with a photographic memory, he committed those precedents to memory, passed the bar, and in 1825 set up a practice in Burlington.

Marsh's practice never truly thrived, however, and he finally abandoned it in 1842. Instead, his mind and interests ran to scholarly and scientific pursuits. He pursued his interests in the natural sciences, purchasing engravings that he eventually donated to the Smithsonian Institution, of which he was an early supporter. After leaving his struggling law practice, he ran for the U.S. House of Representatives, was elected in 1843, and served four terms. Marsh also made forays into a number of businesses, including sheep raising, but success eluded him. By the 1850s, he was on the verge of bankruptcy, but because of his outstanding reputation as a scholar and historian, President Abraham Lincoln appointed him U.S. ambassador to Italy in 1861, a post in which he served until his death in 1882.

During this period, he wrote *Man and Nature*, which he based on his observations of deforestation in Italy and other parts of Europe as well as on his observations in Vermont during his formative years. He visited Italy's forestry school near Florence, where he observed firsthand how forestry had improved the quality of trees. His observations made him an early believer in silviculture: the study and application of how forests are established, what their composition is, and how they develop. Silviculturalists examine both the growth of individual trees and the development of forests as natural systems.

The overriding theme of *Man and Nature* was that nature is a gift, not a package of commodities for humanity to consume. "Man has too long forgotten," Marsh asserted, "that the earth was given to him for usufruct alone, not for consumption, still less for profligate waste."[4] *Usufruct* was the defining concept. It meant that humankind had the right to use nature's gifts to meet its needs, but it did not own those gifts nor have the right to destroy them.

Marsh documented the effects of the heavy logging and widespread fires on soil, water, vegetation, and wildlife, anticipating the twentieth-century view of forests as complex ecosystems. Heavy logging, for example, had destroyed the habitats of numerous birds. He explained, "Birds affect vegetation directly by sowing seeds and by consuming them; they affect it indirectly by destroying insects injurious, or, in some cases, beneficial to vegetable life. Hence, when we kill seed-sowing birds, we check the dissemination of a plant."[5]

In a similar vein, he detailed the effect of deforestation on forest soils. "It is well established," he explained, "that the protection afforded by the forest against the escape of moisture from its soil, insures the permanence and regularity of natural springs, not only within the limits of the wood, but at some distance beyond their borders."[6] Because of the loss of the forest canopy, rains fell directly on the soils, with disastrous effects:

> The soil is bared of its covering of leaves, broken and loosened by the plough, deprived of the fibrous rootlets which held it together, dried and pulverized by sun and wind. . . . The face of the earth is no longer a sponge, but a dust heap, and the floods which the waters of the sky pour over it hurry swiftly along its slopes, carrying in suspension vast quantities of earthy particles.[7]

Marsh piled detail upon detail to buttress his case that forest soils were being depleted and that streams were filling with silt, impeding their natural flow. "Man is at this moment so fast laying waste the face of the earth," he admonished, "that the most serious fears are entertained, not only of the depopulation of those districts, but of enormous mischiefs to the provinces contiguous to them."[8]

In the years after the Civil War, a cohort of new scientific organizations also gathered knowledge about natural systems and the wildlife that inhabited them. In 1870, for example, the American Fisheries Society was founded to advocate for the conservation of fish, and two years later, Congress formed the United States Fish Commission, which investigated declines in fish populations. In 1873, the American Association for the Advancement of Science, which had been founded in 1848, urged the states and Congress to pass laws protecting forests. Birding enthusiasts formed the American Ornithologists Union in 1883 to promote ornithological science. Pennsylvania and New York both passed laws protecting birds. C. Hart Merriam and A. K. Fisher were ornithologists who advocated for the preservation of habitat for birds and other wildlife. In 1885, Merriam founded the Division of Economic Ornithology and Mammalogy in the U.S. Department of Agriculture. It eventually became the U.S. Fish and Wildlife Service and was moved to the U.S. Department of the Interior.[9]

Scientific Forestry and Two Forest Reserve Acts

As the intellectual foundations of forest conservation were being established, practitioners were needed to apply the skills, techniques, and practices that would begin to restore America's forests. Many of those practitioners emerged from the new science of forestry, which spread from Europe to the United States in the 1880s. According to the principles of forestry, timberland owners should manage forests for sustained yield, not allowing the annual timber harvest to amount to more than the annual growth of trees. In addition, forestry prioritized protection against fire and disease and sponsored research into new wood products.[10] By managing forests wisely, timberland owners could ensure a reasonable rate of return on their investments over the long term.

The driving force behind the introduction of forestry to the United States was Bernhard E. Fernow, a Prussian who had trained in forestry in Germany and emigrated to the United States in 1876. In 1886, he became the chief of the Division of Forestry, the precursor of the U.S. Forest Service. His mission, as he saw it, was to educate timberland owners and the public to the economic and scientific benefits of forestry.

Fernow played a central role in the creation of publicly owned forest reserves in the United States. He did so by helping to draft two pioneering laws protecting American forests. In the early 1890s, he, Interior Secretary Carl Schurz, and other conservationists persuaded key members of Congress that the federal government should create forest reserves for two purposes: (1) to protect and manage forests for producing timber over the long term and (2) to introduce and model methods of forestry that private timberland owners could emulate. Together, Fernow and Schurz drafted the Forest Reserve Act, which Congress passed in 1891. It stipulated:

> THAT THE PRESIDENT OF THE UNITED STATES MAY FROM TIME TO TIME SET APART AND RESERVE, IN ANY STATE OR TERRITORY HAVING PUBLIC LANDS BEARING FORESTS, [in] ANY PART OF THE PUBLIC LANDS designated in the act as timber lands, or any lands WHOLLY OR IN PART COVERED WITH TIMBER OR UNDERGROWTH, WHETHER OF COMMERCIAL VALUE OR NOT, AS PUBLIC RESERVATIONS.[11]

On March 30, 1891, President Benjamin Harrison set aside the first reserve, the Yellowstone Timberland Reserves. Soon after, he reserved thirteen million additional acres, all of which were forestlands in the public domain in the West. The law was an important beginning of the national forest system as well as of the concept that government should create and protect forest reserves to be managed using the new science of forestry to ensure timber on a sustainable basis. As will become evident, the creation of forest reserves would gain momentum during the 1890s among forest conservation advocates in the East, South, and Lake states.

Despite the significance of the Forest Reserve Act, though, it failed to clarify what the forest reserves should be used for or how they should be managed. In 1896, the National Academy of Sciences created a Forest Commission, chaired by Charles Sprague Sargent, the director of Harvard's Arnold Arboretum and the publisher of the journal *Garden and Forest*, which had done so much to document the destruction of the nation's forests. Among the other six members of the commission was Gifford Pinchot, the future founding chief forester of the U.S. Forest Service and the only trained forester on the commission.

The commission had the ear of President Grover Cleveland, and on February 22, 1897, with only ten days left in his administration, Cleveland created thirteen additional forest reserves, encompassing 21,279,840 acres in the West. Cleveland's action outraged western congressmen and senators, who resented the outgoing president's failure to consult with them. Indeed, their resentment led them to oppose future efforts by conservationists to create forest reserves in the eastern United States.[12]

Western congressmen tried to nullify the reserves, but when William McKinley won election as U.S. president in 1896, Pinchot and others worked furiously to save the Cleveland reserves. Finally, in late March 1897, Congress passed a bill protecting the reserves and placing the U.S. Geological Survey in charge of surveying them, and McKinley signed the bill into law on June 4. Known both as the Forest Management Act of 1897 and the Organic Act, this law created guidelines for administering the federal forest reserves. "No national forest," it stated, "shall be established except to improve and protect the forest within the [national forest] boundaries or for the purpose of securing favorable conditions of water flow, and to furnish a continuous supply of timber for the use and necessities of citizens of the United States."[13]

The language of the law was far-reaching in three ways. First, it established a stewardship purpose: that the federal government should manage the forests in such a way as to "improve and protect" them. Second, it connected forest protection to water flow, reflecting conservationists' observations that deforestation had caused the silting up of rivers and streams, imperiling fish and amphibians and impeding the use of waterways for navigation and power. Third, the law specified that the forest reserves would deliver a supply of timber that was "continuous." In other words, the government was to manage the forests to supply timber on a sustainable basis. In the Organic Act, the scientific and economic priorities of forest conservation came solidly together, and the law guided management of the national forests until the passage of the Multiple Use–Sustained Yield Act in 1960.

Gifford Pinchot and the U.S. Forest Service

The federal government, however, did not yet have the organizational means to manage these public forests effectively. What was needed was leadership, which Dr. Fernow had failed to provide. He had seen his mission as the dissemination of knowledge rather than the creation of an effective organization. In 1898, he left federal government to assume the directorship of Cornell University's newly established school of forestry.

Soon after Fernow left, Secretary of Agriculture James Wilson offered the position to Gifford Pinchot (figure 4.1). He refused it. Brilliant and strong-willed, Pinchot had by the age of thirty-three built a national reputation as the first American to be trained in forestry. He had been born in 1865 in Connecticut, and his father, James, had made a fortune in lumbering operations and land speculation. Late in life, James Pinchot deeply regretted the destruction that his company's heavy logging had caused, and he embraced conservation. As young Gifford was about to go off to college in 1885, his father asked him, "How would you like to be a forester?"[14] Gifford seized upon the idea and attended Yale College, and after earning his degree, he sailed to Europe to study forestry at the National Forestry School in Nancy, France, as well as in Germany, Austria, and Switzerland.

Upon his return to the United States in 1892, Pinchot undertook the first applications of forestry in the United States, at George W. Vanderbilt's Biltmore Estate in Asheville, North Carolina. The seven-thousand-acre forest had suffered severe damage from overcutting, and Pinchot set out to restore it to health. According to Charles Sprague Sargent, the Biltmore forest was "the first experiment yet undertaken on this continent to restore to a profitable condition a considerable area of what was once forestland."[15] The restoration worked so well that the Biltmore forest has been known ever since as "the cradle of American forestry."

Clearly, Pinchot was the right man to direct the federal government's forestry operations, and Secretary Wilson continued pressuring him. In May 1898, he set up a meeting with Pinchot, and together they lamented the deplorable state of America's forests. Pinchot asserted that U.S. forestry was still in "the Dark Ages," and both agreed that the Division of Forestry

Figure 4.1 *Gifford Pinchot, 1909. Pinchot served as the first chief forester of the U.S. Forest Service. Regarding the stewardship of America's forests, he wrote, "Where conflicting interests must be reconciled, the question shall always be answered from the standpoint of the greatest good of the greatest number in the long run." Library of Congress.*

needed to be larger and more assertive. Pinchot then told Wilson what his conditions would be to take over the Division of Forestry. As he recalled in his autobiography, *Breaking New Ground*, "I could run it [the Division] to suit myself. I could appoint my own assistants, do what kind of work I chose, and not fear any interference from him."[16] Wilson granted him the powers he wanted, and he took the job.

On July 1, 1901, the division became the Bureau of Forestry, beginning a rise in bureaucratic status that reflected Pinchot's organizational ambition. He aggressively expanded the bureau, hiring newly trained foresters from Yale and other colleges that were establishing forestry schools. Pinchot, however, was deeply frustrated that the forest reserves were out of his control and in the Interior Department, which he alleged was riddled with political appointees who lacked any interest or training in forestry.

Pinchot started an aggressive bureaucratic campaign to have the forest reserves moved to the Agriculture Department. In that campaign, he boasted the most important ally possible, President Theodore Roosevelt. Outdoor enthusiast, hunter, angler, Roosevelt came to national attention partly because of his writings about the outdoors. In November 1887, he visited North Dakota's Badlands for five weeks and was shocked to see how bereft of wildlife the Badlands had become because of overhunting. The next year, he cofounded the Boone and Crockett Club, which declared that its mission was to advocate for practices and policies that would preserve populations of large game in the United States.[17]

Roosevelt agreed with Pinchot that the Agriculture Department should administer the forest reserves, and for the next four years, they worked to convince Congress of the wisdom of the transfer. In 1905, Pinchot helped to organize the American Forest Congress, which the American Forestry Association convened in Washington, D.C., from January 2 to 6. The purpose was clear: to consolidate support and work out the details of the transfer of the forest reserves. The pressures brought by the American Forest Congress worked. Later that year, the House and Senate approved and Roosevelt signed the Transfer Act, which moved the forest reserves from the Interior Department to the Agriculture Department, signifying that the national forests would be treated as lands that produced crops.[18] In the Agricultural Appropriation Act of 1905, the name of the bureau was changed to the U.S. Forest Service.

What, though, would the forests be managed for? Pinchot began to answer that question in a letter that Agriculture Secretary Wilson wrote to Pinchot, a letter that historians generally acknowledge was written by Pinchot himself. "Where conflicting interests must be reconciled," he wrote, "the question will always be decided from the standpoint of the greatest

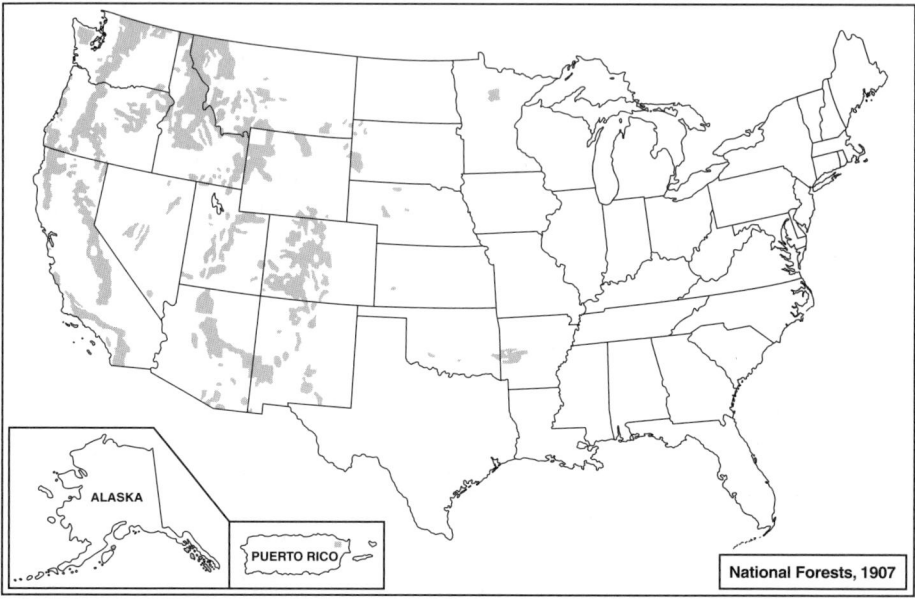

National Forests, 1907

Figure 4.2 *National forests, 1907. Map by Christopher Robinson.*

good of the greatest number in the long run."[19] In Pinchot's utilitarian vision, the Forest Service would manage the national forests for multiple uses, balancing logging, grazing, and mining for long-term production. Despite the maturity of the tourism industry in the White Mountains, recreation was, at this point, a less important motivation for creating national forests, but it would grow in significance as the East, the South, and the Lake states developed their tourism industries in the early twentieth century.

In 1905, the Forest Service published *The Use of the National Forest Service*, which came to be known as the *Use Book*. It detailed rules, regulations, directions, and instructions for managing the national forests.[20] In 1907, the service started to refer to the forest reserves as national forests, creating a brand of the forests as alternatives to the national parks and emphasizing that they existed for the benefit of the entire nation.

Meanwhile, the federal government continued to build its network of national forests, but the vast majority of them were in the West. In 1907, very few national forests existed east of the 100th meridian (figure 4.2).

One of the few was the El Junque National Forest of Puerto Rico, which was originally established by King Alfonso XII of Spain in 1876 and became a national forest in 1906 as a result of the Spanish-American War of 1898.

The reason for a lack of eastern national forests was simple: forests in the West were in the public domain, and the federal government could easily convert those lands to public forests. In the East, South, and Lake states, however, forestlands had all been sold to individuals or corporations. Creating national forests in these regions meant purchasing forestlands from private owners, which the federal government did not yet have the power or the resources to do. For this reason, conservationists began to form the goal of federal legislation that would allow the government to purchase eastern forestlands and create national forests. The specific goals and strategies that conservationists followed in the three regions differed because of divergent histories, cultures, and economic conditions, however.

The Southern Appalachians

In the early years of the twentieth century, a clergyman was awakening southerners to the destruction of their forests. He was Reverend A. E. Brown, director of the Department of Mountain Schools for the Southern Baptist Convention, and as he traveled throughout the South, he decried the widespread deforestation. He told an interviewer in 1910, "When I first started my work in these mountains, 30 years ago, when the forests were untouched, the mountains were full of sparkling brooks and creeks which required a two or three weeks rain to make muddy; today a few hours' rain will muddy them." He continued, "These companies cut practically every tree from 12 inches up, and are utterly indifferent to the interest of the natives."[21]

Brown's warnings reflected that by the 1890s, a conservation movement was growing in the South. Ensuring timber supplies for the future was the overriding goal, but tourism and outdoor recreation were emerging as important considerations in the region's economic revival. Government officials redoubled their efforts to attract investors and travelers from other regions of the country, stressing in particular the ameliorative effects of the mountain air to improve health and combat diseases such as tuberculosis. Many of the travelers were well educated, numbering doctors, lawyers, teachers, and scientists. They formed a pivotal constituency of the southern conservation movement.

As a result, by the 1890s, Asheville, North Carolina, had become an epicenter for conservation. As early as 1885, Dr. Henry O. Marcy of Boston delivered a paper to the American Academy of Sciences in which he called for the creation of a national park in the southern Appalachians. During the 1890s, Asheville physician Chase P. Ambler, who had relocated to Asheville from Ohio, built on Marcy's idea. On November 22 and 23, 1899, he and another Ohio native, Judge William Day, organized a meeting of concerned citizens at the Battery Park Hotel in Asheville. Attending were forty-two political leaders, business executives, and newspaper editors drawn from all parts of the southern Appalachians. Locke Craig, future governor of North Carolina, rallied the audience by asserting, "It would be reckless stupidity, negligence of the grossest kind, if a portion of this grand and picturesque region be not preserved in its original, natural condition for the enjoyment of the people."[22]

The group formed the Appalachian National Park Association (ANPA), with Senator Jeter Pritchard of North Carolina as its director. The goal was to create a national park in the Great Smoky Mountains. On January 2, 1900, the association presented a memorial to Congress urging the creation of national park. Pritchard persuaded Congress to appropriate $5,000 to the Agriculture Department to study the southern forestlands, which culminated in the Wilson Report. (See chapter 3 for an explanation of the report.)

On December 19, 1901, President Theodore Roosevelt forwarded the Wilson Report to Congress, which he called upon to allocate funds for purchasing southern forestlands, but he urged the creation of national forests rather than a national park. "Their management under practical and conservative forestry," he wrote, "will sustain and increase the resources of this region and of the nation at large, will serve as an invaluable object lesson in the advantages and practicability of forest preservation by use, and will soon be self-supporting from the sale of timber."[23]

Lumber industry executives and foresters agreed with the shift of priorities to national forests, for they wanted to keep the mountains open to logging. Leisure industry advocates, on the other hand, were disappointed. The ANPA, however, shifted its goal to the creation of national forests, even renaming itself the Appalachian National Forest Association. The National Lumber Manufacturers' Association and the National Hardwood Lumber Association wholeheartedly supported the call for national forests. On these lands, they would be able to continue logging. In addition, they would not be responsible for paying property taxes because they would no longer own the land.

In 1901, southern legislators introduced a bill into Congress to create a national forest in western North Carolina, but it failed because Congress, led by the outspoken opposition of House Speaker Joseph G. Cannon of Illinois, refused to accept the necessity of purchasing eastern forestlands to protect them. Over the next eight years, several other bills failed. For most senators and representatives, forest protection remained a regional issue rather than a national one.

New England's Conservationists

The forest conservation movement in New England in the late nineteenth century differed markedly from that in the South. For one thing, New England had a mature tourist industry, with resort hotels, campgrounds, hiking trails, bridle paths, and railroads sprawling across the landscape. In addition, New England's conservationists had access to New York's and Boston's newspapers and magazines. During the 1890s, journalists generated a wave of articles about deforestation that appeared in the *Atlantic Monthly*, *Harper's Weekly*, *Scribner's*, and other national magazines. In February 1893, the *Atlantic Monthly* published one of the most comprehensive examinations of the situation: "White Mountains in Peril," by Julius H. Ward. Ward was an Episcopal minister and author in Boston who trekked regularly to the White Mountains. In the *Atlantic Monthly* article, he delved into the knotty question of how to balance the public good of forest protection with the rights of private timberland owners. "In many cases," Ward asserted, "the proprietors have yet to be made to understand that it is quite as profitable to take out the ripe timber and leave the younger trees to grow up to maturity as it is to strip the forests clean and let the future take care of itself."[24]

Ward proposed that New Hampshire offer money to timberland owners who agreed to implement methods of forestry on their lands. He estimated that the plan would cost New Hampshire between $2 million and $3 million. The state would raise additional money by inviting donations from citizens who wanted to preserve specific sections of the forest.[25] Ward's proposal was never implemented, however, for the very simple reason that as a rural state with a low tax base, New Hampshire could not afford such expenditures.

Other figures in the forest conservation movement were Reverend John Johnson, who published *The Constitution of the White Mountains* in 1990, and Joseph B. Walker. As a boy growing up in Concord during the 1820s and 1830s, Walker ranged far and wide through the region's forests and fields. In 1838, he ventured into the White Mountains, which captivated him with their pristine beauty. As he visited the mountains in succeeding years, though, he was an eyewitness to the gradual destruction of the forests. Walker found success as a banker and railroad executive, building enough of a fortune that he could afford to involve himself in the state's fledgling conservation movement. He served as a member of the forestry commissions that New Hampshire created in the 1880s and 1890s. For Walker and others who had studied the state's forests closely, the best solution was to create publicly owned forests, which would be managed for multiple uses: management of timber resources for the long term; protection of the state's tourist industry; and the reduction of flooding, fires, and irregular stream flow.[26]

Walker's support for public forests found a ready audience in New Hampshire among travelers, resort hotel owners, railroad executives, and several of the state's leading political figures. On January 21, 1901, concerned citizens, including Frank W. Rollins, former governor of New Hampshire, convened a meeting at the office of the secretary of the New Hampshire State Board of Agriculture in Concord. Their goal was to create a permanent organization dedicated to saving New Hampshire's forests. On February 6, 1901, that group organized the Society for the Protection of New Hampshire Forests (SPNHF). The society's mission was clear: "To preserve intact the scenic beauty in selected places throughout the state where the forest is an essential element, particularly upon the high and steep slopes in the mountains."[27] In December 1901, Rollins, who had become the organization's first president, made one of the most important moves of the organization's early years. Acting on a hunch, he fired off a telegram to one Philip W. Ayres, inviting him to interview for the position of forester.

Ayres proved to be a visionary who, for the next forty years, would bring wide-ranging curiosity and ample intellect to conservation in New England. He grew up on a farm in southern Illinois, where his father, an avid botanist, had planted five acres of trees native to the Midwest, and Ayres had studied those trees thoroughly. He attended Cornell University, graduated in 1883, and enrolled in graduate school at Johns Hopkins University to pursue a PhD in history. Soon after completing his doctorate, the Charity Organization Society of Brooklyn offered him a position, and for the next sixteen years, he devoted himself to social work. By 1899, he was burned out. His wife, Alice, reminded him that he loved trees and forests, so he cashed in his life insurance and enrolled in the forestry program at Cornell.[28]

He was just finishing his courses in forestry when Rollins contacted him. Ayres had one condition before accepting the position: that the society support the creation of a national forest in the White Mountains. In Ayres's view, only the federal government had the power and the resources to protect the forests. Rollins agreed, and Ayres accepted the job. His contract included language that the SPNHF would support a forest reserve.[29]

Other leaders of the New England conservation movement emerged, including Allen Chamberlain, executive director of the Appalachian Mountain Club; Edwin A. Start, secretary of the Massachusetts Forestry Association; and Thomas E. Will, secretary of the American Forestry Association. One lesson conservationists were learning was that legislators wanted scientific data. Early in 1903, E. Bertram Pike, who represented Haverhill in New Hampshire's House of Representative, introduced a bill appropriating $5,000 for surveying forest conditions in the White Mountains. The New Hampshire Forestry Commission would direct the survey, but the federal Bureau of Forestry was to do the actual survey work.[30] The result was the Chittenden Report, described in chapter 1.

Such studies helped lay a solid foundation of information, yet forest lovers were growing impatient at the slow rate of progress. On January 21, 1903, the *Boston Transcript* published an editorial attacking the SPNHF for fearing "that somebody's feelings might be hurt if it called things by their right names and made practical suggestions bearing upon the immediate dangers of the situation. It roars too much of the sucking-dove tone."[31] Like the southerners, New England's conservationists needed an effective legislative strategy.

The Lake States

In the Lake states, thousands of acres of forestland had been devastated, yet Michigan, Minnesota, and Wisconsin did not yet attract tourists in great enough numbers to raise their voices effectively against heavy logging. According to environmental historian Raleigh Barlowe, "Calls for conservation were not popular in the Lake States region during the 1800s."[32] Those who dared to call for forest conservation were looked upon as "wild theorists." Moreover, lumber industry executives exercised great influence in the legislatures of the three states.

As explained in chapter 2, the efforts to encourage agriculture on cutover lands were largely failures, and the three states shifted their priority to creating state forests and undertaking reforestation to try to reinvigorate their lumber industries. Driving this transition were young foresters—inspired by the vision of Gifford Pinchot—who took positions in state government or at newly established forestry schools at Michigan Agricultural College (later Michigan State University) and the University of Michigan.

Occasionally, a far-sighted political leader would support forestry as a way to make the cutover northern lands productive again. For example, on January 7, 1897, Governor John T. Rich of Michigan delivered his annual message to the state legislature and implored the lawmakers, "A modest appropriation looking to some plan for finally establishing a forestry department of the State is well worthy of your earnest consideration."[33] In 1899, Michigan's state legislature established a State Forestry Commission, and two years later, it began to create a system of state forests.

In the late 1890s, Wisconsin also began to look at ways to rejuvenate its forests. The legislature created a Forestry Commission in 1897, and it examined the status of the state's forests and recommended that the state implement methods of forestry. The election of Robert M. La Follette as governor in 1901 gave the state's conservation movement the shot of adrenaline it needed. La Follette embodied all the qualities of the Progressive movement: belief in science, enthusiasm for modern organization, and use of the nation's resources for the public good. In 1904, La Follette named Edward M. Griffith, who had worked for Pinchot at the Bureau of Forestry, to fill the newly created position of superintendent of state forests. Under Griffith's leadership, the state purchased approximately 250,000 acres of land, starting an extraordinary system that eventually encompassed 1.5 million acres in northern Wisconsin. At Trout Lake in Vilas County, Wisconsin's Board of Forestry also created a tree nursery and gathered seeds for white pines, red pines, Scotch pines, and ponderosa pines. Workers planted seedlings on land that fires and clear-cutting had devastated.[34]

Some business interests and newspapers vehemently opposed what they called the "rejunglizing" of the northern counties because it took land off the tax rolls and, in their opinion, slowed economic development. One newspaper in Rhinelander published a photograph of a farmer next to tall corn with a caption reading, "When corn grows like this picture shows it does in Oneida County, this is too good a country to be given over to reforestation schemes or for state manipulation."[35] The struggle between those favoring reforestation and those favoring agriculture continued into the first decade of the twentieth century.

Minnesota was fortunate to have a native son who provided strong conservation leadership: Brigadier General Christopher C. Andrews, who had risen to prominence during the Civil War. In 1869, President Ulysses S. Grant selected Andrews as the U.S. ambassador to Norway and Sweden. During Andrews's first visit to Sweden, he observed the checkerboard of forests, with trees of various ages, that the Swedish had created by applying methods of forestry.[36] Impressed, Andrews returned to Minnesota and became the state's leading advocate for forest protection.

On September 1, 1894, Minnesotans were shocked to learn that a massive fire had destroyed thousands of acres of forest and killed 413 people in the small town of Hinckley. The tragedy provoked outrage, and in its next session, Minnesota's legislature passed a bill to create a forest commission to enforce "the preservation of forests of this state and for the prevention and suppression of forest and prairie fires."[37] The bill created the new position of state fire warden, and Andrews, who had garnered attention because of his reports on Swedish forestry, won the appointment. In 1899, Judson N. Cross, an attorney in Minneapolis who admired forests, drafted a bill creating a state forestry board to manage state forests, which Minnesota acquired through gifts, takeovers of tax-delinquent properties, and transfers from federal forests.[38]

While Minnesota was founding its system of state forests, conservationists proposed a national forest encompassing some of the most beautiful land in Minnesota. They had their eyes on approximately eight hundred thousand acres surrounding Cass Lake, Leech Lake, and Lake Winnibigoshish, in the north-central part of the state. The land was part of the Ojibwe Indian Reservation, however. Andrews drafted a federal bill to create a commission to negotiate with the tribe to acquire the lands. The government would purchase the lands under the auspices of the Nelson Act of 1889, which Congress had passed to dismantle Indian reservations and distribute land to individual owners.

Andrews proposed, however, that instead of selling the land to private interests, the federal government should acquire major tracts around the three lakes to establish a national forest. Congress, though, ignored the proposal and moved ahead with plans to sell off the lands to corporations and individual entrepreneurs. Minnesota's conservationists were incensed. Maria Sanford (figure 4.3), a professor of rhetoric at the University of Minnesota who was a leading feminist and conservationist, penned a stream of angry editorials for the *Courant*, the journal of the Minnesota Federation of Women's Clubs. Summoning all her rhetorical skills, she attacked Congress for giving the forests away to entrepreneurs who had

Figure 4.3 *Maria Louise Sanford, ca. 1910. Sanford, a professor at the University of Minnesota, wrote letters and lobbied Congress to create a federal forest preserve in Minnesota. Library of Congress.*

made fortunes from timber, and she called for the establishment of a permanent forest reserve in the state to promote healthful living and serve the recreational needs of Minnesotans.[39]

Still, no members of Minnesota's congressional delegation would sponsor a national forest bill. When Sanford heard that the Interior Department had imminent plans to allow logging in the region, she boarded a train to Washington, D.C., with Lydia Phillips Williams, president of the Minnesota Federation of Women's Clubs. There they lobbied Minnesota's congressional delegation to take action to stop the logging, which was put on hold, yet the congressmen still refused to support a national forest.[40]

Florence Bramhall, director of the Forest Reserve Committee of the

Federation of Women's Clubs, then worked with other conservationists in Minnesota to develop a compromise proposal for a smaller national forest of 200,000 acres. Bramhall and her forestry committee barnstormed Minnesota to build public support, and Gifford Pinchot lent his support to the campaign. In 1902, Representative R. Page Morris of Minnesota submitted a bill to create a forest reserve of 225,000 acres, the nucleus of what is today the 650,000-acre Chippewa National Forest, and for the first time, Congress passed legislation creating a national forest.[41] President Theodore Roosevelt signed the bill in June 1902.

With these developments in the Lake states, New England, and the southern Appalachians, the forest conservation movement had become much better defined and had started to persuade the public, government leaders, and progressive executives in the timber products industry that the United States needed to take strong steps to conserve one of its most valuable natural resources, the forests. Yale, Cornell, and other universities were beginning to establish forestry schools, and trained foresters were bringing their expertise to the federal government, state governments, and private corporations. In addition, the federal government had created publicly owned forests in the West from lands that had been in the public domain, and states were creating systems of state forests. Still, a huge dilemma remained: What should be done to protect the eastern forests, which were still being logged heavily? If the answer was to create a system of publicly owned eastern forests, how would the economic and legal obstacles be overcome? Answering these difficult questions required innovative thinking and bold leadership.

CHAPTER 5

Legislation at Last: The Weeks Act

By the first decade of the twentieth century, the situation in the forest-lands of New England, the South, and the Lake states had reached crisis proportions, yet forest conservationists were uncertain about how to proceed with forest protection. Would forest conservation better be pursued by state governments or by the federal government? Arguments in favor of state government were persuasive. New York, for example, had formed the Adirondacks Forest Preserve in 1885, but when logging continued there, the state legislature took a more radical step, sponsoring a convention to update the state's constitution. One new article of the constitution set aside the Adirondacks as a protected preserve, using words that foreshadowed ideas about wilderness: "The lands of the State, now owned or hereafter acquired, constituting the Forest Preserve as now fixed by law, shall be forever kept as wild forest lands."[1] That fall, voters overwhelmingly approved the new constitution, giving the Adirondack forests the strongest protection in the land.

New York, however, had resources for such action that far exceeded those available to agricultural states like New Hampshire and North Carolina. As a result, conservationists looked increasingly to the federal government for solutions, yet knotty questions arose. Did the federal government have the power constitutionally to purchase forestlands from private timberland owners? If so, where would the resources for such purchases come from? And how would the federal government carry out negotiations with private landowners for the purchase of lands?

In New Hampshire, Joseph B. Walker, who had served on several of the state's forestry commissions, was an early voice in favor of public ownership of forests, and he marshaled economic arguments in favor of public stewardship. The New Hampshire Forestry Commission paid special attention to preserving water power because the state's economy depended so heavily on water-powered textile mills on the Merrimack River. In the Forestry Commission's 1891 report, Walker and his co-commissioners, George B. Chandler and J. B. Harrison, spelled out the multiple uses of forests:

1. The first and foremost important function of mountain forests is the preservation of the mountains themselves by clothing them with soil.

2. The second function of mountain forests is the production of a perpetual supply of timber.

3. The third important function of mountain forests is the formation of natural storage reservoirs for the retention and distribution of water.

4. Another important function of mountain forests is the production and maintenance of such conditions of the soil, water, atmosphere, and scenery of the region as are highly favorable to human life, health, and enjoyment.[2]

In the first decade of the twentieth century, forest conservationists turned their attention to Congress in the belief that only the federal government had the powers and resources to purchase privately owned timberlands and create a network of national forests in the East, South, and Lake states, which would be managed for multiple uses. From 1900 to 1905, conservationists and legislators in the South and New England introduced legislation into Congress, but the two regions pursued legislative solutions separately, reinforcing opponents' arguments that forest conservation was a regional issue and not a national one.

Early Legislative Efforts

Much of the early legislative effort was centered in the South. Beginning in 1900, southern conservation organizations and state legislatures petitioned Congress to pass legislation to protect the southern Appalachians. The Appalachian National Park Association, the American Association for the Advancement of Science, and the American Forestry Association all lobbied Congress to take action. In 1901, the North Carolina state legislature went a step further by granting the U.S. government the right to purchase and acquire title to forestlands within the state for the purpose of creating national forests. Later that year, Georgia, Tennessee, and Alabama passed similar enabling legislation.

North Carolina's delegation to the U.S. Congress took further initiatives. After the Wilson Report was issued in 1901 (as explained in chapter 3), Senator Jeter Pritchard of North Carolina, who had written the enabling legislation for the report, submitted a bill that would have appropriated $5 million to purchase two million acres for a forest reserve in the southern Appalachians. (The proposed reserve would later become the Great Smoky Mountains National Park.) Pritchard's bill never came to a vote, however.

Later that year, Senator Joseph R. Burton, a Republican from Kansas who was the chairman of the Committee on Forest Reservations and Game Protection, wrote an eastern forest reserve bill appropriating $10 million for purchasing forestlands in the southern Appalachians. Burton's committee reported the bill to the full Senate, which debated it extensively and then voted in favor of it. The bill won support for a variety of reasons. Some southern senators wanted to protect the forests, but others saw the economic advantage to the region of empowering the federal government to purchase cutover lands of low value and assume the costs of regenerating and protecting the forests.[3] To garner public support for Burton's bill, the Appalachian National Park Association, which changed its name to the Appalachian National Forest Association in 1903, sent out more than a million mailings.[4] In addition, Gifford Pinchot worked closely with and strongly supported the southerners' efforts. At the same time, he kept the New Englanders at arm's length, believing that conservationists should focus on one region at a time and that the South—where he had roots because of his pioneering work at the Biltmore Forest—should take precedence.[5]

In the House of Representatives during the Fifty-Seventh Congress (1901–1902), southern representatives also submitted a number of bills. Representatives Walter Brownlow of Tennessee and Richmond Pearson and James Moody of North Carolina introduced a bill that was reported favorably by the House Committee on Agriculture, but it did not reach the floor for a vote. Over the next few years, representatives from North Carolina and Tennessee introduced other bills, but they all faced the implacable opposition of the Speaker of the House, Joseph G. Cannon of Illinois, a member of the old guard of the Republican Party and a fervid states' rights supporter, about whom more will be said later.[6]

While the southern legislators were introducing legislation, New England's conservationist organizations started to build grassroots support in favor of national forests in the region. The leaders in this effort included Philip Ayres (figure 5.1); Appalachian Mountain Club president Allen Chamberlain, Edwin A. Start of the Massachusetts Forestry Association, and Thomas E. Will of the American Forestry Association, which emerged as a leading force in forest conservation in the early twentieth century. These men and many others traveled extensively around the six states that comprise New England, delivered passionate lectures that showed the devastation of the forests, educated the public to the benefits of forestry, and fired off countless editorials.[7]

Ayres's campaign on behalf of the Society for the Protection of New Hampshire Forests was especially innovative. Convinced that his presentations should be as visual and visceral as possible, he traveled throughout New England in the early years of the twentieth century and used what was known then as a Magic Lantern slide projector, which cast photographic images on glass slides, much like today's slide projectors. Ayres showed images of devastation, such as burned logs scattered up and down mountainsides in the White Mountains. To add to the dramatic effect of the slides, Ayres colorized them. One, for example, showed the black of the

Figure 5.1 *Philip Wheelock Ayres, ca. 1890. As the first forester of the Society for the Protection of New Hampshire Forests, Ayres worked incessantly to inform the public about the plight of New England's forests. Milne Special Collections and Archives Department, University of New Hampshire Library, Durham, New Hampshire.*

charred logs, which stood in stark contrast to a small green patch of trees that had been spared from fire. Ayres journeyed to every corner of New England, speaking to women's clubs, libraries, the Appalachian Mountain Club and other hiking clubs, and Grange halls. In all these presentations, he used his pioneering lantern slide show to convince his audiences of the ruinous aftermath of heavy logging and forest fires.

While Ayres was lecturing in every corner of New England, he also wrote articles for forestry journals and general-interest magazines stressing the severe economic effect of deforestation. In *Commercial Importance of the White Mountain Forests*, a monograph published by the U.S. Department of Agriculture, he noted, "There has been reckless waste of the vast forest wealth of the nation, which is still going on, but changes for the better are being made in important directions. The more thoughtful lumbermen see the issue clearly and have begun to treat the forest more conservatively."[8] He quoted Frederick Weyerhaueser, the leading lumber magnate in the nation, as saying, "The State has interests far beyond those of the individual."[9]

New England's congressional delegation also started to submit bills to protect the White Mountains. Senator Jacob Gallinger and Representative Frank Currier of New Hampshire introduced several bills, but none of them passed, as they faced opposition not only from Speaker Cannon but from the congressional delegations of the western states. By 1905, the repeated failures to pass legislation cast a pall over the conservationists' efforts. By then, southern conservationists and the New Englanders were beginning to realize that although they were separated by physical distance and culture, they would have to unite their efforts and transform forest protection from a regional cause into a national one.

In 1905, a propitious event smoothed the way for unification, and Pinchot was the linchpin. That year, he convened the American Forestry Congress in Washington, D.C., attracting leading conservationists and foresters from around the country to discuss the advancement of forestry. Reverend Edward Everett Hale, the highly respected chaplain of the U.S. Senate who strongly favored forest conservation, attended even though he was in his eighties. At one point, he raised his large frame slowly from his desk, and Ayres and two others helped him to the speaking platform. Ayres recalled admiringly, "With his eloquent voice, he told the story of the White Mountains and offered a resolution that was received with great enthusiasm."[10]

Ayres forged an alliance with Dr. Joseph Trimbel Rothrock, the father of forestry in Pennsylvania. Rothrock then buttonholed Pinchot—a fellow Pennsylvanian—and told him bluntly, "Now, Gifford, your bill for a National Forest in the Southern Mountains has been tried in Congress and failed. It always will fail until you get those Yankees behind it. You have got to have those New England votes and you might just as well agree to a National Forest in the White Mountains." Pinchot reflected for a moment and replied, "All right, I am with you." From then on, he became a stalwart supporter of the White Mountain forests.[11]

In 1906, the American Forestry Association worked closely with Pinchot and a deputy forester, William L. Hall, to draft a "Union Bill" that called for the creation of forest reserves in the southern Appalachians *and* the White Mountains. Proconservation legislators introduced the bills into the House and Senate. Congress held hearings at which Ayres, Edwin Start, Governor R. B. Glenn of North Carolina, and Governor William T. McLane of New Hampshire made powerful presentations. At the same time, Thomas Will lectured in the Lake states in favor of the bills and warned of an impending timber famine.

During the efforts to force a vote on the bills in the House and the Senate, tensions reemerged between the northerners and southerners, partly as the result of lingering resentments from the Civil War. Allen Chamberlain, president of the Appalachian Mountain Club, heard an erroneous report that Pinchot accused the New Englanders of being less than totally committed to the cause of forest legislation. The passionate Chamberlain took offense and attacked Pinchot for "showing the white feather" in the struggle to pass the bills. Pinchot was outraged, but friends managed to calm down both men, and the alliance was preserved.[12]

In May 1906, the House Agricultural Committee voted in favor of the forest legislation, but the bill failed to come to the House floor for a vote. Conservationists accused Speaker Cannon of pressuring the House Rules

Committee to prevent a vote on the bill. Angered by the accusation, Cannon shot back, "What I would consider an insult from an ordinary man I will overlook in your case since I consider men with a forest fad like yours to be nuts!"[13]

"Uncle Joe" Cannon and Representative John Weeks

As Cannon's words amply demonstrated, he was not kindly disposed toward conservationists. Born in 1836 in North Carolina to Quaker parents who had left the state because of slavery and migrated to the town of Danville, Illinois, Cannon had a deep-seated belief in self-reliance born of his rough childhood on the Illinois frontier. After studying law and serving a tenure as a district attorney in Illinois, he won election to the U.S. House of Representatives in 1872, beginning a career in the House that would last until 1923, when he retired at the age of eighty-seven. Cannon won election as Speaker of the House in 1903.

Cannon was a vivid character who was quite popular among his fellow politicos. He came to be known as "Uncle Joe" because of his garrulous personality, his humor, and his colorful stories about the rough-and-ready Illinois frontier. He was a forceful debater who stabbed the air with his pinwheeling arms as he passionately made his points, leading one House member to label him "The Dancing Dervish of Illinois."[14] In his life's story, *Uncle Joe Cannon: The Story of a Pioneer American*, which he dictated to his secretary, L. White Busbey, he proudly proclaimed, "I am one of the great army of mediocrity which constitutes the majority."[15] Woe to those who underestimated Representative Cannon, however. He was a cagey legislator, and when he ascended to the Speaker's chair, he ran the House with an affable but strict hand, rigidly enforcing party unity. When one irate citizen asked his representative for a copy of the House's rules, he received a photograph of Uncle Joe.[16]

As a Quaker and a Republican, Cannon thoroughly despised slavery and had wholeheartedly supported the Civil War, but in the years after the war, he viewed with suspicion the growing power of the federal government. "I have always been inclined," he dictated to Busbey, "to follow the old plan of the beginning of the Federal government and leave much of the Government to the States, and minor political divisions, and but for the slavery question and the civil war, I believe we would still be more devoted to State Rights than we are, and we would continue to look to the State Governments for our domestic laws rather than to Congress." Cannon was also a realist, though, for he continued, "But what has been done cannot be easily undone and Congress has practically taken the place of the State Legislatures as the body to appeal to when any community desires to change the law."[17]

As representatives wrote and introduced various forest bills, Cannon refused even to let them come to the floor of the House for a vote. He apparently found conservationists especially irritating, thundering at one point, "Not one cent for scenery!"[18] Creating eastern national forests, he opined, was an unjustified expansion of federal power. Besides, where would the money come from to purchase privately held forestlands?

In 1907, however, Cannon made a surprising decision: he assigned Representative John Wingate Weeks to the Agricultural Committee in the new Congress. Weeks (figure 5.2) was a native of New Hampshire and a successful Boston businessman, and he had made it known that he favored forest legislation. Born on April 11, 1860, in Lancaster, New Hampshire, he grew up in a family that had been prominent in New Hampshire politics for generations. His great-grandfather, also named John Weeks, had been a captain of the Continental Army during the American Revolution.

After a childhood spent on the family farm in Lancaster, Weeks attended the U.S. Naval Academy in Annapolis, where he won a reputation both for his physical strength and his gregarious personality. He served in the navy until 1883 and then migrated to Florida, where he took a position as a land surveyor. There he met his future wife, Martha A. Sinclair, who had also grown up in New Hampshire. Because she found Florida's climate disagreeable, they moved back to New England, settling in Boston. There, Weeks met Henry A. Hornblower, who owned an investment firm with his father. The senior Hornblower retired, and the two young men decided to go into business together, forming the investment firm of Hornblower & Weeks in 1888. The firm thrived, and by the early twentieth century, it had opened offices in New York, Chicago, and other cities.

Figure 5.2 *John Wingate Weeks, during Weeks's tenure as secretary of the U.S. Department of War, 1921–1925. Weeks, a native of Lancaster, New Hampshire, and U.S. representative from Massachusetts, sponsored the Weeks Act, which permitted the federal government to purchase privately owned forestlands and begin creating eastern national forests. From the collections of the Weeks Memorial Library, courtesy of the White Mountain National Forest.*

Weeks settled in the Boston suburb of West Newton, where he gradually involved himself in local politics. He served as an alderman and then, in 1901, won election as mayor of Newton. He was a staunch Republican but won a reputation as a fair-minded administrator who displayed sound judgment and a mastery of the financial details of city governance. After he had served two terms as Newton's mayor, a group of friends persuaded him to run for the open House seat from Massachusetts's Twelfth Congressional District. He easily won election, and on December 4, 1905, he took his seat in the Fifty-Ninth Congress. Because of his banking background, he was assigned to the Committee on Banking and Currency and the Committee on Expenditures in the State Department.[19]

In a letter to Gifford Pinchot dated June 18, 1912, Weeks recalled when he took his seat in Congress:

> Almost the first thing which attracted my attention was the Forestry Service. I commenced to look it up along general lines and soon ascertained the situation which obtained relating to the White Mountain and Appalachian bills, coming to the conclusion that if any result was to be obtained it must mean cooperation between the Eastern and Southern sections of the country.[20]

Drawing on his experiences growing up in New Hampshire, he proceeded to educate himself on the desperate condition of the eastern and southern forests.

In 1906, Weeks won reelection to the House. As the new Congress started to convene in 1907, Speaker Cannon asked Weeks to come to his office. Despite Cannon's skepticism about the forestry bills, he had come to the realization that forest legislation was inevitable, and he informed Weeks that he wanted to assign the Massachusetts congressman to the Agricultural Committee. Weeks objected that he already had his hands full with his assignments on two other House committees, but Cannon replied that the Agricultural Committee had become increasingly important and would be undertaking initiatives that were "experimental." Cannon undoubtedly was referring to the forest legislation. He then told Weeks that "it was especially essential that trained business men should be on the Committee."[21]

Weeks warned Cannon that he favored certain legislation that the Agricultural Committee would be considering during that session of Congress. Cannon replied, "I suppose you refer to forestry legislation." Weeks acknowledged that he was. Cannon said:

> I think forestry legislation is coming in time, but it has not seemed to me that the time has arrived yet when we ought to commence to purchase lands for forestry purposes. I may be mistaken in this prop-

osition, but my judgment is that it is too early to undertake such a policy. I am not, however, putting you on the Agricultural Committee because I expect you to make my views yours. . . . I want to say this, that if you can frame a forestry bill which you, as a business man, are willing to support, I will do what I can to get an opportunity to get it consideration in the House.[22]

Even while Cannon indicated that he would be open to forest legislation written so as not to alienate business interests, however, nature was about to underscore continued threats to the forests and galvanize public opinion.

The Monongahela Flood of 1907

On March 4, 1907, the Monongahela River rampaged over its banks and inundated lands throughout both Pennsylvania and West Virginia. In his *Fifty Year History of the Monongahela*, C. R. McKim wrote, "Heavy rains brought flood waters down the Monongahela River. . . . The trees and healthy vegetation were no longer there to regulate the rainwater's flow. It devastated all the rich agricultural land in the basin of the Monongahela River, causing some $100 million in damages—a gigantic sum for those times."[23]

The floods visited their full fury on Pittsburgh from March 13 through March 15. On March 13, the Monongahela, Allegheny, and Ohio Rivers rose rapidly and reached the flood stage of twenty-six feet, which was six feet over the danger mark for the city. On Deer Creek in the Pittsburgh suburb of Harmarville, a bridge carrying a freight train collapsed, plunging the train into the roiling waters below and killing three men.[24]

That night, torrents of rain continued to fall, and by the next day the situation was even worse, with floods cresting at thirty-five feet, or thirteen feet above the danger mark, the highest point the rivers had reached in seventy-five years. Most of downtown Pittsburgh was submerged, and streetcar service reached a standstill. Thousands of city residents rushed in a mad frenzy to Union Depot, pushing and fighting one another to force their way onto the few trains that were leaving downtown. Others fled the lower part of downtown for higher ground, overwhelming the few hotels and restaurants that remained open. Police officers tried—valiantly but without success—to restore order to a scene that the *Washington Post* described as "a chaotic mass of humanity." By the end of the day, the floods had claimed fourteen lives.[25]

In Allegheny County, thousands of people were forced to abandon their homes, and steel mills and coal mines temporarily ceased operations, throwing hundreds of people out of work. More than five hundred families climbed to the second stories of their homes and waited to be rescued. On March 15, fires broke out in the cities of Mount Washington, Pennsylvania; Wheeling, West Virginia; and Bridgeport, Ohio, but flood conditions prevented firefighters from reaching the fires.[26]

By then, the waters of the three rivers started to recede. Some streetcar service resumed, the city's many bridges once again became passable, and people who had been stranded downtown were finally able to return home, but pipes set up throughout the city continued to pump water out of the basements of hotels, offices, and homes. In all, ten square miles in and around Pittsburgh were flooded, causing $10 million in damage and leading the *Washington Post* to conclude that the conditions in the city had been "the worst ever recorded."[27]

The Monongahela floods also submerged lands in West Virginia. On March 16, the *Wheeling Daily News* reported that "Old Sol looked down upon a scene of dire desolation."[28] Streetcars had stopped running, telephones and telegraph lines were down, and downtown Wheeling was so flooded that people had to row skiffs to inspect the damage to their homes. Seventeen people died, and another six were missing. In an editorial on March 16, the *Wheeling Daily News* pulled no punches in assigning blame for the floods:

> Again the Ohio River by its conduct forcibly reminds us of the folly of timber destruction. No other cause than the devastation of the forests could have given the Ohio Valley such a deluge following the fall of so comparatively slight a volume of water.
>
> Twenty years ago two inches of rain would have done little else than make a big river. Today it caused the second largest flood in the history of the valley. The barren hillsides are responsible for it. There

is nothing to hold the water back. The river has become little more than a sewer. . . .

The timber is gone; it cannot be replanted and re-grown within the life of the present generation—but for the sake of posterity some action should be taken. France has a law which requires the replanting of a tree for every one cut. If the United States had had such a law Wheeling would have been out of water today.[29]

In 1908, West Virginia's Conservation Commission issued a sternly worded report:

The increase in total discharge of West Virginia rivers, in spite of diminishing rainfall . . . is due solely, so far as available data can be interpreted, to the deforestation of the mountains. There is no reason to doubt that a continuation of timber cutting will increase the fluctuation of the streams.

By keeping the mountains forested, a steady supply of water will be available; but if the woods are destroyed, the water will go down as destructive floods when rain has fallen, and it will quickly disappear when the rains cease.[30]

Legislators in West Virginia responded forcefully to the disaster. Soon after the waters receded, the state legislature passed a law permitting the federal government to purchase lands to create a Monongahela national forest preserve. It was an auspicious step that was calculated to prod the federal government into action.

In their arguments in favor of eastern national forests, conservationists had already begun to emphasize utilitarian arguments, particularly flood control and fire prevention. The Monongahela tragedy reinforced the emphasis on utilitarian arguments and the need to protect entire watersheds. As early as 1902, H. A. Pressey and E. W. Myers of the U.S. Geological Survey had analyzed how deforestation caused an increase in flooding:

To a certain extent, the forest acts as a reservoir, for it keeps the soil porous, allows it to absorb and hold the water for a time, and gradually gives it forth in the form of springs and rivulets. Where the areas have been deforested, however, the rain water forms small but swift-flowing torrents down the sides of the mountains, and quickly reaches the streams below. Deep channels are cut in the mountain sides, and all of the top fertile soil is carried off, leaving only the underlying clays, which are of poor quality and do not yield to cultivation.[31]

The causal connection between deforestation and flooding, however, would emerge as a source of deep disagreement among experts as debate over the forest legislation heated up in 1907 and 1908.

A Bill Is Declared Unconstitutional

With the Monongahela floods rousing public opinion and House Speaker Cannon's apparent acquiescence by appointing Weeks to the Agricultural Committee, the path finally seemed clear to pass an eastern national forest bill. In December 1907, a bill that would use revenues from grazing and logging rights from existing national forests to purchase new forestlands was introduced into the House. Opponents claimed that the bill violated the Constitution, however. In February 1908, the Agricultural Committee referred the bill to the House Judiciary Committee, and two months later, the Judiciary Committee declared the bill unconstitutional because no clause explicitly granted the federal government the power to buy privately held lands. The committee ruled that the government could only purchase lands under the interstate commerce clause: to protect the flow of rivers and streams carrying interstate traffic or providing power to businesses engaged in interstate trade.[32]

Meanwhile, President Theodore Roosevelt was using the powers of the presidency to press Congress for action. On December 7, 1907, he delivered a special message to Congress in which he called for the purchase of forestlands in the southern Appalachians and the White Mountains. Then, from May 12 through May 15, 1908, he convened a governors' conference at the White House to develop strategies for enhancing conservation efforts in the country. In addition to all the state governors, the conference included Roosevelt's cabinet members; the U.S. Supreme Court justices; numerous members of Congress; leaders of scientific and professional societies; reporters and editors from news organizations; and professors of forestry, botany, and other life sciences. The president delivered the keynote speech, commonly referred to as "Conservation as a National Duty," in which he linked conservation and civilization:

> With what we call civilization and the extension of knowledge, more sources come into use, industries are multiplied, and foresight begins to become a necessary and prominent factor in life. Crops are cultivated; animals are domesticated; and metals are mastered.
>
> We cannot do any of these things without foresight, and we can not, when the nation becomes fully civilized and very rich, continue to be civilized and rich unless the nation shows more foresight than we are showing at this moment as a nation. . . .
>
> The wise use of all of our natural resources, which are our national resources as well, is the great material question of today. I have asked you to come together now because the enormous consumption of these resources, and the threat of imminent exhaustion of some of them, due to reckless and wasteful use, once more calls for common effort, common action.[33]

For the next three days, conference attendees strategized how to guide the United States toward more efficient employment of its natural gifts and more scientific knowledge about how best to use those resources. On May 15, the conference ended with a statement by the governors that affirmed the importance of the nation's natural resources, from timber to water to minerals. "This conservation of our natural resources," the statement read, "is a subject of transcendent importance, which should engage unremittingly the attention of the Nation, the States, and the People in earnest cooperation."[34]

The same month as the governors' conference, Representative Charles G. Scott of Kansas introduced an alternative forest bill to the one that had been declared unconstitutional, but the true purpose of this bill was to sidetrack national forests in the East. The chairman of the House Committee on Agriculture, Scott was another member of the Republican Party's old guard who opposed Progressive initiatives and the growth of the federal government. His bill called for the creation of a commission to study the question of publicly owned forests, which was an obvious delaying tactic.[35] On May 21, the Scott Bill passed the House by a vote of 105 to 41, with 124 abstaining. It went to the Senate, where it was referred to the Commerce Committee, and there it languished.

Meanwhile, conservationists undertook an effort to win support for an eastern forest bill from legislators in the West, and Philip Ayres became the designated advocate. On December 28, 1908, Massachusetts's proconservation governor, Curtis Guild Jr., wrote a letter introducing Ayres to western governors and arguing that forest preservation was a national issue. He likened the issue to that of water irrigation, which was extremely important to the western legislators and which congressmen from the East had supported. Armed with this letter of introduction, Ayres undertook a three-month expedition west to win support. In a series of meetings with governors, Ayres argued that the whole nation would suffer economically if eastern timber supplies dwindled. One holdout was John Governor Shafroth of Colorado, who, when he became a U.S. senator, "voted faithfully against every forest measure."[36] All the other western governors, however, agreed to press their congressional delegations to support an eastern forest bill.

The Weeks Bill

On January 22, 1909, three representatives—Weeks, Asbury Lever of South Carolina, and Frank Currier of New Hampshire—introduced a new House bill that they had rewritten to meet the standard for constitutionality by emphasizing the protection of watersheds for interstate commerce. The bill, which now carried Weeks's name to reflect his leadership on the issue, contained significant changes from the 1908 bill:

1. It specified that the federal government could purchase forestlands to protect forests containing the headwaters of rivers and streams used for navigation and water power.

2. It did not mention the southern Appalachians or the White Mountains by name, thus broadening the potential application of the law

to any forests in the country. The references to specific forests were dropped to address opponents' accusations that it was special-interest legislation.

3. It dropped the use of logging and grazing revenues from existing forests and instead appropriated money from the U.S. Department of Treasury for the purchases of forestlands.[37]

On February 3, 1909, the Agricultural Committee reported the bill favorably to the House by a vote of 11 to 7. On March 1, 1909, the full House narrowly approved it by a vote of 157 to 147, with 82 abstaining. The Senate failed to consider the bill during that session, however, and it died.[38]

On July 23, 1909, Weeks submitted a revised bill, which now contained language very similar to that of the final law. It stated that the government could purchase lands "for the protection of the watersheds of navigable streams, and to appoint a commission for the acquisition of lands for the purpose of conserving the navigability of navigable rivers."[39] Senator Jacob Gallinger of New Hampshire introduced an identical bill into the U.S. Senate on December 20, 1909.

At this point, though, controversy erupted over the question of whether conservationists had exaggerated the connection between deforestation and flooding in order to ensure that the Weeks Bill passed constitutional muster. In 1908, Hiram Chittenden of the U.S. Army Corps of Engineers, who supported forest conservation in general, had delivered a report to the annual meeting of the American Society of Civil Engineers in which he accused conservationists of overstating the effect of deforestation on stream flow to justify federal purchases of forestlands.

Upon hearing Chittenden's accusations, Pinchot insisted on the connection between deforestation and stream flow. He testified before Congress that forests did, indeed, prevent flooding, and to make his point, he held up two pieces of paper. One was a blotter that he poured water on, and it absorbed water. The other was a photograph of a deforested mountainside. He poured water on it, and naturally the water ran down onto the floor. The demonstration was of dubious scientific merit, but it had the desired persuasive effect on members of the House.[40]

The issue of the connection between deforestation and stream flow would not go away, though. In early 1910, Willis S. Moore, chief of the U.S. Weather Bureau, distributed *Report on the Influence of Forest on Climate and on Floods*, in which he asserted that "the run-off of our rivers is not materially affected by any other factor than the precipitation" and that "forests should be preserved for themselves alone, or not at all."[41]

On February 23 and March 1 and 2, 1910, the House Agricultural Committee held further hearings on the Weeks Bill, and those hearings focused on the question of whether deforestation was responsible for flooding. The committee called three experts: George F. Swain, professor of civil engineering at Harvard University; L. C. Glenn, professor of geology at Vanderbilt University; and Filibert Roth. In March 1910, the American Forestry Association, which strongly supported the Weeks Bill, published essays in which all three experts attacked Moore's report. According to Glenn:

> This report of Professor Moore is too full of errors to be let pass unchallenged. Some of these errors are due to the statements made by Professor Moore being too broad and sweeping; some are due, either to Professor Moore's failure to grasp what the advocates of reforestation really propose to do, or to a failure on his part to make an adequate statement of their proposals; some are due to his confusing conditions on mountain head-waters with conditions on the lower navigable portions of river systems.[42]

Roth asserted as well that Moore's report was not based on scientific observation and evidence. At one point, for example, Moore had written, "On the whole, it is probable that forests have little to do with the heights of floods in main tributaries and principal streams, etc."[43] Such sentences, Roth claimed, were far too vague and ambivalent to qualify as science. Indeed, Roth asserted, Moore had failed even to consider whether land adjoining rivers was flat, hilly, or gullied. Roth explained:

> In this very matter of run-off Mr. Moore fails entirely to connect run-off with erosion, the gullying or development of the innumerable drain lines due to clearing of land, and aggravated by plowing.
>   That every furrow, every rod of gully, acts as a drain and hastens

run-off and prevents water storage does not seem to be of importance to Mr. Moore's position.[44]

With their thorough refutations, Roth, Swain, and Glenn ably persuaded members of the House Agricultural Committee of the connection among deforestation, irregular river flow, and flooding like that in Pennsylvania and West Virginia in 1907, but the issue would continue to be a controversial one during the implementation of the Weeks Act. The House Agricultural Committee issued a favorable report on the Weeks Bill, and, as promised, Cannon did not stand in the way of the bill coming to the House floor for consideration. In the last week of June 1910, the House scheduled floor debate, and Philip Ayres rushed to Washington to watch. He himself described the scene:

> John Weeks marshal[ed] his forces on the floor of the House. Joseph Cannon left the Speaker's chair to oppose the bill, and John Sharp Williams, leader of the Democrats, made a brilliant antagonistic charge of unnecessary extravagance. But the work of John Weeks was thorough; the bill passed, . . . and our ten year struggle was over.[45]

The historic vote, on June 24, 1910, was 130 to 111. In the Senate, though, passage was delayed for several months as opponents made one more last-gasp effort to kill the bill, with Senator Theodore Burton of Ohio leading a filibuster and Senator W. B. Heyburn of Idaho ranting against it as the "most radical piece of fancy legislation that has ever been proposed in the Congress of the United States."[46]

Once again, though, nature intervened to show dramatically why a new direction was imperative for the management of the nation's forests. On April 29, 1910, fire burst out on the Blackfeet National Forest in northwestern Montana, beginning one of the most incendiary summers the United States has ever had. On August 16, reports streamed into the U.S. Forest Service's district office in Missoula, Montana, about conflagrations on forests throughout northern Idaho and western Montana. President William Howard Taft called out four thousand federal troops to help combat the fires.

By August 19, officials thought they had the fires under control, but on

August 20, gale-force winds swept through the region and whipped the fires into a frenzy. For the next two days, the big blowup—as the fires came to be known—devastated forests throughout Idaho, Montana, and eastern Washington, killing more than eighty people and devastating three million acres of land. The fires galvanized the public and gave supporters of the Weeks Bill in the Senate a real-life demonstration of the desperate need for public forests and forest conservation in the East. In addition, Henry S. Graves, who had succeeded Gifford Pinchot as chief forester of the U.S. Forest Service, supported the idea of firefighting cooperation between the federal government and the states, a principle that was written into the Weeks Bill.[47]

Girded by the devastating events in the West, conservation organizations and the public turned up the pressure on the Senate, and on February 15, 1911, the Senate finally approved the bill by a vote of 58 to 9. Cannon bowed to the legislators whom he had fought so long and exclaimed, "Gentlemen, you have my scalp!"[48] On March 1, 1911, President Taft signed the Weeks Bill into law.

Provisions of the Weeks Act

The Weeks Act stated its purpose forthrightly: "To enable any State to cooperate with any other State or States, or with the United States, for the protection of the watershed of navigable streams, and to appoint a commission for the acquisition of lands for the purpose of conserving the navigability of navigable rivers." The general nature of the law—it named neither the White Mountains nor the southern Appalachians—proved to be a key element in its flexibility and success. The act established a clear procedure for purchasing forestlands:

- The act designated $1 million for fiscal year 1911 and $2 million for each year after, through 1915, for the purpose of surveying and acquiring lands containing the headwaters of navigable rivers. A total of $11 million was designated.

- It established a National Forest Reservation Commission to examine and recommend purchases by Congress. The commission consisted of the secretary of war, the secretary of agriculture, the secretary of the interior, two members of the Senate, and two members of the House.

- The secretary of agriculture was responsible for surveying and recommending lands for purchase to the commission. In time, the U.S. Forest Service and the chief forester gained significant new powers through these responsibilities.

- The commission could grant rights for cutting timber and mining the land even after the federal government had purchased it. In such cases, however, the law specified that "such rights of way, easements, and reservations . . . shall be subject to the rules and regulations prescribed by the Secretary of Agriculture for their occupation, use, operation, protection, and administration." Over the years, the granting of mineral and timber rights led to perhaps the greatest controversies in applying the law.

- The commission could approve the sale of certain lands appropriate for agriculture as long as the sale of these lands and their use for agriculture would not do "injury to the forests or to stream flow and which are not needed for public purposes."

- The federal government could provide grants to the states for the purpose of fighting forest fires. To receive the federal money, each state had to appropriate an equal amount for fighting fires. The act allotted a total of $200,000 for this program, which led to steady improvement in the firefighting capacities of states.[49]

In an editorial in *American Forestry* in March 1911, the American Forestry Association hailed the law, stating that although it was "greatly circumscribed," it represented a major step forward because "it makes our national forest policy really national" and "is a notable triumph of enlightened public sentiment over political obstruction."[50] The law was an enormous step forward for the country's young conservation movement. It affirmed the desirability of federal stewardship of forests and established a legislative precedent that led to the eventual passage of the Wilderness Act and other wilderness-protection initiatives. Equally important, the public gained a major and continuing voice in policies affecting the environment. Before the Weeks Act, the public's voice struggled to be heard; after the law, its voice would never again be ignored. The law's passage was a triumph by citizens who did not reside in the halls of power but who forced their way into those halls and created a living legacy that we renew each time we follow a trail in one of the eastern national forests.

On January 8, 1913, Joseph B. Walker, who had lived long enough to see the Weeks Act become law, passed away. He was ninety years old. In the obituary the *Concord Monitor* ran on January 9, 1913, Philip Ayres hailed Walker's leadership in introducing forestry to New Hampshire.[51] Ayres knew very well that he and conservationists around the nation had carried on Walker's work to its triumphant conclusion, yet they would face a new set of daunting challenges in implementing the law and creating a system of national forests in the East, the South, and the Lake states.

CHAPTER 6

Creating the Eastern National Forests

On May 11, 1911, a *New York Times* article fired off a sharp complaint: "Why has not Director [George Otis] Smith of the Geological Survey at Washington waked [*sic*] up sooner to the fact that he has not yet compiled data, showing whether the control of the 'strategic areas' of watersheds in the White Mountains and in the southern Appalachians will promote or protect the navigation of the streams that spring therefrom?"[1] The item accused Director Smith and the U.S. Geological Survey (USGS) of moving too slowly to identify possible tracts for purchase under the Weeks Act as well as of jeopardizing the $1 million appropriated under the law for purchases during the first year of its implementation. "How long," the *Times* demanded, "will Director Smith's new investigations take?"[2]

That government officials were somewhat slow to implement the Weeks Act was perhaps not surprising. The ambition of the law was momentous: to restore millions of acres and create a well-managed network of public forests that would serve a wide variety of uses. Over the next thirty years, from 1911 to 1941, the federal government would purchase a vast amount of acreage, plant millions of trees, and implement sound principles of forestry. The effort was marked by scientific methods, and a new class of scientists emerged to take leadership in the field of conservation. As environmental historian Samuel P. Hays has argued in *Con-*

servation and the Gospel of Efficiency, "These leaders brought the ideals and practices of their crafts into federal resource policy."[3]

These leaders' priorities were primarily utilitarian. They, and the political leaders who sponsored forest legislation and the development of forestry policies, regarded the protection of the eastern national forests as a project with tangible economic and social benefits. The forests would provide timber, minerals, and grazing lands. They would also provide places to hunt, fish, hike, and camp for the practical purpose of rejuvenating people's bodies and spirits so that they could serve as more productive members of society.

Implementing the Weeks Act

The processes that paved the way to create and expand the eastern national forests were notable for their thorough and methodical nature. Indeed, the legislators who devised the Weeks Act had required that officials from several branches of government be involved. The U.S. Forest Service was charged with identifying lands for potential purchase. The USGS had the responsibility for conducting surveys to make sure that proposed tracts met the legal requirement of containing headwaters of navigable rivers and streams. The Forest Service then recommended purchases to the National Forest Reservation Commission (NFRC), which comprised the secretaries of interior, agriculture, and war; two members of the House of Representatives; and two senators. The NFRC had to approve the purchases.

For the process to work smoothly, the Department of Interior and the Department of Agriculture had to cooperate, but tensions had been building between the two since 1898 because of Pinchot's criticism of the Interior Department's management of the national forests and his campaign to have the forests transferred to the Agriculture Department. These tensions reached the breaking point during the Ballinger-Pinchot affair in 1910. When William Howard Taft became president, he named Richard A. Ballinger, the former mayor of Seattle, to become secretary of the interior. Ballinger favored unimpeded development over conservation, and soon after taking office, he approved and granted Alaskan coal-mining rights to Seattle businesspeople, several of whom had been clients of his law practice. A whistleblower in the General Land Office complained to Chief Forester Pinchot that, in approving the coal-mining rights, Ballinger was clearly involved in a conflict of interest. When Pinchot took the complaint to Taft, however, the president fired the whistleblower. Pinchot then went public with his accusations against Ballinger. In early 1910, Taft fired Pinchot for insubordination, raising the fury of conservationists.

The Forest Service, however, was fortunate to find a highly competent successor as chief forester: Henry Solon Graves. Born on May 3, 1871, Graves earned his master's degree from Yale. In 1900, Graves told Pinchot that if his family put up the money to found the Yale School of Forestry, Graves would serve as the founding dean. The family did, and Graves assumed the position. After Pinchot was fired, several trusted associates urged Taft to select Graves as the replacement, and on January 12, 1910, the president named him chief forester. Graves's first priority was to prove that the new agency could wisely manage the nation's forests.

Identifying Forestlands for Purchase

The Weeks Act had appropriated $11 million: $1 million for purchasing forestlands for 1910 and $2 million for each year after through 1915, when the law would come up for renewal. The Forest Service considered a variety of lands for purchase, including (1) fully timbered lands in the watersheds of navigable rivers and streams, (2) cutover lands, (3) lands with enough brush to protect watersheds, (4) burned-over lands that had enough soil to nurture young trees, and (5) abandoned farmland that was not of sufficient quality to grow crops.[4]

The immediate goal was to purchase five million acres in the southern Appalachians and six hundred thousand acres in the White Mountains. Other regions identified for early purchase were the watershed of the Youghiogheny River in Maryland, the Potomac area in Virginia and West Virginia, the watershed of the Monongahela River in West Virginia, the Shenandoah area in Virginia, Natural Bridge in Virginia, the Iron Mountains in Tennessee and Virginia, the watershed of the Yadkin River in North Carolina, Mount Mitchell in North Carolina, the Great Smoky Mountains in North Carolina and Tennessee, Mount Pisgah and the surrounding area in North Carolina, the Nantahala area in North Carolina and Tennessee, and the Savannah area of Georgia and South Carolina.[5]

Purchase agents worked on behalf of the NFRC to identify blocks of land in a "purchase unit" that would be reviewed by the NFRC to ensure that the land met the requirements of the Weeks Act and that money was available to acquire it. Once the NFRC approved, the land in a purchase unit could be acquired from willing sellers. When enough land was acquired, the president would sign a proclamation designating the purchase unit as a national forest. Sometimes several purchase units were combined to form a national forest, and sometimes the Forest Service disbanded a purchase unit when agents had trouble acquiring enough land.[6]

To oversee the process for the Forest Service, Graves selected William L. Hall, an assistant forester who had established the Forest Service's new Forest Products Laboratory in 1910 in Madison, Wisconsin. Hall moved quickly, establishing an office in Washington, D.C., by May 1911 and hiring thirty-five men to go into the field and identify forestlands for possible purchase. That same month, Maine, Maryland, Virginia, West Virginia, North Carolina, South Carolina, Tennessee, and Georgia passed laws enabling the federal government to purchase forestlands within their borders. Offers to sell forestland came in from private timberland owners in the southern Appalachians and the White Mountains. The American Forestry Association urged private owners to consider the public good, asking, "Will those who hold the lands recognize the public necessity, as Congress has somewhat reluctantly done, and meet the government halfway? Or will they hold their property for impossible prices and thereby delay and obstruct the development of this great enterprise?"[7]

On March 27, 1911, the Forest Service issued a circular that explained the multiple uses for which it would manage the new national forests. The circular, which included blank forms that timberland owners could use to submit proposals to sell forestlands, stated that although the primary purpose of the law was to improve the flow of rivers and streams, it also had other objectives and benefits:

Table 6.1 Weeks Law Expenditures for Cooperative Forest-Fire Management, 1911

State	State Expenditures	Allotment to States	Balance of Allotment Not Yet Spent	Balance of $200,000 Fund, January 1, 1912
Maine	$23,557.07	$10,000	$8.20	—
New Hampshire	13,876.21	7,200	980.50	—
Vermont	2,243.90	2,000	782.00	—
Massachusetts	400.12	1,800	1,435.00	—
Connecticut	513.96	1,000	994.00	—
New York	3,837.59	2,000	—	—
New Jersey	1,241.51	1,000	10.00	—
Maryland	262.85	600	339.00	—
Wisconsin	20,841.87	5,000	562.75	—
Minnesota	25,675.77	10,000	—	—
Oregon	8,758.89	5,000	1,695.00	—
Total	$101,209.74	$45,600	$6,806.45	$161,206.45

Source: Robbins, *American Forestry: A History of National, State, and Private Cooperation*, 56.

1. Prevent soil erosion on the sides of mountains

2. Prevent forest fires from destroying soil

3. Preserve water power by making stream flow more even, securing water sources for municipalities

4. Manage and protect the supply of timber for the needs of U.S. industry

5. Protect the beauty of forestlands for people who use them for recreation and other leisure pursuits[8]

As it was identifying lands for purchase, the Forest Service also upgraded the ability to fight forest fires. Section 2 of the law allotted $200,000 for the federal government to match state appropriations for improving firefighting capacities, and states took advantage of the matching grants to hire fire marshals and other staff, purchase equipment, and build fire towers. By January 1, 1912, a number of states had moved ahead aggressively to qualify for the matching funds, as table 6.1 shows.

Controversy over Stream Flow

As the *New York Times* article of May 11, 1911, indicated, though, the USGS moved slowly to certify lands for purchase. The agency's lack of progress resulted partly from the difficulty of proving definitively the connection between deforestation and regular stream flow. Otis Smith directed the surveyors to conduct rigorous examinations, and after the USGS conducted preliminary surveys in early 1911, Smith wrote in *American Forestry* that "forests are *not* everywhere essential to the regulation of stream flow"[9] [italics added]. Outraged conservationists accused Smith and the USGS of two mistakes: establishing an impossibly high standard for showing the connection between deforestation and stream flow and prejudging the appropriateness of tracts for purchase.

Proconservation forces put heavy pressure on Smith to hurry the process along. On June 22, 1911, for example, the *New York Times* reported, "Dr. Smith has been severely criticized for thwarting the will of Congress by holding up the required approval from his bureau to the tracts selected by the Forest Service."[10] The pressure had the desired effect; by July 1912, the USGS had certified 1,962,800 acres for purchase in the southern Appalachians.

Surveys progressed more quickly in the South because in that region, surveyors found clear evidence of erosion due to deforestation. In the White Mountains, though, it was not as obvious, but in April 1912, the USGS finally found the evidence it sought, on the east branch of the Pemigewasset River, which drains much of the White Mountain region. Surveyors identified one tract, Shoal Pond Brook, that had virgin timber. They compared the tract with an area of about equal size around Burnt Brook, which had been heavily logged. During the time of the survey, both areas still had snow. The surveyors set up rain and snow gauges to measure precipitation and hydrometric stations to measure stream flow. During a seventeen-day period in April, three storms moved through the region. During the storms, the runoff of water into Shoal Pond Brook measured 6.48 inches, whereas the runoff from Burnt Brook measured 12.87 inches. The forested tract adjacent to Shoal Pond held the precipitation more effectively. In contrast, on the deforested tract, solar radiation caused the snow to melt and water to run off more rapidly. In its preliminary report, Director Smith concluded, "The results of the Burnt Brook–Shoal Pond Brook studies are held to show that throughout the White Mountains the removal of forest growth must be expected to decrease the natural steadiness of dependent streams during the spring months at least."[11] The findings were sufficient to allow the NFRC to approve White Mountain tracts. It was largely political pressure that pushed the USGS to issue a preliminary finding that allowed the acquisition of lands for this forest. The stream-flow controversy lives on a century later.

Challenges in Creating Eastern National Forests

Because the eastern forests had almost all been privately owned, the federal government had to go through a prolonged process of identifying potential purchases, surveying lands, clearing titles, and negotiating prices. The Forest Service realized that it would not be able to reach agreements on purchasing all forestlands within a purchase unit. The president designated a Proclamation Boundary to indicate where land could be acquired for a national forest. As a result, the federal government sometimes owns less than half of the lands inside a national forest proclamation boundary. For example, the federal government holds title to only one-fifth of the lands in Wayne National Forest in Ohio.[12] As a result, many eastern national forests consist of a patchwork of ownership that combines federally owned lands with private inholdings. This ownership pattern complicates management of a forest.

At times, state legislators, U.S. representatives and senators, and other political figures encouraged federal forest purchases because the infusion of federal funds could help revitalize a local—and usually rural—economy by restoring forests and building infrastructure. David K. Rice, chairman of the board of commissioners in Warren, Pennsylvania, noted:

> I fully realize that when the Allegheny National Forest was created, there was nothing here. Most of this land was nonproductive because it had been logged off. It was worthless. Everybody was exploiting

everything. The Allegheny National Forest took over this land and brought it up to the point where they could make one or two million dollars off the acreage during the course of a year.[13]

Surveying lands to be acquired could present challenges. Surveyors were examining lands that were mountainous, hard to get to, and often overgrown. Thomas Cox, who was the survey examiner in Georgia, described the difficulties in a report that he wrote in 1914: "Tracts difficult to locate as owners do not know anything definate [sic] of corners."[14] During the same year, James Denman attempted to survey former Vanderbilt estate lands near Asheville as part of the Pisgah Purchase Unit. He complained, "No one either in Vanderbilt employ or otherwise seems to know much about the location of their lands on the ground."[15]

After the Forest Service created a purchase unit and the NFRC approved the purchase, lawyers and Forest Service title specialists began to clear the titles for purchase. The boundary descriptions in old deeds could be exceedingly vague, however, referring to trees and other landmarks that had long since disappeared. William L. Hall described one grant in which the deed described a boundary as the point at which a white cow stood on a hillside. "Needless to say," Hall noted wryly, "the cow cannot be located."[16] Despite such challenges, the Forest Service proceeded methodically to survey, acquire, and restore forestlands in the East, South, and Lake states. Regional differences, though, led the process to unfold somewhat differently in each region.

The Southern Appalachians

In the southern Appalachians, the process of acquiring forestlands brought an infusion of federal dollars into the region, which was fervently trying to develop its economy. The South had suffered extensive deforestation (figure 6.1), but federal money would help start the process of forest restoration so that the region could take advantage of its geographical advantages, such as a long growing season and ample rainfall, to provide timber on a sustainable basis. In 1907, Agriculture Secretary James Wilson proposed acquiring five million acres of land in the southern Appalachians, and by

Figure 6.1 *Devastation on Mount Mitchell, North Carolina, 1923. Because of the Weeks Act, federal foresters in the U.S. Forest Service were able to restore forestlands that had been devastated by heavy logging and forest fires. National Archives (95G-176379).*

1912, the Forest Service had identified eleven purchase units (table 6.2). In December 1911, the government completed its first purchase: 18,500 acres in McDowell County, North Carolina.

One of the first eastern national forests created under the auspices of the Weeks Act was the Pisgah National Forest in North Carolina, and the first purchase in what would become the Pisgah was a tract of 8,100 acres along Curtis Creek, near Asheville.[17] The heart of the forest came from the Pisagh Forest, called the cradle of forestry. After the passage of the Weeks Act, George Vanderbilt, who believed that the forest was a public trust, started negotiations to sell 86,700 acres from his estate to the government. In 1913, Henry Graves and members of the NFRC toured the forest and stayed with Vanderbilt at his hunting lodge on Mount Pisgah.

Table 6.2 Eleven Original National Forest Purchase Units in
the Southern Appalachians

Name	Location	Initial Acreage
Mount Mitchell	North Carolina	214,992
Nantahala	North Carolina and Tennessee	595,419
Pisgah	North Carolina	358,577
Savannah	Georgia and South Carolina	367,760
Smoky Mountains	North Carolina and Tennessee	604,934
White Top	Tennessee and Virginia	255,027
Yadkin	North Carolina	194,496
Boone	North Carolina	241,462
Cherokee	Tennessee	222,058
Georgia	Georgia and North Carolina	475,899
Unaka	North Carolina and Tennessee	473,533
Total		4,004,157

Source: Mastran and Lowerre, *Mountaineers and Rangers: A History of Federal Forest Management in the Southern Appalachians, 1900–1981*, 50–51.

The visit and tour helped smooth the way for the negotiations, which were proceeding well when, on March 6, 1914, Vanderbilt died suddenly. His widow, Edith, shared her husband's belief that the Pisgah Forest was a public trust and resumed the negotiations with the Forest Service. On May 21, 1914, the two parties finalized the sale for $433,500, which was $200,000 less than the initial price that Vanderbilt had quoted to the purchasing agents.[18] On October 17, 1916, President Woodrow Wilson issued a proclamation creating the Pisgah National Forest as well as the Pisgah National Game Preserve.

Acquisitions continued in the South from 1912 to 1915, and as the Forest Service took on the management of these lands, it began to spread the methods of forestry to privately owned timberlands. Collaboration and persuasion were the watchwords. The Forest Service would not force private timberland owners to change how they managed their forests, but it could offer advice. In October 1912, William B. Greeley, who was the assistant chief forester in charge of silviculture and would take the reins as chief forester from 1920 to 1928, counseled southern timberland owners on the long-term economic advantages of selection cutting:

> The aim of [conservative forestry] should be to restrict the trees removed to those which are mature, leaving on the ground the younger, thriftier trees which are still making a fair rate of growth. Ordinarily this would mean probably the leaving of a quarter or a third of the merchantable stand per acre which is usually removed. . . . By leaving such trees standing and restricting the cut to the older growth which contains the best quality of lumber, it is my judgment that operators would often find the results beneficial from a business and manufacturing standpoint.[19]

Federal foresters did not always receive the warmest of welcomes. Inman Eldredge, who worked for the Forest Service in the South, recalled a "murky atmosphere of animosity" between federal foresters and some southern logging operators. Eldredge lamented, "All the rest of the people didn't know and didn't give a damn. Forestry was as odd and strange to them as chiropody or ceramics."[20] Such negative reactions were not universal, however. In February 1912, Verne Rhoades, who later became the first supervisor of the Pisgah National Forest, reported that people in North Carolina and Tennessee "in general regard most favorably the movement on the part of the government to purchase these mountain lands."[21]

The NFRC focused on purchasing tracts of land that were two thousand acres or larger. In forming the Nantahala National Forest in North Carolina and Tennessee, for example, the federal government purchased 60 percent of the lands from only twenty-two owners.[22] By focusing on large landowners, the Forest Service was able to move more quickly. After the creation of the Pisgah National Forest in 1916, other national forests followed in quick succession (table 6.3).

Progress in fire protection in the South was more inconsistent, though. In 1912, Kentucky wrote new fire-protection laws and created the position of state forester, steps that qualified the state to receive matching funds from the federal government. In 1914 and 1915, Virginia also created the position of state forester and initiated fire patrols in the western counties,

Table 6.3 Southern National Forests, 1916–1920

Year	Forest	Location	Changes
1916	Pisgah	North Carolina	
1918	Alabama	Alabama	Name changed to Black Warrior and then William B. Bankhead National Forest
	Shenandoah	Virginia	Merged with Natural Bridge and name changed to George Washington National Forest
	Natural Bridge	Virginia	Merged with Shenandoah and name changed to George Washington National Forest
1920	Boone	North Carolina	Merged with Pisgah National Forest in 1921
	Nantahala	North Carolina	
	Monongahela	West Virginia	
	Cherokee	Tennessee	
	Unaka	North Carolina	Split among Pisgah, Jefferson, and Cherokee National Forests in 1923 and 1936

Sources: Shands, *The Lands Nobody Wanted: The Legacy of the Eastern National Forests*, 9–10; Mastran and Lowerre, *Mountaineers and Rangers: A History of Federal Forest Management in the Southern Appalachians*, 59–60.

where the Jefferson National Forest was eventually formed. Not until 1921, however—ten years after the passage of the Weeks Act—did North Carolina appoint a state forester and create a fire-protection association. Georgia, Tennessee, and South Carolina were also slow to create fire-protection associations. Finally, by the mid-1920s, the federal government and the southern states had made substantial progress in improving fire-protection measures.

The White Mountains

In contrast to the South, land acquisitions in the White Mountains proceeded slowly at first. Agriculture Secretary Wilson had targeted about 600,000 acres for purchase, but by July 1915, the federal government had acquired only 265,000 acres in the White Mountains. Philip Ayres complained bitterly, "To friends of the White Mountains, this is disappointing . . . especially to those who realize the close relationship between the

mountain forests and the extensive water-power upon which New England's industry largely depends."[23] The Weeks Act had stipulated that $1
million must be spent in fiscal year 1910 and $2 million in 1911, but by
the time federal agencies worked out procedures for making purchases,
the time for spending those appropriations had passed, costing the federal
government $3 million of the original allotment of $11 million.[24]

In the White Mountains, the first purchase was the 7,000-acre Bertram
Pike in Benton, New Hampshire, on January 2, 1914. A few weeks later, the
Tract government made a second purchase, of 30,365 acres north of the
Presidential Range. One of the largest tracts consisted of 29,570 acres that
the Berlin (N.H.) Timberland Company owned and for which the government paid $8.00 an acre. According to the American Forestry Association,
"The tract has been carefully protected from fire for a number of years so
that the ground where the mature timber was removed a number of years
ago is now fully restocked with a good quality of young growth."[25]

In September 1914, the NFRC announced the purchase of another
tract: 85,000 acres on Mounts Washington, Adams, and Jefferson. Negotiations for the land had dragged on for three years, and conservationists feared that private landowners would harvest timber from the forests
before the federal government completed the purchase. The owners had
offered the land at a starting price of $28.60 an acre, which the Forest Ser-
vice regarded as too expensive. As the public clamored for the mountains
to be protected, the federal foresters conducted an inventory of the standing timber and, armed with data, negotiated a final price of $8.50 an acre.[26]

Private timberland owners, outdoor recreation buffs, and the federal
government often cooperated to ease the way for many of the purchases
in the White Mountains. For example, in 1914, proconservation residents
in the Sandwich Range, on the southern apron of the White Mountains,
urged the government to acquire a scenic area known as the Bowl. Kather-
ine Sleeper Walden, a Boston native who had opened a country inn in the
area and was a cofounder of the Wonalancet Out Door Club, reacted with
horror when she heard that loggers were about to start harvesting timber
in the Bowl. She approached Louis Tainter, the president of the Publishers
Paper Company, which owned the land, and persuaded him to give her

an option to buy 3,000 acres in the Bowl for $50,000.[27] Negotiations over these and other tracts stretched out over the next several years, but eventually the Publishers Paper Company and the Conway Lumber Company, of which Tainter was a vice-president, sold 204,000 acres to the federal government, nearly a fourth of what became the White Mountain National Forest.[28]

The federal government bought lands in three purchase units in 1911: the White Mountain Purchase Unit, the Androscoggin Purchase Unit, and the Kilkenny Purchase Unit. On May 16, 1918, the Forest Service merged those purchase units to form the White Mountain National Forest (WMNF), which encompassed an area of 950,114 acres in northern New Hampshire and western Maine. In 1928, the Forest Service dropped the Androscoggin Purchase Unit from the WMNF, reducing the forest to 801,900 acres in New Hampshire and 53,300 acres in Maine.[29]

As these land acquisitions proceeded, Assistant Forester Hall attended a conference on forestry at Lake Sunapee in New Hampshire in July 1913 and delivered an address titled "The White Mountain Forest and How It Is to Be Made Useful." The address, which was reported in the September 1913 issue of *American Forests*, provided a clear view of the Forest Service's management objectives as it organized the network of eastern national forests. The prominence of the word *useful* in Hall's title echoed the gospel of efficiency and utilitarianism. Hall asserted that the central principle guiding the Forest Service was "the right use of land" for the practical purposes of "giving a living" and "making life better and more enjoyable."[30] Hall discussed the importance of improving fire protection and restoring forests to regulate stream flow. The rejuvenated forests would produce timber on a sustainable basis to meet the needs of a rapidly industrializing nation.

Hall, however, also devoted as much time to recreation as he did to resource extraction. "Conditions here are so delightful," he rhapsodized, "as to attract each year increasing thousands, not only from New England, but from all over the country."[31] These delights, he continued, would have utilitarian benefits, as the thousands who sojourned into the forests would "rest and renew themselves for their labors in their own cities and towns. Used in this way, the White Mountain Forest is an intangible, but nevertheless, a real asset, and possible one of vastly underappreciated importance in our national life."[32] Hall's speech reflected the Forest Service's belief that it could balance recreational needs with resource uses by providing timber, mineral, and grazing lands to support the country's economy.

The Forest Service also made extensive efforts to educate the public in forestry. In February 1915, Hall published an article in which he explained in detail how the Forest Service would apply methods of forestry. After a bidding process, the agency had recently awarded the sale of half a million board feet of timber in the White Mountains. The purpose of the sale was to "clear away a large amount of timber and liberate a fine stand of young trees."[33] The logging operation would generate revenue for local municipalities, as the Weeks Act had stipulated that 5 percent of revenues from federal timber sales must go to the towns and counties to compensate them for lost property taxes. (The amount of revenues reverting to local municipalities soon rose to 25 percent.)

The area to be harvested encompassed 162 acres in the notch west of Mount Moosilauke, some thirty miles southwest of Mount Washington. The loggers would use selection cutting, in which the loggers selected the trees to be harvested by considering numerous factors, including the age of the trees, the mix of species, and past methods of logging on the tract. The largest tract, consisting of 118 acres, had been heavily logged fifteen years before. The remaining hardwoods were, in Hall's words, "defective" and "on the decline." In addition, Hall explained, "The previous cutting left several big holes in the forest where all the trees were removed."[34] The Forest Service would direct the logging company to remove maples and yellow birches more than ten inches in diameter and paper birches and other species more than eight inches in diameter. The loggers would harvest four thousand board feet of timber per acre but would leave old-growth timber on the summit of Mount Moosilauke. Hall was careful to emphasize that the Forest Service would attempt to preserve and enhance the aesthetic values of the forest. For example, on either side of a state highway that ran through the White Mountains, the forest supervisor would direct loggers to remove all brush along the highway and cut dead trees and snags.[35] By 1920, the federal government, local conservationists, and private timberland owners had engaged in an extensive amount of collaboration to create the White Mountain National Forest and to begin restoring its forests.

The Lake State Forests

In Michigan, Wisconsin, and Minnesota, the creation of national forests moved more slowly, particularly compared to the South. The geography of the region made it harder to fulfill the provision in the Weeks Act that lands must contain the headwaters of navigable streams. In addition, the focus during passage of the Weeks Act had been entirely on New England and the South.[36] Even so, the depletion of forests in the Lake states was acute, and the states faced economic devastation as one of their most important natural resources dwindled. As a result, the three states were proactive in creating state forests and introducing methods of forestry to private timberland owners.

One economic concern that drove conservation efforts in the Lake states was the fear of a timber famine. In February 1920, Senator Arthur Capper of Kansas requested a report from the Forest Service on the timber situation in the United States, and the resulting Capper Report, issued in 1920, documented the dwindling supplies of timber. In 1925, the U.S. Senate Select Committee on Reforestation reported that the Lake states contained fifty-seven million acres suitable for growing trees. Of that vast acreage, ten million acres still had virgin timber, twenty-six million acres had second growth that one writer termed "haphazard," and twenty-one million acres were completely barren of trees.[37]

The cut-and-run approach had also created major tax headaches for the Lake states. After a company finished harvesting a tract, it often stopped paying the property taxes, and the land reverted to the state or county. In 1925, for example, the state of Michigan owned seven hundred thousand acres of delinquent properties. All three states tried to divest of the land by encouraging agriculture in the north country, but as chapter 5 explained, such efforts failed because of climate and soil conditions. For example, although Michigan had 15,500,000 acres of logged-out lands, the amount of land being used for agriculture in the state increased by only 93,000 acres between 1910 and 1920.[38]

Figure 6.2 *Planting crew in Minnesota National Forest, 1921. After the passage of the Weeks Act, workers planted millions of trees on lands that had been cut over. The cost of planting was less than $3 an acre. USDA Forest Service, Chippewa National Forest.*

From the early 1900s to 1924, the federal government created national forests in Michigan and Minnesota but not in Wisconsin, using lands that were still in the public domain. Starting in 1902, the General Land Office began to identify potential national forests in Michigan's Upper Peninsula, and it held back from sale several thousand acres that were still in the public domain. On February 11, 1909, the government officially created the Michigan and Marquette National Forests, both in the Upper Peninsula. The forests had only one ranger and no roads, fire towers, or offices. In 1912, the Forest Service built the Norway Ranger Station, and in 1915, the government consolidated the two national forests. During the 1930s, these forests would undergo rapid expansion and a number of administration changes.[39] The agency also undertook reforestation efforts (figure 6.2).

In Minnesota, the federal government had acquired lands formerly belonging to the Ojibwa Indians and created the Minnesota Federal Forest Reserve in 1902. In 1908, this reserve became the Minnesota National Forest. Soon after the passage of the Weeks Act, the federal government acquired 37,135 more acres of former Ojibwa Indian lands.[40] For several years, the federal government and the Ojibwa tribe negotiated over compensation for the lands, and in 1923, the government finally paid the tribe $14,091,976. In 1928, the Forest Service changed the name to the Chippewa National Forest, "Chippewa" being an inaccurate Anglicization of "Ojibwa."

General Christopher C. Andrews, the forestry advocate who spearheaded the creation of the Minnesota National Forest and served as the state's forestry commissioner, then turned his sights to protecting the region of pristine lakes and rivers known as the Boundary Waters Canoe Area. Andrews urged state legislators to set aside 500,000 acres for forestry purposes rather than sell the land to farmers, and in 1902, the state withdrew the acreage from sale. The state also set aside another 141,000 acres along the Canadian border in 1905 and more than 500,000 acres in 1908. On February 13, 1909, in the waning months of his presidency, Theodore Roosevelt formally approved the Superior National Forest, which then totaled 1,018,638 acres, including Boundary Waters.

Legislation After the Weeks Act

The Weeks Act came up for renewal in 1915, and that year, conservation groups, including the American Forestry Association, the Appalachian Park Association, the North Carolina Forestry Association, and the Appalachian Mountain Club, launched an intensive lobbying effort to persuade Congress to extend it. In July 1915, Philip Ayres wrote in *American Forestry*, "This [the law] was an experiment. It has been worked out successfully."[41] On September 23, 1915, conservationists met with the secretary of agriculture, David F. Houston, and convinced him that although the law had led to important acquisitions, it urgently needed to be extended. Houston supported the extension, and in 1916, Congress approved the expenditure of $2 million per year for another five years.

During these years, the Forest Service, state forestry agencies, and volunteers throughout the East, the South, and the Lake states undertook the restoration of the forests. In 1915, Philip Ayres reported, "The method of logging pursued by the Federal Government on the National Forests provides for removing the mature trees, clearing up the debris, protection from fire, and preservation of the crown-cover of the forest so that the sun does not beat upon and dry out the soil."[42]

As the U.S. economy boomed during the 1920s, the demand for forest products soared. Prodevelopment forces complained that the Forest Service had made too much forestland off-limits for timber harvesting, raising once again the specter of a timber famine. William B. Greeley, who became chief forester in 1920 and believed strongly in cooperating with the private sector, implemented a policy of assisting private owners to adopt sustainable forestry practices as part of a strategy to produce adequate quantities of timber. In addition, Congress strengthened federal efforts to restore eastern forests and increase timber supply by passing three new land-acquisition laws: the General Exchange Act of 1922, the Clarke-McNary Act of 1924, and the McNary-Woodruff Act of 1928. All three laws dramatically expanded the eastern national forests.

To create national forests of contiguous lands, the Forest Service wanted to use land-exchange programs. The General Exchange Act allowed the Forest Service to exchange lands with private owners, but under tight restrictions. The lands to be exchanged had to be located within the same state, carry comparable monetary value, and not contain valuable mineral resources. The Forest Service used land exchanges sparingly, but the strategy proved to be an important tool in building the eastern forests.

The Clarke-McNary Act of 1924 expanded the types of lands that the federal government could purchase. This act removed the Weeks Act requirement on protecting only the headwaters of navigable streams and instead permitted the purchase of "lands within the watersheds of navigable streams" and to produce timber.[43] Under the law, the Forest Service could purchase almost any land from willing sellers because nearly all land lies within the watershed of a river or stream. It also expanded the ability of the federal government to work with the states on forest fire protection and reforestation issues.

In 1927, destructive floods occurred along the Mississippi River, and conservationists blamed the flooding on cutover forestlands. In response, the Forest Service proposed a more aggressive program of land acquisition, and Congress responded with the McNary-Woodruff Act of 1928. The law stipulated that the federal government could spend up to $8 million between 1928 and 1931 to purchase forestlands. Using the funds from this law, the government purchased 4.6 million acres east of the Mississippi River.[44]

The passage of the Clarke-McNary Act opened the way for purchasing lands to create national forests in the Lake states. Wisconsin had no national forests, but in December 1928, the federal government purchased more than four hundred thousand acres in six counties in northern Wisconsin, and on March 2, 1933, Nicolet National Forest was created. The Forest Service divided the two forests into Nicolet East and Nicolet West, and in November 1933, Nicolet West became the Chequamegon National Forest. At last, Wisconsin had joined Minnesota and Michigan in having national forests.

As a result of the General Exchange Act, the Clarke-McNary Act, and the McNary-Woodruff Act, national forests now spread throughout the states east of the Mississippi. Those lands were in varying states of health. Some boasted stands of virgin timber, whereas others had to be replanted and restored. In 1930, Congress passed the Knutson-Vandenberg Act, which appropriated funds for the federal government to restore national forests, improve timber stands, and establish nurseries to provide seedlings for restoration. In the years since the passage of the Weeks Act, Congress and the Forest Service had taken major steps in creating a well-managed eastern national forest system. The stage was set for the far-reaching activism of the New Deal.

New Deal National Forests and the Civilian Conservation Corps

On Tuesday, October 29, 1929, the stock market in the United States plummeted, wiping out the life's savings of millions of people and sending the U.S. economy into a sickening spiral. As the Great Depression engulfed the country, farmers, who had never benefited from the economic boom of the 1920s, found that they could not sell crops for any kind of a profit. Then, in 1930, drought spread across the Great Plains, and thousands of farmers abandoned the environmental disaster known as the Dust Bowl. By 1934, banks were foreclosing on nearly forty of every one thousand farms.

When he became president in 1933, Franklin D. Roosevelt recognized that lands throughout the country were in desperate need of conservation and restoration. He himself took great pride in having restored the woodlands on his estate in Hyde Park, New York. Under his leadership, the federal government embarked on an ambitious and far-reaching program of soil and forest conservation. From 1933 to 1942, the Roosevelt administration oversaw the creation of twenty-six new eastern national forests—stretching from Texas to Georgia and from Florida to Minnesota—forests that have been known ever since as the New Deal forests.[45] The federal government also developed comprehensive forest policies for the first time. In 1932, Senator Royal Copeland of New York submitted a resolution to the Senate calling for the Forest Service to provide a comprehensive report on the state of the nation's forests. In 1933, the Forest Service turned over the report to the Senate, which issued it on March 13, 1933, under the title *A National Plan for American Forestry*. Coming in at 1,677 pages, the Copeland Report provided an unprecedented look at fire protection, timber harvesting, water quality, grazing policies, mining practices, wildlife, and federal-state relations.[46]

The Copeland Report criticized forest management practices by private timberland owners and argued for greater public ownership and more assertive management of the nation's forests by the federal government. In response, proconservation forces in Congress submitted the Omnibus Forestry Bill, which would have expanded public ownership and regulation of privately owned forests. The bill never passed Congress, but its twin goals of expanding public ownership and managing national forests guided federal policy throughout the rest of the New Deal.

To add new eastern national forests, the Roosevelt administration used the powers of the Clarke-McNary Act, the McNary-Woodruff Act, and $20 million in funding for special emergencies.[47] Because of falling land prices and foreclosures, the government purchased countless small tracts of land. Forest historian David E. Conrad writes, "The purchase units of this period looked like crazy quilts."[48] They comprised a blend of privately owned and publicly owned lands. Adding to the complication was that owners often retained the mineral rights to the land they sold. (Chapter 12 will explain the effect that mineral rights continue to have on eastern national forests, particularly regarding oil and natural-gas extraction.) The resulting patchwork ownership, which the Forest Service referred to as "land ownership adjustments," became a permanent challenge in managing eastern national forests.[49]

Another challenge for the federal government was finding thousands of people to do the work of forest restoration. Only five days after his inauguration, Roosevelt convened a meeting at the White House in which he outlined to advisors his plan to employ half a million jobless men to work on conservation projects. The White House collaborated with the secretaries of war, labor, agriculture, and the interior to draft a bill forming the Civilian Conservation Corps (CCC), which practically overnight would create an army to restore devastated forests and farmlands. The bill passed Congress easily, and the president signed it on March 31, 1933, only twenty-eight days after he had been inaugurated.

The law spelled out a plan that combined the simplicity of life lived close to nature with the efficiency of an army camp. Enrollees in the CCC were to be paid $30 a month. They could be single or married, although the majority turned out to be single men in their early twenties. If they had dependent children or parents, they were required to send a portion of their earnings to their families.

The director of the CCC was former labor leader Robert Fechner. Under Fechner's leadership, the Forest Service identified forest planting and soil erosion projects and supervised the work. To guide environmental restoration projects, the agency hired approximately twenty-five thousand local foresters, spreading the benefits of the CCC to rural communities near national forests. The foresters taught the young recruits the rudiments of forestry and sound techniques of forest restoration.[50] The CCC's efforts to restore forests were nothing less than astonishing. In 1933, fewer than 25 million trees had been planted in national forests. By 1938, the CCC had planted 190 million trees. Its workers also built 3,470 fire towers, connected them with 65,100 miles of telephone lines, and constructed 97,000 miles of roads for firefighting.[51] They destroyed gypsy moths and cut trees that had been infested by beetles. Even though it existed for only nine years, the CCC played a pivotal role in restoring the eastern national forests.

On July 22, 1937, Congress also passed the Bankhead-Jones Farm Tenant Act, which authorized the federal government to acquire damaged lands to rehabilitate them for conservation purposes. The acquisitions helped stabilize rural economies and provided opportunities to put people back to work restoring the damaged lands. Several eastern forests came into the system as a result of Bankhead-Jones authorization, including the Tombigbee National Forest in Mississippi (1959) and the Finger Lakes National Forest in New York (1985).

By 1941, with the United States poised on the edge of war, three decades had passed since the passage of the Weeks Act. In that short period, the law and its descendants—the General Exchange Act, the Clarke-McNary Act, and the McNary-Woodruff Act—had wrought nothing less than a revolution in preserving and restoring the forests of the East, the South, and the Lake states. Forests that had been cut over and burnt now sprouted millions of saplings, and land that had been clotted with ugly debris and barren ground was now carpeted with millions of acres of trees. The years had also seen the development of a Forest Service that was highly professional, efficient, and committed to Gifford Pinchot's vision of managing the nation's forests for multiple uses. Moreover, the period had seen a high level of cooperation among federal agencies, state agencies, and, to a growing extent, private timberland owners in implementing sound forestry practices to ensure timber uses for the future.

These years of success also churned with unanswered questions, though. How could harvesting the ample resources of the forests be balanced with competing claims for outdoor recreation and wilderness? Who would make decisions about priorities, and how would the public be included in the decision-making dialogues? Even as such questions began to emerge, the success of the thirty years from 1911 to 1941 stood as a momentous accomplishment. During these years, the United States established itself as a world leader in protecting and managing a magnificent resource: its forests.

PART II

ISSUES FACING THE EASTERN NATIONAL FORESTS TODAY

By the end of World War II, the United States had created more than forty-five eastern national forests as the result of the Weeks Act, the Clarke-McNary Act, and other actions by the federal government. The map on page 143, "U.S. National Forests, 2012," shows just how extensive the eastern national forests have become. The U.S. Forest Service, the Civilian Conservation Corps, other agencies, and countless volunteers worked to plant trees and repair forests that had been cut over and degraded by erosion. This remarkable effort placed the country in the forefront of forest restoration in the world. The forests gradually regained their health, providing timber and other resources to the country and attracting outdoor lovers who flocked to the restored forests to hike, camp, climb, hunt, fish, and engage in other leisure pursuits.

Starting in the 1950s, societal changes, such as a rapidly growing economy and the development of the environmental movement, presented the managers of the eastern national forests with a new set of issues and challenge. The purpose of part II is to examine those issues, which range from the amount of timber harvesting to the protection of wilderness. We take a case-study approach, exploring important issues through specific national forests.

Tying these case studies together are four trends that emerged in the development of the forest conservation movement and the passage of the

Weeks Act: (1) the expansion of scientific knowledge about forest ecology; (2) the associated understanding of the interrelationships of flora, fauna, water, soil, and air in forest ecosystems; (3) changing attitudes among the American public toward forests and other natural systems; and (4) citizens' involvement in policies affecting the eastern national forests. Citizens' opinions about the eastern national forests range widely, including those who favor a continuation of the traditional multiple-use policies, those who do not want to see resource extraction from their back windows (the so-called not-in-my-back-yard, or NIMBY, constituency), and those who oppose all resource extraction from national forests. It is important to listen to all these voices and understand how the Forest Service has responded to them in managing the eastern national forests.

These trends have led to significant changes in how the eastern national forests are managed, how the public involves itself in decisions affecting the forests, and how timber companies and other private companies operate on the forests. All these trends will be important as the forests continue to evolve during the twenty-first century.

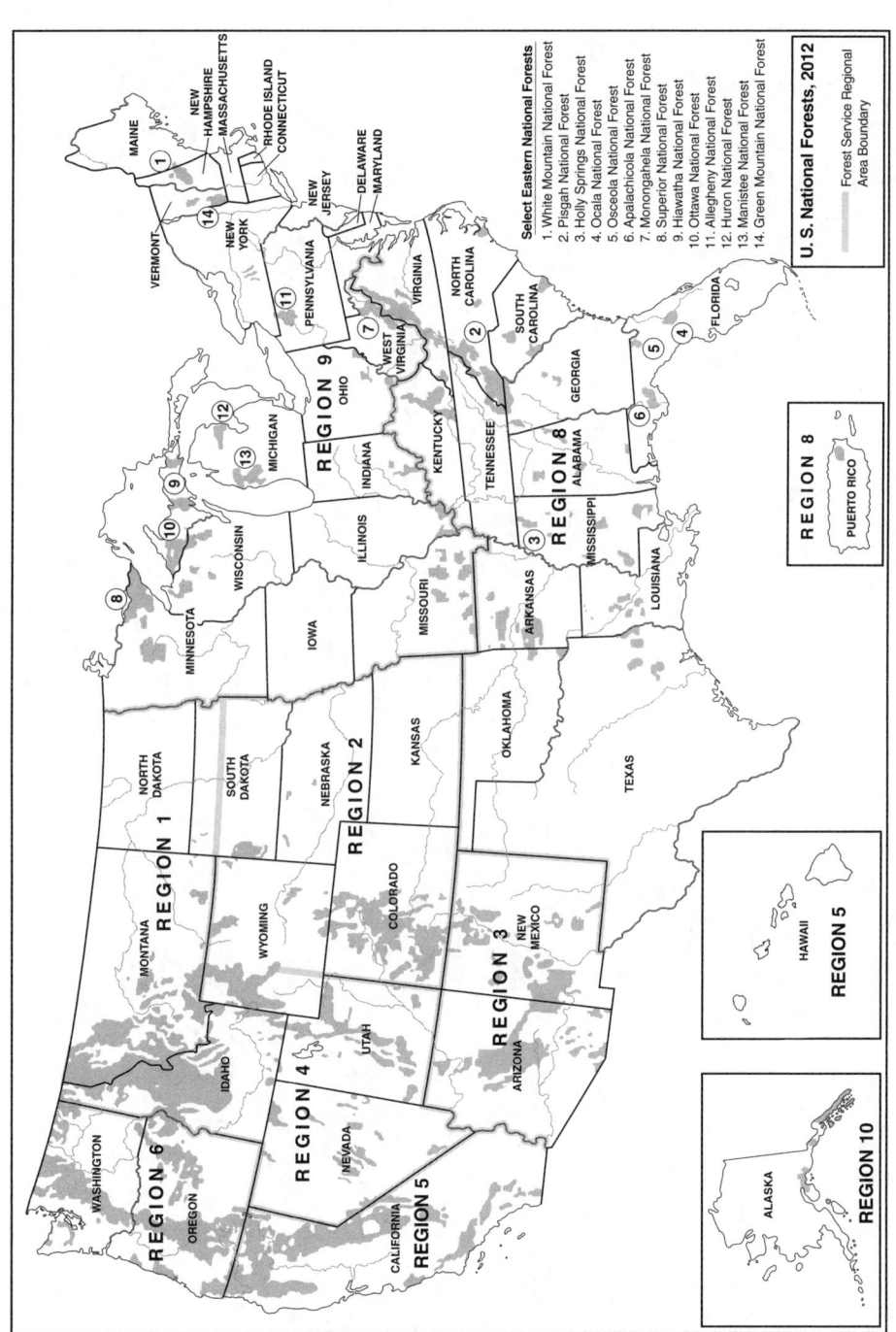

Select Eastern National Forests

1. White Mountain National Forest
2. Pisgah National Forest
3. Holly Springs National Forest
4. Ocala National Forest
5. Osceola National Forest
6. Apalachicola National Forest
7. Monongahela National Forest
8. Superior National Forest
9. Hiawatha National Forest
10. Ottawa National Forest
11. Allegheny National Forest
12. Huron National Forest
13. Manistee National Forest
14. Green Mountain National Forest

U. S. National Forests, 2012

Forest Service Regional
Area Boundary

REGION 8

PUERTO RICO

REGION 5

HAWAII

REGION 10

ALASKA

National Forests, 2012. Map created by Christopher Robinson.

CHAPTER 7

Holly Springs National Forest: A Study in Forest Management Reform

Holly Springs National Forest, which sits just outside the graceful city of Oxford, Mississippi, spreads like a gentle sea of green across 155,000 acres in northern Mississippi. The forest is graced by sparkling streams and rivers, including the Tallahatchie River and the Wolf River, and by crystalline lakes, including the Chewalla, the Sardis, and the Puskus. The locals embrace the forest for hiking, bird-watching, horseback riding, hunting, and fishing. During Mississippi's lovely springs, the blossoming dogwoods attract visitors from Memphis and other parts of the mid-South (figure 7.1). Thousands of hunters trek there every year to hunt deer, turkey, and quail, and anglers flock to the rivers and lakes to catch bass and trout.

In May 2010, the forest's tranquility was shattered when a tornado ripped through the forest, leaving an ugly two-hundred-yard-wide swath that looked as if Paul Bunyan had dragged his axe through the forest. By early 2011, salvage trucks were carrying away fallen pines from 312 acres in the northern part of the forest. The district ranger, Joel Gardner, a native Louisianan with a wry sense of humor, commented that once the salvage operation was completed, the U.S. Forest Service would plant shortleaf pines, which are native to the region.[1]

Gardner explained that Holly Springs had a more diverse forest than it did twenty years before. "We don't *want* a single-species forest," he said. "In recent years, the forest has gotten more hardwood, and the composition

Figure 7.1 *Blossoming dogwoods in Mississippi. Every spring, tourists travel to Holly Springs National Forest to observe and take photographs of the flowering dogwoods that grace the restored forest. © Clint Farlinger/Alamy.*

of the forest is now 35 percent hardwoods and 65 percent pine."[2] The pine species consisted mostly of loblolly and shortleaf pine, whereas the hardwoods included oak, hickory, dogwood, green briar, sassafras, sweet gum, and beech trees. The Forest Service was even planting chestnut trees, using a seedling that scientists hoped would resist the chestnut blight that destroyed millions of those magnificent specimens in the early twentieth century.

After the tornado, the Forest Service bid out the job of salvaging the fallen timber and awarded the contract to a private timber operator, all in about six months. Gardner and Rives "Buddy" Lowery, a longtime forester at Holly Springs, pulled out a four-inch-thick file on the sale of the salvage timber that included an environmental impact assessment, detailed maps showing the topography of the forest, tables showing the number of trees on the site, and contracts for the salvage company.

In recent years, the number of standing trees logged at Holly Springs has declined dramatically, whereas the amount of salvage timber has risen because of weather events and insect infestations. Gardner said, "In the 1980s, we were harvesting 22 million board feet of standing timber a year."[3] By 2010, however, only 1.6 million board feet of standing timber were harvested from the forest. Table 7.1 shows timber sales on Holly Springs National Forest for the fiscal years 2007 through 2011. (A fiscal year starts in October, so fiscal year 2011 started on October 1, 2010.) It also shows the amount of road construction, which inevitably accompanies timber harvesting. This work was primarily to rebuild existing roads, although in 2007, about half the work was to construct new roads.

The Forest Service had developed a detailed process to determine the potential effect of proposed timber harvesting. Loggers could not clear-cut a tract more than forty acres in size, and the Forest Service had to provide a rationale for even those clear-cuts by showing the effects on wildlife, recreation, and the scenic beauty of the forest. The Holly Springs National Forest staff emphasized that in making decisions about timber harvesting, they had moved toward an interdisciplinary approach that included soil and water specialists, wildlife biologists, and landscape architects, all of whom could weigh in on timber-harvesting plans. Caren Briscoe, the recreation land environmental coordinator at Holly Springs, said, "We have visual quality objectives. For example, we limit harvesting around streams, rivers, and lakes. A harvesting inspector observes the operation to make sure that aesthetic guidelines and other contractual obligations are followed."[4]

As recently as the mid-1990s, timber-harvesting policies at Holly Springs were far different from what they had become by 2011. For decades, the primary emphasis had been on the production of timber, but in 1997, controversy over timber harvesting broke out, beginning a ten-year debate over how the forest was being managed. The controversy forced the U.S. Forest Service in Mississippi, which manages six national forests, to examine how it made timber-harvesting decisions as well as how the public was involved in those decisions.

The controversy at Holly Springs was regional, but it had national implications, raising issues that have emerged in similar disagreements on

Table 7.1 Timber Harvest in Holly Springs National Forest, 2007–2011

Fiscal Year	Green[a]/ Salvage[b]	Road Work (Miles)	Sawtimber[c] (Million Board Feet)	Roundwood[d] (Million Board Feet)	Total (Million Board Feet)
2011	Green	2.4	4.880	1.385	6.265
	Salvage	0.1	2.100	3.860	5.960
	Total	2.5	6.980	5.245	12.225
2010	Green	0	0.170	1.407	1.577
	Salvage	0	14.535	7.265	21.800
	Total	0	14.705	8.672	23.377
2009	Green	0	0.670	3.625	4.295
	Salvage	0	3.350	0.600	3.950
	Total	0	4.020	4.225	8.245
2008	Green	3.5	2.325	2.270	4.595
	Salvage	0	8.218	1.660	9.878
	Total	3.5	10.543	3.930	14.473
2007	Green	5.6	1.785	10.140	11.925
	Salvage	0	0.300	0.055	0.355
	Total	5.6	2.085	10.195	12.280
Five-Year Totals	Green	11.5	9.830	18.827	28.657
	Salvage	0.1	28.503	13.440	41.943
	Total	11.6	38.333	32.267	70.600

Source: USDA Forest Service, Holly Springs National Forest, table given to Christopher Johnson on March 11, 2011.
[a]Standing timber
[b]Trees damaged by wind, insects, ice, or other natural causes
[c]Trees suitable for production of sawlogs
[d] A length of a cut tree having a round cross section, such as a log

other eastern national forests. To what degree should the national forests be managed? How much emphasis should be placed on resource uses, such as timber production, as opposed to recreational, aesthetic, and scientific uses? What role should the public have in the management of the national forests? How could the Forest Service be more proactive in communicating its forest management decisions to that public? In addition to raising these questions, the Holly Springs controversy reflected two other trends: the increasing involvement in forest issues by environmentally conscious citizens and reforms in the Forest Service that have affected timber-harvesting practices.

Roots of the Holly Springs Controversy

The debate over timber harvesting had its roots in national economic, societal, and environmental trends that emerged after World War II. After the passage of the Weeks Act, the Forest Service undertook a comprehensive effort to restore the eastern national forests by planting trees and instituting principles of forestry that would ensure a steady supply of timber for the future. In the early 1950s, however, two things happened that greatly affected the forests. First, many of the second-growth trees that had been planted during the 1920s and 1930s began to reach maturity and were available for harvesting. Second, the U.S. economy boomed, and demand for wood products skyrocketed.

To meet congressionally mandated timber-production targets and ensure a sustainable supply of timber, the Forest Service used—and continues to use—a variety of logging methods. In weeding, foresters remove unwanted species of trees to encourage the growth of desirable species. Thinning refers to the removal of overstocked, diseased, or damaged trees to improve the quality of remaining trees.[5] Shelterwood cuts are used to harvest all the trees in a stand of trees over a period of time. Shelterwood cuts occur in two stages. In the first stage, loggers remove approximately 50 percent of the trees, allowing sunlight to penetrate through the forest canopy and reach seedlings, which take hold and begin to grow. After the new trees have reached heights of five to ten feet, loggers remove the remaining trees. In shelterwood cuts, foresters direct loggers to leave some mature trees to improve wildlife habitats for certain birds and other animals.[6]

Clear-cutting is also part of a forester's repertoire in managing a forest. Contemporary forestry practice usually calls for limiting the size of a clear-cut to forty acres or fewer. From a forester's and wildlife biologist's point of view, the technique is appropriate to encourage the growth of

149

sun-favoring species, such as pines and aspens, and to increase wildlife diversity. In 1997, Clemson University's Department of Natural Resources published a comprehensive review of scientific literature showing that clear-cutting did, in fact, improve the availability and quality of food and cover for white-tailed deer, black bears, moose, rabbits, and many early successional–habitat songbirds. Slash and snags also provided habitat for certain birds and reptiles. Researchers also found, however, that clear-cuts could have a negative effect on amphibian populations because sunlight hit the soil directly, creating drier and warmer conditions on the ground. If clear-cutting had positive ecological effects, why did the practice stir such anger among conservationists? Clemson's report cut to the heart of the issue:

> The temporary absence of merchantable trees after cutting, the presence of logging slash and soil disturbance made clearcuts seem uglier than areas harvested by other cutting methods. . . . Indeed, over the years, clearcutting's unsightly appearance caused a general lack of public acceptance.[7]

From the 1920s through the 1950s, clear-cutting largely subsided as an issue, but in 1964 it shot into prominence again when the Forest Service clear-cut several stands of timber in the Monongahela National Forest in West Virginia. Hikers, wild turkey hunters, and campers were shocked to see favorite areas shorn of timber. Extensive clear-cuts and the creation of terraces in the Bitterroot National Forest in the late 1960s further fueled opposition to Forest Service policies. Wilderness advocates and environmentalists joined forces with the Izaak Walton League, the Sierra Club, and other groups to file lawsuits against the Department of Agriculture and the Forest Service, asserting that clear-cutting violated the Organic Act of 1897. A district court found in favor of the plaintiffs, putting a temporary halt to logging on national forests in West Virginia, Virginia, North Carolina, and South Carolina.[8]

Because of these crises and to create new processes for managing the national forests, Congress passed several laws that attempted to define more clearly the function and role of the national forests. The Multiple

Use–Sustained Yield Act of 1960 stated that the national forests should be managed "for outdoor recreation, range, timber, watershed, and wildlife and fish purposes," signaling a shift away from the traditional emphasis on timber production.[9] The Forest and Rangeland Renewable Resources Planning Act (RPA) of 1974 and the National Forest Management Act (NFMA) of 1976 required the Forest Service to develop a forest management plan for each national forest, including plans for timber sales, mining, grazing, extraction of oil and natural gas, and other resource uses. A central feature of the legislation was that members of the public must have an opportunity to comment on the plans and request administrative reviews if they objected to portions of a forest management plan. If plans included clear-cutting, the Forest Service had to demonstrate that this method was the most appropriate silvicultural practice for a particular stand of trees. In the Monongahela Forest Plan of 1986, for example, foresters could use clear-cuts that were no larger than twenty-five acres, and such tracts had to be separated from one another by at least one-eighth of a mile. In addition, after a tract was clear-cut, an adjacent tract could not be clear-cut until the previous tract had grown to one-fifth the height of the nearby forest.[10]

Two other laws also had a far-reaching effect on the eastern national forests. One was the National Environmental Policy Act (NEPA) of 1969, which established federal environmental policies and created the Council on Environmental Quality. To fulfill NEPA requirements, the Forest Service assessed the environmental impact of timber harvesting, road construction, and other activities on the forest ecosystem. Another important law was the Endangered Species Act (ESA) of 1973, which protected species threatened with extinction. Under the auspices of this act, the Forest Service had to detail in its forest management plans how it would protect rare and endangered species of flora and fauna. The law led the Forest Service, for example, to take action to protect habitat for the red-cockaded woodpecker in Florida and other forests in the southeastern United States. With these new laws, Gifford Pinchot's concept of multiple uses expanded to include the protection of biodiversity and of recreational and aesthetic values on the national forests. The laws also initiated gradual changes in the staff of the Forest Service, which began to hire wildlife biologists, hydrologists, botanists, and ecologists to ensure that it was complying with NFMA, NEPA, and ESA regulations.

Holly Springs and Timber Harvesting

In many ways, however, Mississippi's national forests remained immune to these trends. Even as late as the mid-1990s, timber harvesting in Holly Springs and Mississippi's five other national forests—the Bienville, the De Soto, the Homochitto, the Tombigbee, and the Delta—generated little controversy. Indeed, it was a point of pride in the state that the forests had been restored and made a valuable contribution to the economy. During the 1920s and 1930s, Holly Springs had fallen into deplorable condition because of the region's overreliance on cotton farming and lack of crop rotation (figure 7.2). Forester Buddy Lowery commented, "The gullies were so big that you could drop in a building as big as our office. The region was very badly eroded, and loblolly pine was the tree of choice for erosion control."[11] Wind and rain had worn deep crevasses in the earth, creating a landscape that bore a startling resemblance to craters on the moon. Holly Springs became a national forest in 1936, and the Forest Service and the CCC planted hundreds of thousands of loblolly pines, gradually restoring the forest to health.

By the 1950s, the replanted loblollies were thriving, and the forest had recovered substantially from the earlier degradation. One ongoing challenge for the Forest Service, though, was the forest's scattered ownership pattern. District Ranger Gardner explained:

> There are approximately 155,000 acres of national forest land within the Holly Springs proclamation boundary, but the national forest exists within a mosaic of 530,000 additional acres of private lands. This national forest is very scattered, and it is interspersed with privately owned lands, which are primarily owned by investment companies and insurance companies. Because of the scattered ownership pattern, the Forest Service has no access to some of the tracts.[12]

Despite the scattered ownership pattern, Mississippi's national forests were highly productive. It was estimated that loggers routinely harvested

Figure 7.2 *Erosion gullies in Mississippi. One-crop farming and a lack of crop rotation caused heavy erosion and enormous gullies during the 1920s and 1930s in Mississippi. Photo courtesy of the Forest History Society, Durham, North Carolina. Photograph by Louis A. Rowland Jr. for the Mississippi Forest Commission.*

between twenty million and forty million board feet a year, enough lumber to build thirteen hundred to twenty-six hundred average-sized houses. Gene Sirmon, who worked for the Forest Service for many years and then formed his own timber-consulting firm in Mississippi, estimated that in the late 1990s, the timber industry was pumping $40 million into the state's economy, making Mississippi the top timber producer in the Southeast.[13] Local communities depended on the resulting revenues, as 25 percent of timber sales and other revenues from national forests went back to local counties and towns to fund schools and roads.

Consequently, when the Forest Service announced in 1997 that it would harvest timber on a 1,251-acre tract in the national forest and that two stands would be clear-cut, it expected no opposition or controversy. What happened over the next several months caught the agency by surprise. In March 1998, the *Memphis Commercial Appeal* announced, with the headline "Chain Saws Buzzing in Miss. Forest Sound Like Massacre to Ex-Urbanites," that the timber-harvesting debate had landed in full force in northern Mississippi. Scott Banbury, then president of the Memphis Audubon Society, told the newspaper, "I believe the emphasis has been on timber production. The environmental assessment they've produced on their timber production has not adequately addressed the other values of the forest."[14] Local critics claimed that the proposed logging project and other timber harvesting at Holly Springs would threaten wildlife habitat for songbirds and cause soil erosion and compaction. They also accused the Forest Service of gradually converting the forest to pine monoculture—a forest consisting only of pine trees—by killing hardwood seedlings through the use of herbicides. The result, environmentalists claimed, would be single-species tree farms that would not support a rich variety of birds, mammals, and reptiles. Furthermore, critics asserted, the interests of the lumber and paper industries were pushing aside those of recreationists, who were increasingly using the forest for hiking, horseback riding, fishing, and hunting.

Gary Yeck, who was the forest supervisor at Holly Springs at the time, believed that one factor bringing the forest under closer scrutiny was demographics. Former urbanites from Memphis and other cities were snapping up properties near the forest, and they wanted to preserve the forest landscape that had attracted them. Yeck told the *Memphis Commercial Appeal*:

> I think they look at the woods differently than I look at the woods. They have different values, different ideas of how their land should be managed, and they're expressing that. They moved from the city to live near the forest, and they don't want the forest cut.[15]

Yeck responded by taking critics to the forest to show them the Forest Service's strategies and plans, emphasizing that the amount of land to be clear-cut was very limited and that the method was appropriate for regenerating the forest. "If we want to get that stand of trees back and we don't have a seed source," he said, "then we'll clearcut and plant."[16] Yeck also maintained that the Forest Service had converted very little of the forest from hardwood to pine.

The Holly Springs controversy reflected very important changes that were occurring in attitudes toward nature among a certain portion of the American public, changes that help explain the opposition to the Forest Service's timber-harvesting policies. In the early twentieth century, people like Gifford Pinchot who had led the forest conservation movement had been Progressives who adhered to the utilitarianism that helped define that political movement. For practical economic and scientific reasons, they supported the conservation of trees and other forest resources, the improvement of water and soil quality, and the creation of recreational opportunities that had practical benefits. Overall, they achieved a consensus about multiple uses of the national forests that held together for forty years.

During the 1950s and 1960s, however, this consensus began to fall apart. These decades saw the evolution of the environmental movement, which placed a greater emphasis on preserving and protecting natural environments from harm or decay. Writers like Aldo Leopold, Rachel Carson, Edward Abbey, Barry Commoner, and Edwin Way Teale helped create the philosophical foundation of the environmental movement, and activists absorbed Leopold's proviso that "the land ethic simply enlarges the boundaries of the community to include soils, waters, plants, and animals, or collectively: the land."[17] To the growing cohort of environmentalists, forests and other natural systems had inherent scientific, recreational, aesthetic, and spiritual value, and protection of forests became a moral obligation. They wanted less timber harvesting, mining, oil drilling, and grazing in public forests and more protection of wildlife habitat, species diversity, and wilderness.

In northern Mississippi, a good representative of this point of view was Dr. Ann Phillippi, a biology instructor at the University of Mississippi who became involved in the local movement to change how Holly Springs National Forest was being managed. She explained:

> Philosophically, I would like to see logging, mining, and any other extraction completely eliminated from the national forests in order to promote biological diversity. I think that the philosophy of the

management of the national forests needs to change. They are places with important biodiversity in our country because of old growth and because the harvesting has been limited.[18]

In the late 1990s, environmental activists in Mississippi began to form grassroots groups to pressure the Forest Service to reduce timber harvesting. In 1998, a group of residents in the Oxford area, including Phillippi and Ruthann Ray, who lived near the forest boundaries and had been upset by timber harvesting near her home, formed Citizens for Holly Springs National Forest. For the next several years, the organization worked with other environmental groups to file appeals and lawsuits to attempt to change how the Forest Service was managing Holly Springs and Mississippi's five other national forests. Phillippi recalled, "I wrote letter after letter to the Forest Service to appeal against timber sales by the Forest Service. I wrote enough letters to fill a foot and a half in my files."[19] Environmentalists argued that change was imperative to protect the biodiversity of the forest and to enhance the hunting, fishing, hiking, and equestrian experiences. They also believed that the Forest Service was converting the forest into a pine monoculture by using herbicides to kill hardwood species. Such practices, they claimed, violated the National Forest Management Act of 1976.[20]

For technical and legal assistance, Citizens for Holly Springs turned to Heartwood, an Indiana-based consortium of environmental groups seeking the kinds of changes in national forest policy that Friends of Holly Springs wanted. A brief history of Heartwood is essential to understanding the attitudes that led environmental groups to urge reform in how national forests were managed. Heartwood had grown out of the activism of Andy Mahler, a farmer in southern Indiana who has been involved in national forest issues since the 1980s. Mahler explained that in the mid-1980s, "People in Bloomington and Indianapolis who were in the environmental movement didn't like how the Forest Service was manipulating the composition of Hoosier National Forest. Every plan that the Forest Service developed reduced the population of oaks."[21]

Mahler hosted a gathering of activists, and in 1986, they formed the group Protect Our Woods (POW).[22] Their first objective was to overturn plans to build trails for off-road vehicles in Hoosier National Forest. Soon after, they began to protest against logging operations. Mahler recalled, "We opposed the timber sales program, and we made a three-pronged argument against the Forest Service's approach to timber sales. One leg was environmental. Another was economic, and the third was political."[23]

POW questioned whether the Forest Service was actually making a profit on timber sales, claiming that the agency did not factor in the price of building roads to remove timber. The organization also commissioned a poll, which showed that a majority of the people living in and around Hoosier National Forest wanted to reduce the amount of timber harvesting. In 1991, Mahler and others who had been involved in POW formed Heartwood, which then coordinated efforts and provided expertise to environmental groups throughout the Midwest that were pressuring the Forest Service to reduce logging on forests in Illinois, Indiana, Kentucky, Ohio, and Missouri.[24]

When Citizens for Holly Springs asked for help, Heartwood sent Joe Glisson, the organization's Forest Coordinator. Glisson was a criminal justice professor at Southern Illinois University who had participated in antilogging protests at Shawnee National Forest. He recalled, "I was at the center of the Holly Springs controversy. I traveled around the Southeast for Heartwood in the 1990s and filed numerous lawsuits to stop clear-cutting, among other practices."[25] In Mississippi, Glisson worked closely with Phillippi; her husband, Mel Warren, who was a research scientist for the Forest Service; and other local activists. According to Glisson, Heartwood claimed that local Forest Service officials had not filed adequate statements on the environmental impact of timber-sales programs. Glisson said, "We met with Citizens for Holly Springs, and Heartwood hired a lawyer to write and file the appeals."[26]

A pivotal issue for Citizens for Holly Springs and Heartwood was their perception that the forest was being converted from hardwoods to pine monoculture. The Forest Service gathered extensive data on the composition of the forests, and some of the data gathered in the late 1980s and early 1990s supported the claims that pine plantations on national forests were increasing. For example, in 1988, the Department of Agriculture and the Forest Service issued a comprehensive study titled *The South's Fourth Forest: Alternatives for the Future*, which examined the region's public forests and privately owned forests. The study included table 7.2, which shows how the forests had changed by type from 1952 to 1985.[27]

This table shows that from 1952 to 1985, the number of acres with pine plantations on southern national forests more than doubled, from 239,000 acres to 583,000 acres. On privately owned industrial forests in the South, acreage with pine plantations increased even more, from 660,000 acres in 1952 to 13,153,000 acres in 1985. In 2007, an updated Forest Service study titled *Forest Resources of the United States, 2007* presented reliable data showing that the trend toward pine plantations continued, as there were 294,000 acres of longleaf pine and 703,000 acres of loblolly and shortleaf pine plantations.[28] At the same time, the data did not support the contention that hardwoods were declining. From 1952 to 1985, the acreage of upland hardwood forests actually increased, from 4,121,000 acres to 4,430,000 acres. By 2007, the southern national forests had 6,093,000 acres of oak and hickory forests.[29] These studies demonstrated that pine plantations *and* hardwood were both increasing.

Another issue on which Heartwood and other environmental groups criticized the Forest Service was road building, which has been a controversial issue on national forests since the 1960s. Indeed, one of the stipulations of the Wilderness Act of 1964 was the prevention of road construction in those parts of national forests that were designated as wilderness. Every year, the Forest Service was building logging roads in nonwilderness areas to reach timber-harvesting sites. According to ecologists, roads destroy wildflowers and archaeological sites and cause soil erosion because silt washes from roads into streams and rivers. Furthermore, roads fragment forests, give access to motorized vehicles, and can take land in national forests out of consideration as wilderness.

To prepare administrative appeals, Citizens for Holly Springs and Heartwood worked with Ray Vaughan, an attorney in Alabama who had been involved for several years in environmental causes. Vaughan grew up in Montgomery, Alabama, where he became an avid outdoorsman. As

Table 7.2 Area of Timberland by Ownership
and Forest Management Type in Twelve Southern States[a]

(Thousands of Acres)

Forest Type	1952	1962	1970	1977	1985
National Forests					
Pine plantation	239	468	473	450	583
Natural pine	3,765	3,844	3,878	3,814	3,460
Mixed pine and hardwood	1,536	1,530	1,651	1,740	1,841
Upland hardwoods	4,121	4,258	4,244	4,446	4,430
Bottomland hardwoods	708	612	489	460	459
Total	10,369	10,712	10,735	10,910	10,773
Forest Industry					
Pine plantation	660	3,127	5,714	8,723	13,153
Natural pine	14,576	13,646	13,357	11,340	8,818
Mixed pine and hardwood	4,955	5,170	6,342	6,395	6,025
Upland hardwoods	5,814	6,469	6,229	6,655	7,118
Bottomland hardwoods	7,379	7,386	6,774	6,996	7,148
Total	33,384	35,798	38,416	40,109	42,262

Source: *The South's Fourth Forest*, 113.

[a]Alabama, Arkansas, Florida, Georgia, Lousiana, Mississippi, North Carolina, Oklahoma, South Carolina, Tennessee, Texas, and Virginia

a young man, he had considered becoming a forester but decided instead to study law. After receiving his law degree, he focused on environmental law and formed an organization called Wild Alabama, which sought to preserve wild nature in the state. Wild Alabama evolved into WildLaw, which still exists. Vaughan helped Citizens for Holly Springs National Forest and Heartwood prepare the appeals, which sought a new forest management plan that would assess the effect of timber harvesting and protect the biodiversity of the forest.[30] WildLaw also challenged the management of Mississippi's five other national forests. In December 1998, for example, the organization appealed plans by the Forest Service to harvest timber on the DeSoto National Forest, near Hattiesburg.[31]

In response to the administrative appeals and the negative public reaction that the logging plans had catalyzed, the Forest Service management at Holly Springs revised its forest management plan to eliminate clear-cutting, lower the volume of timber harvesting by 19 percent, and reduce the use of herbicides.[32] Even so, the Memphis Audubon Society, Heartwood, and Citizens for Holly Springs appealed to the regional forester in Atlanta to set even lower targets for timber harvesting. In October 1998, local residents and environmental groups brought a federal lawsuit, seeking a termination of logging until the Forest Service had written a new forest management plan.[33]

At the same time, some local residents had serious concerns about the revenue that would be lost if timber harvesting were reduced on Holly Springs and Mississippi's other national forests. Local school boards filed countersuits to allow the timber harvesting to go forward.[34] Communities were also concerned about losing jobs in the forest products industry. The Forest Service was caught in the middle, between people opposed to harvesting timber and the local communities and businesses that needed the products and revenues the national forests provide.

Reforms in the Forest Service

In response to these outside pressures and because of developments in forestry, wildlife biology, environmental science, and other academic fields, the Forest Service initiated a period of reform in the 1980s that led to a greater emphasis on forest ecosystem management and changes in the composition and training of Forest Service personnel. These reforms would ultimately affect how the Forest Service responded to the Holly Springs controversy.

Between 1985 and 1993, the proportion of Forest Service employees with training in forest management, range management, and engineering declined from 66 percent to 51 percent, whereas the proportion who had been educated in landscape architecture, hydrology, botany, and biology rose from 10 percent to 15 percent. Specialists with the latter kind of training were more likely to emphasize ecological values, such as preservation of biodiversity, over timber harvesting and other forms of commodity production.[35] In addition, a small group of current and former employees of the Forest Service formed the Association of Forest Service Employees for Environmental Ethics (AFSEEE) in 1989. Its mission was to change the agency's culture so that the national forests were managed more for ecological values than for resource extraction. In his book *Mississippi Forests and Forestry*, University of Memphis professor James E. Fickle observed that this group "is far more likely to identify with [Aldo] Leopold's land-ethic philosophy than are their fellow employees, who tend to be more attuned to [Gifford] Pinchot's values," which were utilitarian.[36]

In 2002, Greg Brown, assistant professor of environmental science at Alaska Pacific University, and Chuck Harris, associate professor of forestry at the University of Idaho, published the results of a survey of environmental attitudes among Forest Service employees, 939 of whom responded, including 231 members of AFSEEE.[37] The survey respondents included foresters, engineers, hydrologists, range specialists, ecologists, wildlife biologists, geologists, recreational specialists, fire management officers, and administrative personnel. Most respondents fell between the extremes of preserving forests in a pristine ecological condition and allowing maximum economic development.

AFSEEE members were more aligned with the forest preservation movement, however, with 13 percent calling themselves preservationists, 17 percent identifying with the segment labeled "John Muir," and 49 percent associating with the segment identified "Aldo Leopold," or land ethic values. Wildlife biologists, geologists, and ecologists were more likely to agree with the land ethic values of Aldo Leopold than were foresters, engineers, and hydrologists, who tended to favor forest management and multiple uses. Harris and Brown also reported that 29 percent of supervisors and managers had backgrounds other than forestry or range management, indicating that wildlife biologists, ecologists, geologists, and hydrologists were rising to positions as district rangers and forest supervisors.[38]

By the early 1990s, this new cohort of managers and supervisors was beginning to move Forest Service policies away from the utilitarian emphasis on commodity production. In 1991, Tom Mills, acting associate deputy chief for programs and legislation for the Forest Service, said, "The essence of the future direction of the Forest Service programs is described in four themes: enhancement of recreation, wildlife, and fisheries resources; a commitment to environmentally acceptable commodity production; expanding scientific knowledge about natural resource systems; and responding to global resource issues."[39] In 2000, the Forest Service issued a new RPA Assessment, which established the overall goal of sustainable forest management. The report established five criteria for managing and conserving forests and rangelands:

1. Conservation of biological diversity

2. Maintenance of productive capacity of forest and range ecosystems

3. Maintenance of forest ecosystem health and vitality

4. Maintenance of forest contribution to global carbon cycles (i.e., contribution to carbon sequestration)

5. Enhancement of long-term multiple socioeconomic benefits to meet the needs of societies[40]

The 2000 RPA reflected that nontimber uses of the national forests were receiving a higher priority in the Forest Service's strategies for managing national forests than they once had. For example, the document highlighted the significance of the nation's forests as carbon sequesters to mitigate the effects of greenhouse gases. In the report, the Forest Service estimated that forests stored approximately 24 percent of the carbon dioxide emitted by the United States every year.[41] The RPA also warned that fragmentation of forests, particularly through road construction, presented a severe threat to wildlife. (Chapter 14 examines the issue of forest fragmentation in depth.) Consequently, the Forest Service itself was raising concerns about forest fragmentation and loss of biodiversity, indicating that the agency was trying to embrace a new paradigm, that of the forest as a complex ecosystem that should be managed for a balance of scientific, economic, recreational, and aesthetic interests.

Reforms in the Management of Holly Springs National Forest

In the meantime, the administrative appeals and legal actions brought by Heartwood, WildLaw, and Citizens for Holly Springs National Forest had mixed results. Appeals and lawsuits slowed the rate of logging in Mississippi, and the state fell behind Arkansas in the amount of board feet harvested.[42] WildLaw successfully appealed plans by the Forest Service to harvest timber on twelve thousand acres of the Homochitto National Forest, which is southwest of the state capital of Jackson. When Citizens for the Holly Springs National Forest and Heartwood brought lawsuits against the Forest Service for its alleged mismanagement of the forest in 1999, the cases took several years to be resolved. In 2003, the courts decided that the timber sales could proceed.

Despite these mixed results, the pressures brought by environmentalists *and* the changes in the staff of the Forest Service had a substantial effect on how the Holly Springs National Forest was managed. These reforms were reflected in the careful approach that the forest managers took to the salvage operation that introduced this chapter. The forest managers conducted detailed environmental studies, shared information, and created opportunities for public input. Ray Vaughan commented, "The management of the forest at Holly Springs improved greatly. The change occurred with new leadership. Tony Dixon [in 2011, the acting regional forester of the Rocky Mountain region] came in as the forest supervisor in the early 2000s, and he changed logging policies and the overall management of the forest."[43] Joe Glisson added, "Heartwood had a real impact on the approach to timber harvesting that the Forest Service took. They became much more careful and did far greater analyses of the environmental impact."[44] Andy Mahler agreed. "The Holly Springs National Forest is better managed now than it was in the 1990s," he said.[45]

In the aftermath of the battles over logging in Mississippi, another powerful idea—the idea of ecological restoration—also modified how Holly Springs was managed. In 2009, Tom Vilsack, secretary of agriculture during the Obama administration, placed the goal of restoration front and center in a policy speech in Seattle. "Our shared vision begins with restoration," he said. "Restoration means managing forest lands first and foremost to protect our water resources, while making our forests more resilient to climate change."[46] In Mississippi, some of the key players who had locked horns with the Forest Service ten years before began to cooperate with the agency on forest restoration projects in Mississippi and Alabama. For example, Vaughan and WildLaw participated in restoration projects on the DeSoto and Homochitto National Forests. Vaughan explained, "We submitted a proposal under the Collaborative Forest Landscape Restoration Program."[47] Congress passed this legislation in 2009, authorizing the secretary of agriculture to grant up to $40 million annually for implementing and monitoring forest restoration projects. Stakeholders, who could include nongovernmental organizations and Forest Service staff, prepared proposals that regional foresters then submitted to the secretary of agriculture for evaluation.

Vaughan explained that in the DeSoto and Homochitto National Forests:

> We have been doing longleaf pine restoration. It took a long time to develop seedlings that would grow. Longleaf pines are the native species in southern Mississippi, whereas shortleaf pines are native to northern Mississippi. The Forest Service also engaged in an open and collaborative process and provided choices for how to deal with salvage timber. In deciding what to do after Hurricane Katrina, they listened to the scientists. As a result, they left three-quarters of the trees on the ground.[48]

This approach had the benefit of returning nutrients to the soil as the trees decayed and providing habitat for wildlife. According to Vaughan, the William B. Bankhead National Forest in Alabama undertook the largest hardwood restoration project in the eastern national forests. WildLaw was working with the Forest Service to restore hardwoods—primarily oak and hickory—on seventy-five thousand acres of the forest, and they were experimenting with planting American chestnuts. Looking back at the Holly Springs controversy, Vaughan commented, "Now, there are plenty of places at Holly Springs that have hardwoods, and in fact, I believe that the percentage of hardwoods might be more than 35 percent."[49]

In 2011, as part of preparing a new Forest Plan, the Forest Service in Mississippi issued an overview that identified several forest management goals, including the restoration of 23,000 acres of shortleaf pine, longleaf pine, and hardwood forests; improvement of forest quality on another 150,000 acres; and use of prescribed burns on 200,000 acres every year. The agency also proposed to enhance recreation by upgrading trails and campsites and restoring water quality on rivers and streams throughout the six national forests. At the same time, the Forest Service expected to maintain a flow of timber of approximately 94 million board feet a year. In short, the agency was proposing a balance among the three Rs: restoration, recreation, and resource use.[50]

Lessons Learned

In the Holly Springs controversy over timber harvesting, two very important trends emerged that will continue through several other case studies: (1) the pressure by environmental groups to deemphasize commodity production on the eastern national forests and (2) the capacity of the Forest Service to undertake reforms in the management of those forests. The debate over timber harvesting in the Holly Springs National Forest was a highly productive one, bringing to the surface different assumptions about the purposes of national forests and catalyzing the following important changes in how the Forest Service managed Holly Springs and Mississippi's other national forests.

1. *Considerations of the interrelationships among the different aspects of the forest ecosystem are increasingly affecting decisions on eastern national forests.* After years of unopposed timber harvesting in the Holly Springs National Forest, environmentalists urged change and cited the effect that logging and road construction had on birds, mammals, reptiles, vegetation, water, and soil.

2. *Ecological restoration is an emerging goal in managing the eastern national forests.* The experiments in longleaf pine and hardwood restoration that Vaughan described highlight the growing desire to return forests to a mix of tree species that approximates the original condition of the forests. The expectation is that through such restoration, forests will provide habitat for birds, wildlife, and vegetation, preserving some

of the biodiversity that is being diminished by the growth of human populations in suburbs and exurbs.

3. *Despite these trends, economic factors are still very important, and the economic effect of changes in timber harvesting must be considered.* In the decades following the passage of the Weeks Act, rural areas close to national forests benefited from the timber revenues and jobs created by the wood products industry. As critics call for a reduction or elimination of timber harvesting in eastern national forests, they must also take into account the economic effect of these changes. Certainly an increase in tourism can bring new jobs, as has been true in communities like Asheville, North Carolina, and North Conway, New Hampshire, both of which border national forests. If commodity production in the eastern national forests declines, local communities, counties, and states must mitigate the effect by improving regional planning, with strategies for reeducating workers and attracting low-impact industries to rural areas.

The timber-harvesting controversy at Holly Springs National Forest underscores the dramatic changes that have shaped the eastern national forests in the years since the passage of the Weeks Act. The emphasis during the Pinchot era was on the concept of "the greatest good, for the greatest number, for the long run." In the twenty-first century, defining the greatest good for the greatest number has become highly complex. Recreationists demand more facilities, and ecologists and biologists urge the protection of endangered species of vegetation and wildlife. At the same time, the wood products industry and some local communities argue for fewer regulations and more harvesting of forest products. In the Holly Springs National Forest, the Forest Service undertook important reforms in response to the criticisms that it received. The controversy—and the changes that grew out of that controversy—underscore the continuing legacy of the Weeks Act and other conservation legislation in creating a network of public forests to which citizens feel invested and committed.

CHAPTER 8

Florida's National Forests: A Revolution in Prescribed Burning

On a fine day in March 2011, the longleaf pines stood at attention like sentries on Apalachicola National Forest, which radiates south from Tallahassee to the Gulf of Mexico in Florida's panhandle. Those stately pines, spaced twenty to twenty-five feet apart, created a park-like landscape in which three-feet-long blades of wiregrass waved gently in the breeze (figure 8.1). The grass blades were mostly brown, but some of them were speckled with the green of approaching spring. Steve Parrish, the zone fire management officer whose friendly, even-keeled temperament drew other people to him like a magnet, made his way through the forest of pine and remarked with obvious pride, "It's beautiful, isn't it? We did a prescribed burn in this part of the forest about a year ago."[1]

Sometimes referred to as controlled burns, prescribed burns are fires that forest managers intentionally set for a variety of purposes: to clear away underbrush; stimulate the growth of trees, wildflowers, and other vegetation; create habitat for wildlife; and reduce hazardous fuel accumulations. Parrish and other trained fire officers had carried out the burn to reduce the forest understory, which had become clogged with a mass of undesirable vegetation. Saw palmettos, with stems as sharp as the blades of serrated knives, dotted the landscape, but their numbers were relatively few, and they lived in a kind of peaceful coexistence with the wiregrass

Figure 8.1 *Pine forest in Apalachicola National Forest in Florida. The use of prescribed burns here prevents the leafy palmettos from taking over the forest floor and allows the wiregrass to thrive. Photograph by Christopher Johnson.*

and the pines. Without the prescribed burns, though, the aggressive palmettos would soon have crowded out the wiregrass and taken over the forest understory.

To Parrish and his fellow forest managers at Apalachicola, fire was a powerful ally in maintaining the ecological balance in this forest, preventing the dominance of the palmetto, and preserving the habitat of the prize bird that calls the forest home: the beautiful but endangered red-cockaded woodpecker (RCW). The RCW is about seven inches long and has black and white stripes running horizontally on its back. It features a black cap and nape that form a circle around white patches on the cheek, and on either side of the black cap is a small streak of red called a cockade, which is difficult to see except when the bird is breeding or defending its territory. By using prescribed burns at Apalachicola, Parrish and his Forest Service colleagues were helping protect the bird's habitat as well as restoring important components of the forest's biodiversity.

In addition to Apalachicola, Florida contains two other national forests: Ocala National Forest, west of Daytona Beach; and Osceola National Forest, just south of the Georgia border on the southernmost fringes of the Okefenokee Swamp. The story of prescribed burning in these three national forests is similar to the story of timber harvesting in Holly Springs National Forest: a story of reform. From its inception in the early 1900s, the Forest Service emphasized fire suppression to the point that Smokey Bear and his slogan, "Only you can prevent forest fires," imprinted themselves on the consciousness of just about every American who visited a national forest. Today, though, the Forest Service, state foresters, and private timberland owners embrace prescribed burns as an ally. According to the Wildland Fires Lessons Learned Center, based in Tucson, Arizona, land management agencies in the United States conduct four thousand to five thousand prescribed burns every year in managing forestlands and grasslands.[2] For eons, Nature has used fire to manage forests. For the last few decades, humanity has been doing the same thing.

A Brief History of Fire

To understand why fire has become a positive tool to manage forests, it is essential to take a short look at how attitudes and policies toward fire have evolved since the early twentieth century. If heavy logging was one factor that led to the passage of the Weeks Act, forest fires were the other major factor. Not only did the Weeks Act help create eastern national forests, but it provided federal aid for establishing state forest-fire agencies, acquiring equipment, and building fire towers. Under the leadership of Gifford Pinchot and Henry S. Graves, the U.S. Forest Service established a policy of fire suppression.[3]

In 1924, the Clarke-McNary Act further supported firefighting efforts by allocating funds to the states for fighting fires and requiring annual fire reports from regional foresters. Two years later, the Forest Service directed the staff on national forests to fight fires aggressively before they reached ten acres in size. The American Forestry Association sponsored lectures and showed films that reinforced the federal policy of fire suppression. During the New Deal, the Civilian Conservation Corps built roads, communication networks, and fire towers throughout the national forests.

The Forest Service escalated the war on fire in 1935, when it issued the "10:00 a.m. policy," which stated that if a fire were larger than ten acres, it should be controlled by ten o'clock the next morning, when rising temperatures created the most dangerous period for the spread of a fire. During the 1940s, the Forest Service developed the Smokey Bear mascot and campaign, which was extraordinarily successful in making Americans aware of the dangers of forest fires. The initiatives of the Forest Service and state forest agencies to suppress fire were miracles of modern organization, and the rapid responses undoubtedly saved millions of acres of trees from being devastated.[4]

In various small corners of the forestry establishment, however, individuals began to look more closely at fire as a natural tool in managing forests and to question the policy of total fire suppression. One of the earliest voices favoring prescribed burns was Herman Haupt Chapman, who worked for the Forest Service from 1904 to 1906 and then became a professor at Yale's School of Forestry. In an article in *American Forestry* in 1912, he asserted, "There is abundant evidence that the attempt to keep fire entirely out of southern pine lands might finally result in complete destruction of the forests."[5] Chapman explained the importance of regular fires to the longleaf pine, for the tree "has adapted its whole structure and growth as a seedling to the primary object of surviving ground fires."[6] Moreover, if fire were suppressed, then the needles from the longleaf dropped to the ground, dried out, and became highly combustible fuel. To regenerate longleaf pines, Chapman recommended that fire be suppressed in an area with seedlings for five years, by which time the young trees would have attained diameters of about an inch and would be able to resist low-intensity fires. After that period, he wrote, "Fire must be introduced into this young stand, and the ground burned over to get rid of the fire trap formed by the accumulated litter."[7]

Other federal agencies were also looking at the positive uses of fire.

During the 1920s, the U.S. Biological Survey sponsored research to determine ways to increase the population of the bobwhite quail, a desirable game species. The lead researcher, a wildlife biologist named Herbert L. Stoddard, reported the findings in an influential book, *The Bobwhite Quail: Its Habits, Preservation, and Increase*, published in 1931. The quail, Stoddard asserted in a key chapter titled "The Use and Abuse of Fire on Southern Quail Preserves," needed open woodlands for food and nesting, but the suppression of fire allowed forest understories and pastures to become overgrown with broomsedge and other thick vegetation, preventing quails from locating seeds and fruits to eat. Fire suppression also inhibited the germination of legumes such as partridge peas and bush clovers, which were part of the quail's diet.[8] To increase quail populations, Stoddard supported occasional controlled burns:

> If burned over in February, after the seeds have matured and fallen and before new growth has started, by fires of the "creeping" type, the ground is cleaned of accumulated grasses, and the legumes, especially the perennials, appear stimulated, grow thriftily, and seed well and abundantly.[9]

Controlled burns had the further advantage of preventing the spread of parasites and organisms carrying diseases, both of which harmed quail. After Stoddard submitted his manuscript for review by the Biological Survey, the chapter about fire stirred up a hornet's nest within the agency. Angered, Stoddard threatened to resign and write a book that would take a no-holds-barred approach in attacking the policy of fire suppression. The agency relented and allowed publication of the book.[10]

Pioneering research like that by Chapman and Stoddard laid the foundation for experiments with controlled burns, most notably on privately owned forestlands. For example, Superior Pine Products, which was based in Fargo, Georgia, owned two million acres, and the company's forest managers burned between thirty thousand and fifty thousand acres every year. In addition, the Carolina Fiber Company, based in South Carolina, had been using controlled burns since 1920 to reduce fuel.[11]

Some government foresters in the South also started to question the orthodoxy of fire suppression. In 1933 and 1934, the Southern Research Station of the U.S. Forest Service, located in Asheville, North Carolina, prepared several articles recommending periodic light burning to meet specific objectives such as clearing away fuel. The Forest Service delayed publication of the papers, however, reflecting the agency's desire to forestall the movement in favor of prescribed burning.[12] The Society of American Foresters had traditionally favored fire suppression, but in a sign that the consensus in favor of suppression was beginning to disintegrate, Chapman, who had become the president of the organization, put controlled burning on the agenda for the society's annual conference in January 1935. Elwood L. Demmon, one of the participating foresters, read a paper stating that forest managers should use controlled burns, but only after carefully examining forest conditions and establishing clear objectives for the use of fire, but he cautioned, "Let no one infer from these statements that protection of forests from fire is not essential to the practice of forestry in the longleaf region."[13]

During the following discussion, several speakers expressed concerns that changing policies toward fire would prove unpopular among the public. One speaker, Shirley Allen, commented, "I hope you people from the South will be careful in talking to southern visitors from the Lake States. They may become enthusiastic about the merits of fire and come back home dangerous citizens. Forestry and fire won't mix in the Lake States."[14] At the same time, Ed Komarek, who had worked with Stoddard on the bobwhite quail research, asserted, "The whole program up to this time has been one-sided. This is the first time that censorship on the subject has been removed and we have been told the facts."[15]

Given the extensive destruction that forest fires had caused in the late nineteenth and early twentieth centuries, however, senior officers in the Forest Service continued to favor fire suppression. H. N. Wheeler, the director of public relations for the Denver office of the Forest Service, wrote to Chapman that although fire might do some good in limited circumstances, its overall effect was to harm forests.[16] By the late 1930s, though, cracks appeared in Forest Service opposition to controlled burning as the agency released the experimental results that the Southern Research Station had compiled earlier in the decade.

One part of the country where foresters started to look more closely at fire was Florida. Fire has always been an essential part of the Sunshine State's ecological menu. Nature rattles the region with more thunderstorms than any other part of the United States, and those storms arrive like Zeus, throwing down approximately two million lightning bolts a year, or about 10 percent of the total in the country.[17] According to biologist Jerome A. Jackson, "The lightning associated with these storms, coupled with the fast-draining sandy soils of the coastal plain, resulted in frequent natural fires that blanketed upland areas."[18]

During the 1940s, Florida's climate and several natural disasters conspired to drive changes in the traditional fire policy. In 1941, the state suffered through a long drought, and fierce wildfires attacked Osceola National Forest, devastating twelve thousand acres, or 7 percent of the forest. Because fires had been suppressed, the dense understory of saw palmettos and the accumulation of fallen leaves and branches fueled the fire. As the situation grew increasingly dangerous, the Forest Service faced two challenges: (1) whether to modify the policy of fire suppression to fit the conditions in the forests and (2) how to incorporate growing scientific knowledge about the role of fire in forest ecosystems.

L. S. Newcomb, an Osceola district ranger, made the bold decision to initiate controlled burns, but to avoid violating Forest Service policy, he described the burns as experiments intended to collect data about the effect of fire on forests. The Osceola forest managers referred to the burning as a *prescription*, or a method to remedy a condition that harmed the forest, and from then on, the staff called the process *prescribed burning*.[19] In 1941 and 1942, the managers continued their experiments, which showed that the burns did no harm to the growth of trees and improved forage for wildlife.

Another wave of fires devastated eleven thousand more acres in 1943. Because the Osceola forest managers' experiments had covered only a small portion of the forest, they blamed the wildfires on the accumulation of large amounts of fuel in the rest of the forest and urged immediate expansion of their experiments. On August 3, 1943, Lyle Watts, chief forester of the Forest Service, agreed to a change of policy in the South: controlled burning would be used in forests that had stands of slash pine and longleaf pine. District Ranger Newcomb initiated a forest-wide program of prescribed burns, which the staff agreed to study and document for the Forest Service's national office. Osceola adopted an officially sanctioned policy of prescribed burns in 1944 and 1945, burning about 20 percent of the forest every year.

The policy of allowing prescribed burns quickly expanded to Florida's two other national forests. In 1943, Ocala National Forest faced a crisis even more severe than the one at Osceola. Its location in central Florida made it the center of thunderstorm activity, with lightning fires occurring at the rate of 1.7 a year per ten thousand acres.[20] That year, fires swept across thirty thousand acres and burned thousands of trees. In December 1943, Ocala National Forest officials issued a directive allowing controlled burning, which insiders referred to as the "Treaty of Lake City" (Lake City being located just outside the boundaries of the forest).[21] Apalachicola also started prescribed burns during the winter dormant season of 1943–1944. Through the 1940s and 1950s, forest managers at Apalachicola burned an average of 14,820 acres every year.[22]

Prescribed Burns Become Accepted Practice

Certainly the Forest Service continued to fight wildfires, but the experiments in Florida represented a first step toward a more balanced approach to fire management on the eastern national forests. During the post–World War II era, prescribed burning gradually spread throughout the national forest system. The growing use of the practice has been accompanied by a fund of knowledge about the significance of prescribed burning in promoting biodiversity by encouraging vegetation growth and creating habitat for wildlife.

The Tall Timbers Research Station and Land Conservancy, located in Tallahassee, Florida, played a pivotal role in understanding fire. In 1958, Henry L. Beadal, an avid hunter in Florida's panhandle who had come to

believe in the value of fire in improving habitats for wildlife and vegetation, willed his inheritance to found Tall Timbers "to create a fire type
nature preserve . . . to conduct research on the effects of fire on quail, turkey, and other wildlife as well as on vegetation of value as cover and food
for wildlife, and experiments on burning for said objectives."[23] Since its
founding, Tall Timbers has conducted and disseminated extensive information on fire ecology.

In addition, the Forest Service's Southern Research Station continued
to gather data about the frequency and beneficial effects of prescribed
burns. In the mid-1990s, the agency issued *Influences on Prescribed Burning
Activity and Costs in the National Forest System*, by Forest Service researchers
David A. Cleaves, Jorge Martinez, and Terry K. Haines, which underscored
the degree to which prescribed burning had become an integral tool in
managing public forestlands. Excluding Alaska, which did not respond to
the survey, the researchers found that all the Forest Service's regions were
using prescribed burning, but the practice was most common in the Southern Region, which accounted for 48 percent of the total acreage burned.[24]
The frequency of prescribed burning in the region reflected geographic
factors like the long growing season, the rapid growth of vegetation, the
frequency of thunderstorms, and the protection of habitats for endangered
species like the RCW.

Since that survey in 1997, the National Interagency Fire Center, which
provides logistical support for fighting forest fires from its center in Boise,
Idaho, has conducted annual surveys of prescribed burning across the
entire national forest system. As shown in table 8.1, their data clearly
showed the degree to which the Forest Service employed prescribed burning as an ecosystem management tool and how much the use of prescribed
burning has increased since the late 1990s.

Prescribed burns were used for a variety of purposes: to lower fire haz-
ard, stimulate reforestation, reduce unwanted vegetation, control pests
such as pine beetles, improve habitat for wildlife, upgrade grasslands for
grazing, and reintroduce fire as part of the natural management of forestlands.[25] In surveying the burning activities on the national forests, David
Cleaves and the other researchers from the Forest Service asked fuels

Table 8.1 Prescribed Fires and Acres Burned by the U.S. Forest Service, 1998–2011

Year	Number of Prescribed Fires	Acres Burned
2011	2,890	960,992
2010	3,766	1,224,638
2009	3,795	1,244,342
2008	3,193	955,016
2007	4,771	1,291,889
2006	5,138	1,091,714
2005	3,782	1,329,439
2004	4,859	1,501,697
2003	4,134	1,275,310
2002	4,339	1,076,811
2001	4,058	1,071,473
2000	2,954	728,237
1999	4,021	1,239,429
1998	2,938	505,103

Source: National Interagency Coordination Center, http://www.nifc.gov/fireInfo/fireInfo_stats _prescribed.html, accessed September 13, 2012.

management officers to rate the importance of each purpose for prescribed burning. Table 8.2 shows the mean rating of the importance of each purpose for prescribed burns for each national forest region.

The statistics reveal that managers of national forests used prescribed burns for a variety of purposes, and these differences were explained by climate, soil conditions, dominant forest type, and other geographic factors. For example, forest managers in the East were less likely to use fire for fuel reduction because the region's relatively high levels of precipitation led to a lower incidence of wildfires than other areas of the country. The South placed a higher priority on prescribed burns for protecting habitat for threatened and endangered species because that region's subtropical climate supported a large number of such species.

Table 8.2 Purposes for Prescribed Burns

0 = "no importance"; 5 = "most important"

Resource Objective	R1 Northern	R2 Rocky Mountain	R3 Southwestern	R4 Intermountain	R5 Pacific Southwest	R6 Pacific Northwest	R8 Southern	R9 Eastern	All Regions
Fire control hazard reduction	4.58	3.87	4.82	4.18	4.71	4.47	4.32	2.10	4.21
Silviculture									
Reforestation	3.73	1.87	1.68	2.80	3.15	3.31	3.21	2.30	2.85
Vegetation control	1.82	1.73	2.95	2.30	2.33	1.88	2.89	1.80	2.26
Pest protection	1.91	1.70	1.95	2.60	1.92	1.53	1.44	0.90	1.71
Wildlife									
Nongame	1.36	2.10	2.86	2.36	2.08	1.47	3.33	3.30	2.37
Threatened and endangered species	2.18	1.30	2.36	1.36	1.83	1.24	4.37	1.30	2.15
Game birds and animals	2.82	3.30	3.14	2.91	3.07	1.94	4.00	4.20	3.15
Range									
Grazing	2.18	2.30	2.91	2.27	1.25	1.12	1.42	0.70	1.70
Ecosystem									
Reintroduction of fire ecosystem management	4.42	3.90	4.45	4.00	4.39	2.53	3.47	2.40	3.65

Source: Cleaves, Martin, and Haines, *Influences on Prescribed Burning Activity and Costs in the National Forest System*, 6–7.

Effect on Vegetation and Wildlife

All these benefits emerged in the six decades since the Forest Service began using prescribed burns, and the effects became apparent for both vegetation and wildlife. The forest managers in Apalachicola National Forest gradually expanded their burning to control palmettos and encourage wiregrass and other vegetation. The expanded use of controlled burns occurred almost by accident. In 1985, the state suffered from a winter drought, and forest managers curtailed dormant-season burning because of the threat that the fires would spread. They decided instead to set fires during the growing season of late summer and found that the low-intensity fires promoted the flowering of endangered, threatened, and sensitive plants. For example, only three skullcaps flowered before prescribed burns, but one hundred flowered after the burns. Growing-season fires also helped control the titi, three varieties of brushy trees that grew naturally in wetlands but invaded upland forests, where their waxy leaves made wildfires burn hotter. Prescribed burns pushed the titi back to their native wetlands.[26] Since then, growing-season burns have become part of the regular rotation.

Southeast from Apalachicola—and not far from the fairy-tale setting of Disney World—the Ocala National Forest resembled a tropical rain forest, as gauzy-fronded cypress trees danced alongside tall pines, which towered over an understory bursting with a bountiful mixture of grass and palmetto. Mike Herrin, district ranger, who had recently moved to Florida after a lengthy stint in national forests in Montana, said, "We have extensive stands of sand pine and longleaf pine. The trees benefit from prescribed burns because the undergrowth can become so thick that it chokes off nutrients to the trees. The burning also facilitates seeding by the trees."[27] The sand pine is found almost exclusively in Florida, with the Choctahatchee variety located primarily in northwest Florida and the Ocala variety ranging from south Florida to the northeast part of the state. Herrin compared the sand pine to the lodgepole pine, the dominant species in Yellowstone National Park. Both pines have seritonous cones, which require heat to open up, drop their seeds, and reproduce. Consequently, prescribed burns are essential to the continued health of sand pine forests.[28] The tree provides habitat for more than twenty endangered or threatened species of wildlife, including the Florida scrub jay, which builds its nest in the tree's branches. Songbirds, woodpeckers, and squirrels also use the trees for nesting and cover.

One of the great success stories in Florida has been the resurgence of the red-cockaded woodpecker, and fire has played a critical role in the recovery of the bird's population. According to Jerome Jackson, an RCW habitat must have four characteristics: (1) a pine forest that is burned every three to five years to prevent a thick understory; (2) old growth, with longleaf pines that are at least ninety-five years old and loblollies that are more than seventy-five years old; (3) a minimum of two hundred acres for each cluster to use for foraging; and (4) the presence of many clans to stabilize the population and provide variety in the genetic pool.[29] In addition, the RCW has adapted to Florida's natural fires by inhabiting the cavities of living pines rather than dead pines, which are more likely to burn during fires.

In the late 1960s, the population of the RCW was declining rapidly because of a loss of these native habitats due to deforestation, economic development, and the policy of fire suppression, and it was listed as an endangered species in 1970. As the bird's population plummeted, an RCW Recovery Team was formed in 1975 that included members from government, academia, and private industry. The U.S. Fish and Wildlife Service accepted the recovery plan in 1979, but the plan was never implemented because of disagreements between wildlife advocates and timber-industry interests. In 1983, the Fish and Wildlife Service and a representative of the Forest Service revised the plan, which both agencies then accepted. The RCW population continued to decrease, however, largely because continued timber harvesting left only 2.5 percent of forests in the South as suitable habitats for the bird. The American Ornithological Union (AOU) then recommended more aggressive preservation efforts, particularly reintroducing fire and protecting old pines from timber harvesting.[30]

The AOU explicitly criticized fire suppression in national forests. If fire was suppressed, then understories developed, fueling wildfires that were catastrophic to the RCW. Southern pine beetles were also a problem. According to biologist Jerome Jackson, "These tiny insects are natural components of the southern pine forest ecosystem."[31] When a forest was in ecological balance, lightning struck a tree down, and pine beetles entered the tree. The RCW, in turn, ate the beetles. Fire suppression, though, led to an overpopulation of pines and subsequent epidemics of pine beetles. Faced with this situation, private forestland owners and the Forest Service burned entire stands of the diseased trees, eliminating habitat for the RCW. In the 1970s, the Forest Service logged about 40 percent of what is now the Kisatchie Hills Wilderness Area in Kisatchie National Forest in Louisiana. As a result, one RCW colony in the forest was surrounded by an area of the forest that had been clear-cut, forcing the birds to fly half a mile for forage. Eventually, the birds abandoned the habitat island.[32]

In 1995, the U.S. Supreme Court decided in the case of *Sweet Home v. Babbitt* that the Endangered Species Act required private forestland owners and the federal government to maintain the habitats that endangered species used for shelter, feeding, and breeding. The court found that modifying habitat constituted "harm" against such species. The decision forced far-reaching forest management changes, including the use of prescribed burns to preserve the RCW's habitat.[33]

In the Apalachicola National Forest, restoring the RCW's population has become an enormous point of pride for forest managers. During that beautiful day in March 2011, Steve Parrish waded through the sea of wiregrass until he came to a pine with a broad horizontal white stripe painted on it. With an evident sense of satisfaction, he said, "This is one of the trees that the red-cockaded woodpecker uses for nesting, so this tree is protected from logging."[34] About twenty feet up, a woodpecker had carved out a cavity as perfectly round as if a skilled carpenter had used a compass to trace its circumference. Hanging just below the cavity was a medallion carrying the initials "RCW" and a number indicating the identity of the tree in the Forest Service's database and telling loggers that the tree was off-limits. "The RCW wants an open, mature forest," Parrish emphasized, "and prescribed burns keep the forest understory open."[35]

Scattered throughout the Apalachicola forest were eight thousand known woodpecker homes. Parrish explained, "The forest has 576,000 acres, of which 300,000 acres are burnable, meaning that they are not swamp. We aim to burn 100,000 acres a year to maintain wildlife habitat and ameliorate unplanned ignitions [i.e., wildfires]."[36] Chuck Hess, a wildlife biologist on the Apalachicola National Forest, added, "The prescribed burns are driven by the RCW. It's the best management indicator of the health of the forest."[37] The Apalachicola managers were also conducting experiments to find out how best to control the growth of palmetto and other woody plants. They found that the most effective strategy was to alternate dormant-season and growing-season burns, leading to a 24 percent decrease in the number of undesirable plants over a ten-year period.[38]

On the Osceola National Forest, wildlife biologist David Dorman explained, "The population [of the RCW] has been increasing by 10 to 12 percent a year. We have 140 active clusters, which is up from 75 active clusters ten years ago."[39] The RCW lives in clusters, or clans, of six to ten birds, with one breeding female and several males that help raise the young. Every member of the clan occupies a separate cavity in a pine, which it uses for nesting and protection from predators and the weather.[40]

Forest managers on Osceola initiated a program in which they have burned thirty-five thousand acres to forty thousand acres each year. A recent burn on the Osceola had scorched the palmettos and blackened the trunks of pine trees, although green tufts of palmetto were already sprouting through the charred floor of the forest. The burns were effective, as wiregrass dominated the saw palmetto, creating an open understory and nurturing a greater number of insect species as forage for the bird.[41] Before a prescribed burn, workers mowed around the base of each tree that had an RCW nest. The mowing prevented fire from reaching RCW trees and igniting the flammable sap that ran down the trunks from woodpecker cavities. The Forest Service's logging program at Osceola also protected RCW habitat. According to Dorman, "We take out slash pine and loblolly and create one- and two-acre openings, which the woodpecker likes. We then replant with longleaf pines, which are native to the region."[42]

Forest managers at Ocala, Osceola, and Apalachicola have found that their prescribed fire programs benefit other species, including deer, wild turkeys, bobwhite quail, and the gopher tortoise. In 2009, for example,

Robin J. Innes of the Forest Service reported on the benefits of fire for the gopher tortoise, a foot-long reptile that inhabits Florida, Georgia, South Carolina, Mississippi, Alabama, and Louisiana. This endangered species carves tunnels that can run to thirty, forty, or even fifty feet in length, and there, in the depths of the earth, it lays its eggs. Researchers discovered that the tortoise benefits from one- to three-year burning rotations, which stimulate the growth of plants the tortoise uses for forage and opens up the understory so that the reptile can dig its subterranean home without being impeded by thick vegetation.[43]

The efforts to use prescribed burning to improve habitat for the RCW and the gopher tortoise point to a larger insight: that managers of the eastern national forests are increasingly managing forests as ecosystems comprising highly sophisticated relationships among climate, wildlife, vegetation, soil, water, and air. In addition, forest managers are integrating the growing body of knowledge about how nature's tools, such as fire, create the conditions in which a forest thrives. A one-size-fits-all approach to forest management cannot work because the characteristics of a forest are so related to the unique geographic characteristics of the region in which a forest is found.

The Technology of Prescribed Burns

Despite the ecological benefits, though, prescribed burning does not come without hazards. Mike Drayton, the fire management officer for the Ocala National Forest, said, "In Florida, one of the dangers in the prescribed burning program is the wildland-urban interface. There are many inholdings and cities nearby, including Orlando and Daytona Beach. We have to be very careful about the fires."[44] As of 2011, Ocala's Forest Service staff was conducting prescribed burns seventy-five to eighty days a year, covering upwards of 60,000 acres, or more than 20 percent of the 383,000-acre forest. Before they can burn on a particular day, they have to receive permission from Florida's Division of Forestry, which monitors and approves all prescribed burns in the state.

In addition, a staff meteorologist closely monitors atmospheric conditions for days on which burns are scheduled. If the humidity is too low, burns are postponed because the air is too dry and fires can burn out of control. Naturally, fire officers do not burn on days when winds are expected, not only because wind can whip fire out of control but because it can blow smoke off the forest. Drayton remarked, "I've had people curse at me because of fires and smoke, but people generally accept what we're doing. This is especially true of long-time Floridians, but newcomers can be alarmed at the fires."[45] Drayton added that the smoke had once drifted over to Daytona during the running of the Daytona 500, upsetting race officials. After that unfortunate incident, the Forest Service made a point of avoiding burns during the race. The Forest Service also informs the public about the schedule for prescribed burns by issuing press releases and posting schedules on the Forest Service Web site.

Such efforts to explain the importance of prescribed burning has paid dividends in increased public acceptance. In 2001, John B. Loomis and Lucas S. Bair of Colorado State University teamed with Armando González-Cabán of the Forest Service to conduct research on changes in public attitudes toward prescribed burning programs. The survey asked, "Do you think forest managers should periodically burn underbrush and debris in pine forests?" Before receiving educational materials about prescribed burns, 68.6 percent of respondents agreed that prescribed burns should be used. The researchers then sent out an illustrated booklet explaining the importance of prescribed burning in preventing catastrophic wildfires and improving wildlife habitat. After reading the materials, 84.9 percent of the respondents agreed with the value of prescribed burning.[46] The Floridians were particularly supportive of using prescribed burning to eliminate fuel and reduce the effect of wildfires, particularly because many of them had vivid memories of wildfires that caused widespread damage in 1988, 1994, and 1998.

In the Ocala National Forest, fire management officers like Mike Drayton play a pivotal role in educating the public about prescribed burning. As he explained the benefits of fire, Drayton also touted the ways in which technology has made its use more efficient. He stepped out of his mud-splattered four-by-four pickup truck and led the way through the forest to a recently burned tract. Drayton, who had started his career in the Youth

Conservation Corps and then steadily moved up the ranks of the Forest Service, walked quickly but purposefully through the charred underbrush. The lower trunks of most pines were blackened, but the trees, Drayton emphasized, were unharmed. He bent over and showed where green was already sprouting in the midst of the charred stem of a palmetto. "We'll burn this stand of the forest twice more, once during dormant season and once during growing season," he explained. "There's still a lot of vegetation on the forest floor because it's been a while since we burned this tract. But two more prescribed fires will burn away the fuel."[47]

From there, he drove to the nerve center of Ocala's prescribed burning program, a flat, low building painted the beige and dark green colors of the Forest Service. About a hundred yards away was a helicopter pad with a gleaming Bell Jet Ranger helicopter. During the 1940s, 1950s, and 1960s, the fire officers did all the prescribed burns by hand, using torches that they lit the old-fashioned way, with kitchen matches. In the early 1970s, however, in a clever adaptation of existing technology, they started using helicopters. Drayton introduced John Vinson, the helicopter manager, who supervised the ignition of the forest below. Vinson explained that he would fly low over the forest, usually around fifty feet, and a fire specialist would release aerial ignition spheres, which looked like Ping-Pong balls. When the spheres hit the ground, they ignited the surrounding vegetation.[48]

The machine that ignited and released the spheres was a device swathed in bright, shiny aluminum. It was the Premo MK III Plastic Sphere Dispenser, manufactured and distributed by SEI Industries, a company based in British Columbia, Canada (figure 8.2) The device injected ethylene glycol—or automotive antifreeze—into the plastic spheres, which already had potassium permanganate. When the two chemicals met, they began to ignite. The MK III dropped the spheres through a chute, and by the time the spheres reached the ground, they had burst into flame and started small fires on the forest floor. In case the MK III malfunctioned, it had a built-in fire-extinguishing system, and if worst came to worst, the operator could jettison the MK III in seconds, which Drayton and Vinson emphasized they had never had to do.[49] Once the fires had been ignited, fire management officers on the ground ensured that they stayed under

Figure 8.2 *Premo MK III Plastic Sphere Dispenser. Pilots fly helicopters over an area of Florida's national forests scheduled for a prescribed burn, and fire officers dispense plastic spheres from this dispenser. The spheres ignite and set small fires, which are controlled by officers on the ground. Photograph by Christopher Johnson.*

control. In addition, the fire officers carefully planned the fire so that it burned toward natural firebreaks, such as roads, trails, lakes, and streams.

Lessons Learned

Since the end of World War II, the Forest Service, state forestry agencies, and private owners of forestlands have instituted a major shift in the uses of fire. As evidenced by the actions of forest managers on the Apalachicola, Osceola, and Ocala National Forests, fire has evolved into an important tool in maintaining healthy forests. Forest managers on all three forests took several steps to use fire to improve the ecological health of forests.

1. *Forest managers were open to experimentation.* The managers of the Osceola National Forest were going against established Forest Service policy when they started prescribed burns in the early 1940s. District

Ranger L. S. Newcomb listened to his staff and allowed the experimentation to go forward, however.

2. *The forest managers supported experimentation with data.* The Forest Service did not just allow the advocates of prescribed burning to move ahead with their experiments. The managers insisted on data to support the experimental burns, and the resulting statistics showed that prescribed fires reduced fuel and the severity of wildfires, encouraged the growth of desirable species like longleaf pines, created habitat for endangered species like the red-cockaded woodpecker, and helped control invasive species.

3. *The forest managers changed their stewardship of the forest to suit geographical conditions of the ecosystem.* When Florida's forest managers began their experiments in prescribed burning, they demonstrated the wisdom of modifying the policies to fit local conditions. In this case, the area had frequent thunderstorms with heavy lightning, a long growing season that nurtured a thick understory, and high humidity. It turned out that frequent prescribed burns during dormant and growing seasons were appropriate for those conditions.

The changing attitudes and policies toward the use of fire reinforce a larger point that emerged in the discussion of reforms in timber harvesting: that the managers of national forests are increasingly prioritizing ecosystem management based on the findings of ecology and related sciences. The use of prescribed burns has had benefits for wildlife and for trees and other vegetation. Through an understanding of the role of fire in the countless interrelationships in a forest ecosystem, forest managers can more closely approximate the natural conditions in which a forest thrives. Yet even as conservation thinkers have changed their attitudes, a knotty question remains: When and to what extent should prescribed burning be used in wilderness? In turn, this question raises the larger issue of the extent to which humanity should manage wilderness. It is an issue that requires a close examination of the ways in which the wilderness movement has affected the eastern national forests.

CHAPTER 9

Monongahela National Forest: Wilderness at Heart

The reforms that unfolded at Holly Springs National Forest in timber harvesting and at Florida's three national forests in the use of prescribed burning reflected important changes in the management and purposes of the
eastern national forests. Driving these reforms were developments in the
biological sciences, public pressure by citizens, and forces within the U.S.
Forest Service and other government agencies. The values of ecosystem
management, wildlife protection, recreation, and aesthetics were gradually
assuming major roles in the management of the eastern national forests.

The development of the wilderness movement, which was born in the
1930s and reached its maturity in the 1960s, was an outgrowth of these
trends. Wilderness advocates sought to preserve the primeval characteristics of forests and other landscapes, enhance primitive outdoor recreation,
and protect scenery of extraordinary value. Some even found profound
spiritual truths in wilderness. The development of this movement has been
especially important in the East, the South, and the Lakes states because
these regions have a scarcity of public recreational lands compared with
the western United States.

Wilderness supporters, however, have also clashed with economic
interests, such as logging and mining corporations, which argue that
resource extraction must remain an important use of the national forests.
Exacerbating these tensions has been that many eastern national forests

are located near major metropolitan areas. Because of the proximity of the forests, corporate interests and their allies have argued against making timber and mining resources unavailable through wilderness designation. Proximity fuels the opposing argument as well: millions of nature lovers in the New Yorks and Bostons and Philadelphias want the types of recreation that wilderness promises: backpacking and primitive camping in areas that offer solitude and respite from the stresses of modern life.

To examine the knotty issues surrounding wilderness in the eastern national forests, we have chosen two national forests: the Monongahela National Forest (MNF) in West Virginia and the Superior National Forest in Minnesota, which contains the Boundary Waters Canoe Area Wilderness (covered in chapter 10). In both West Virginia and Minnesota, wilderness protection grew out of citizen involvement, and campaigns for wilderness involved working closely with legislators. The two states are also very different in important ways. Geographically, the MNF is mountainous, as it is part of the Allegheny Range, whereas Boundary Waters is a pristine water ecosystem. Each region also has a unique culture, political environment, and economic mix.

The MNF is pivotal because it was the epicenter of the movement to protect wilderness in the East. In 1964, controversy erupted on the Monongahela when the Forest Service allowed clear-cutting on several tracts of the forest. The agency based its decision on research showing that clear-cuts, when used judiciously, produced more volume of timber, stimulated the growth of desirable species of trees, and created better habitat for certain wildlife than non-clear-cut land. The decision did, however, represent a change of policy for the eastern national forests. As forest historian David Conrad has written, "It was contrary to common practices of the timber industry in the East, and it was foreign to the management of eastern National Forests."[1]

Fueling the controversy on the MNF was the humble turkey. In the mid-1960s, when gobbler-seeking hunters went to favorite spots such as the Cranberry Backcountry, they were appalled to see so many trees—and the accompanying wildlife habitat—gone.[2] Ed Cliff, who was chief forester of the Forest Service at the time, recalled:

We had a visitation from two or three people from the Monongahela National Forest, who came in to protest against the clear-cutting on the Monongahela. They were very upset because of some timber harvesting in the area that they had been using for years for turkey hunting. They felt that the cutting that was being done was not compatible with hunting. They demanded that this be stopped.[3]

During the same period, the Forest Service had allowed clear-cutting in several other national forests, including the Bitterroot in Idaho. One of the agency's goals was to increase timber production. During the 1960s, the U.S. economy was booming, and builders were erecting new homes at a feverish pace. In 1940, the MNF produced nine million board feet of timber. By 1960, the amount of timber harvested on the forest soared to twenty-two million board feet.[4]

The negative reaction to the expansion of timber harvesting clearly caught the Forest Service by surprise. Cliff remembered, "Personally, I didn't pay as much attention to that protest as perhaps it deserved. I didn't recognize the strength of the opposition that could be developed over the issue."[5] The controversy revealed a deep division between conservationists and national forest managers over the direction of the eastern national forests. The Monongahela controversy erupted just at the time that the U.S. Congress was debating the Wilderness Act of 1964, which was destined to bring far-reaching changes in how the federal government managed the forests.

Origins of the Wilderness Movement

To understand how a constituency developed to protect wilderness on the MNF, it is necessary to step back in time to the origins of the wilderness movement. During the 1920s, three young conservationists—Aldo Leopold, Arthur Carhart, and Robert (Bob) Marshall—started to build the intellectual underpinning for setting aside natural areas of extraordinary value as wildernesses. Leopold, who had grown up in Iowa in the late 1890s and earned a degree in forestry from Yale University, started a rapid ascent in the Forest Service when he became an assistant district forester in the Gila National Forest in New Mexico. There he fell in love with the wild peaks and valleys of the forestlands surrounding the Gila River. He lobbied for and, in 1924, won approval from the Forest Service to set aside part of the Gila National Forest as a primitive area. It was the first wilderness in the national forest system.

Carhart, who, like Leopold, had grown up in Iowa, joined the Forest Service in 1919 in the newly created job of "recreation engineer." He soon became a strong advocate of the recreational possibilities of the national forests. After he took a position in District 2 (later designated as Region 2) in the West, Carl J. Stahl, the assistant district forester, assigned him the task of assessing the recreational potential of Trappers Lake, a crystal-blue body of water in northwestern Colorado. Specifically, Stahl wanted Carhart to identify a route for a loop road around the lake and sites for one hundred summer homes. Carhart, though, had a different vision: of Trappers Lake as a protected wilderness area. "There are a number of places with scenic values of such great worth that they are rightfully [the] property of all people," Carhart wrote. "They should be preserved for all time for the people of the Nation and the world. Trappers Lake is unquestionably a candidate for that classification."[6] At Carhart's urging, the Forest Service decided to stop development at Trappers Lake and manage it as a primitive area.

Marshall, the third founder of wilderness thought, was the son of a prominent lawyer in New York. Marshall embraced the outdoors while spending summers at his parents' summer camp in the Adirondacks, where he and his brother eventually climbed all the range's four-thousand-foot mountains. After completing his doctorate in plant physiology from Johns Hopkins University, he published one of the seminal early works in the wilderness movement, "The Problem of the Wilderness," which appeared in *Scientific Monthly* in 1930. In the essay, Marshall celebrated the physical, mental, and aesthetic benefits of wilderness.

In 1934, Marshall attended the annual meeting of the American Forestry Association in Knoxville, Tennessee. On October 19, he took a field trip to visit a nearby Civilian Conservation Corps camp and was accompanied by several conservationist friends, including Harvey Broome, the leader of the Smoky Mountains Hiking Club; and Benton MacKaye, a New Englander who had, in the early 1920s, made the original proposal for what became the Appalachian Trail. Along the way, the group started talking excitedly about forming an association that would dedicate itself to preserving wilderness in the United States. The organization blossomed into the Wilderness Society, and they quickly invited several other prominent conservationists to join, including Leopold; Robert Sterling Yard, who had been a founding member of the National Parks Association; Ernest Oberholtzer, a Minnesotan who would be instrumental in protecting the Boundary Waters Canoe Area; and Harold Anderson, who was a leader of the Potomac Appalachian Trail Club. Yard served as the organization's first president.

Under the prodding of the Wilderness Society and other conservationist groups, the Forest Service and the National Park Service set aside 14.2 million acres in the West as primitive areas before World War II. Marshall; Leon Kneipp, who was an assistant chief forester; and other wilderness advocates in the Forest Service also succeeded in changing the designation of primitive areas to wilderness areas and strengthening their protection. These areas were classified into two categories: wilderness areas, which comprised more than one hundred thousand acres; and wild areas, which encompassed five thousand to ninety-nine thousand acres.

When the economic boom that followed World War II created a massive demand for wood, the Forest Service gradually began to rearrange the boundaries of the wilderness areas to reduce their size and open up more forestlands to logging as well as mining. Alarmed by what they perceived as the Forest Service's lagging commitment to wilderness, the Wilderness Society, the Sierra Club, the Izaak Walton League, and other conservation organizations decided to press Congress to pass a law that would protect wilderness.[7]

In 1945, Yard passed away, and taking his place as the leader of the Wilderness Society was Howard Zahniser, who proved to be an indefatigable fighter for wilderness. Zahniser drafted an early version of a wilderness bill in 1955 and persuaded Senator Hubert H. Humphrey of Minnesota and Representative John P. Saylor of Pennsylvania to introduce the bill into their respective houses of Congress, which they did in 1956. The mining, timber, and cattle industries vehemently opposed the bill, but over the next eight years, it steadily gained public backing and the support of legislators, and in the summer of 1964, both houses of Congress passed the Wilderness Bill. President Lyndon B. Johnson signed it into law on September 3, 1964.[8]

According to the law, "A wilderness, in contrast with those areas where man and his own works dominate the landscape, is hereby recognized as an area where the earth and its community of life are untrammeled by man, where man himself is a visitor and does not remain." It did not have "permanent improvements or human habitation." It also provided "outstanding opportunities for solitude or a primitive and unconfined type of recreation."[9] The law established a National Wilderness Preservation System, which would comprise wild areas with at least five thousand acres of land. (The five-thousand-acre rule is no longer in effect.) The law also stipulated that within ten years, the secretary of the interior would review roadless areas for the purpose of assessing their suitability as protected wildernesses.

In wildernesses, there were to be no roads and "no commercial enterprise," such as logging, grazing, farming, mining, or other economic uses. All motorized vehicles were banned, including motorboats and aircraft. The law did, however, include a compromise that allowed continued use of motorboats in the Boundary Waters Canoe Area Wilderness of Minnesota, an exception that will be examined in chapter 10. Another compromise allowed mining for minerals and drilling for natural gas until December 31, 1983.[10] The Wilderness Act, however, applied only to federally owned lands, not to state forests or privately owned forests.

Wilderness in the Eastern National Forests

The Wilderness Act specified 9.1 million acres to become wilderness, but protecting more wilderness would require additional acts of Congress. Wilderness turned out to be wildly popular among American outdoor recreationists, yet almost immediately, serious problems surfaced in creating wildernesses in the eastern national forests. From 1964 to 1974, a mere four areas east of the 100th meridian received wilderness designation: Boundary Waters in Minnesota, the Great Gulf in New Hampshire's White Mountains, and Linville Gorge and Shining Rock in North Carolina.[11] The Forest Service took a strict constructionist approach to interpreting the law, refusing to propose as wilderness any areas that had been previously logged or contained roads or other improvements. The agency's approach came to be known as the "purity policy." Because nearly all the eastern national forests had been logged at one time or another, the purity policy meant that the Forest Service recommended very little forestland in the East, the South, or the Lake states as wilderness.

In the early 1970s, another key piece of legislation further affected how the Forest Service evaluated potential wilderness. In January 1970, President Richard M. Nixon signed the National Environmental Policy Act (NEPA), which established the Council on Environmental Quality and required federal agencies, including the Forest Service, to consider alternative strategies when making policy decisions that affected the environment. As part of its implementation of NEPA—and to get in front of the rapidly growing wilderness movement—the Forest Service undertook, from 1971 to 1973, an inventory called Roadless Area Review and Evaluation (RARE). The RARE process identified 235 areas of potential wilderness, encompassing eleven million acres. Few eastern forests met the criteria established in RARE, though, and the public had no opportunity to provide input into the process. In the eyes of many conservationists, RARE failed to solve the problem of wilderness in the eastern national forests.[12]

West Virginia and the Eastern Wilderness Act of 1975

Because of the failure of RARE, eastern wilderness advocates grew increasingly frustrated by the lack of wilderness, and state by state, they began banding together to persuade Congress to create new eastern wildernesses. One of the epicenters of this blossoming grassroots activism was West Virginia. Over the span of thirty years, West Virginian conservationists organized three campaigns—from 1969 to 1975, during the early 1980s, and from 2001 to 2009—to protect wilderness in the MNF. The stories of these campaigns show the challenges that conservationists faced in protecting wilderness in the East and the strategies they used to win wilderness protection.

Examining these campaigns in some depth reveals several insights about the wilderness movement and how it has affected the eastern national forests and the Forest Service. First, the three West Virginia campaigns involved the strategy of building broad-based coalitions that encompassed environmental organizations, hunters and anglers, advocates for preserving rural culture, and business interests. In addition, winning the active support of the state's congressional delegation was critical. Finally, wilderness advocates developed and employed increasingly sophisticated economic arguments, asserting that wilderness protection boosted tourism to West Virginia.

Wilderness advocacy in West Virginia had its roots in the late 1960s when conservationists from a number of different organizations, including the Sierra Club, came together to form the West Virginia Highlands Conservancy (WVHC). The organization would play a leading role in all three wilderness campaigns and, along the way, establish itself as one of the premier environmental organizations in the state. Its newsletter, *The Highlands Voice*, provided a rich accounting of environmental protection in the state.

Two issues drove the conservancy members' concerns: increased timber harvesting and two proposals for highways that would have bisected some of the most gorgeous vistas in the state. In 1968, Helen McGinnis, a transplanted Californian who had taken a job at the Smithsonian Institution in Washington, D.C., went hiking for the first time through Dolly Sods, one of the most scenic areas in the MNF. During the mid-nineteenth century, the Dahle family had grazed sheep there on grass fields that they called "sods," giving the area its name. It was one of the most rugged and challenging expanses in the East, sitting high on the Allegheny Plateau, with elevations ranging from twenty-seven hundred feet to forty-seven hundred feet. McGinnis, who had been an outdoor enthusiast while growing up in California, recalled, "This area reminded me of the West—open, sweeping vistas, just much smaller."[13] She wondered why the area was not protected as wilderness, as similarly pristine areas in the West were protected.

At meetings of the WVHC, McGinnis met other wilderness enthusiasts, including Rupert Cutler and Stewart Brandborg of the Wilderness Society; Michael Frome, who had written extensively about wilderness in the United States; and Ernie Dickerman, who had been a stalwart supporter of wilderness in the East. They suggested to McGinnis that she consider advocating for Dolly Sods as a possible wilderness. The WVHC organized a meeting to identify potential wilderness in the MNF and to persuade the state's congressional delegation to bring a West Virginia wilderness bill before Congress.[14]

In addition to Dolly Sods, the WVHC proposed two other wildernesses: Otter Creek and the Cranberry Backcountry. Otter Creek (figure 9.1), which was an enormous bowl of approximately twenty thousand acres formed by Shavers Mountain on the east and McGowan Mountain on the west, had been owned by the Otter Creek Boom and Lumber Company, which had harvested timber there early in the twentieth century.[15] Otter Creek's forests had recovered extremely well, prompting Dr. Thomas King, a local dentist with a bent toward the outdoors, to rave, "I was almost overwhelmed by the wilderness beauty of the Otter Creek Valley." He called it "the most beautiful area in the state."[16]

In 1970, though, the area faced a new threat when the Island Creek Coal Company proposed to conduct exploratory drilling for underground coal reserves in five sites. In what became a landmark case for environmental protection, the WVHC filed suit to stop the company from building roads to provide access to the drilling sites, and Judge Robert E. Maxwell found in favor of the conservancy. Although the judgment was favorable, the WVHC realized that proposals to mine or log in the area would continue because the area's resources were just too tempting. The only way to prevent those efforts, the group realized, was to have Otter Creek designated as wilderness. WVHC members made presentations throughout West Virginia to build public support for protecting Otter Creek as wilderness.[17]

The third area the WVHC proposed as wilderness was the Cranberry Backcountry, which encompassed more than thirty-five thousand primeval acres that were crisscrossed by ridges and deep valleys and dominated

Figure 9.1 *Otter Creek Wilderness, Monongahela National Forest, West Virginia.*
In 1974, Congress created the Otter Creek Wilderness, which ranges over approximately
20,698 acres in the northern sector of the Monongahela National Forest. Photograph
by Christopher Johnson.

by a variety of hardwoods and red spruce. It was also the habitat for a sizable population of black bears. Because the Forest Service had managed
the area as a back country, motor vehicles had been prohibited. Then, in
October 1970, *The Highlands Voice* reported that the Forest Service was
planning to increase timber harvesting, including the use of clear-cuts.
In addition, the Princess Coal Company proposed a deep-mine project for
extracting coal.[18] The developments added a sharp note of urgency to the
WVHC's efforts.

Meanwhile, events on the national scene were affecting the wilderness campaign in West Virginia. In an attempt to forestall the creation of
eastern wildernesses, the Forest Service proposed in the early 1970s a new
category called "wild areas," which would include areas that had previously been logged or mined. In December 1972, however, Ernie Dickerman
and Doug Scott of the Wilderness Society opposed the creation of this new

category and argued instead that eastern wildernesses should be added to the National Wilderness Preservation System under the auspices of the 1964 Wilderness Act. Their efforts proved successful in heading off the creation of the second category.

To press for the protection of Dolly Sods, Otter Creek, and the Cranberry Backcountry, the WVHC worked assiduously to build public support, writing letters to editors of local newspapers, convening public meetings, and working to gain the support of West Virginia's congressional delegation, most notably, Senator Robert W. Byrd. Meanwhile, the Nixon administration, which recognized the growing electoral power of environmentalists, directed the Forest Service to analyze potential wildernesses in the East. While the WVHC was pressing for wilderness in West Virginia, conservationists in other areas of the East, the South, and the Lake states testified to Congress in 1973 and 1974 on the importance of creating eastern wildernesses.[19]

On December 16, 1974, the House Interior Committee approved a bill that created new eastern wildernesses, and two days later, the entire House approved the bill. After the Senate approved quickly of the bill, President Gerald R. Ford signed it on January 3, 1975. Through a clerical error, the law had no title, but it has been known ever since as the Eastern Wilderness Act of 1975. The law immediately created fifteen eastern wildernesses, including Dolly Sods and Otter Creek. It also identified seventeen Wilderness Study Areas, including the Cranberry Backcountry. Amidst the jubilation shared by members of the WVHC was disappointment that the Cranberry Backcountry had not been listed as wilderness, but as *The Highlands Voice* noted, "It is not a wilderness area, but it will be managed as one just the same while it is being studied for possible inclusion in the Wilderness Preservation System."[20]

RARE II and the 1983 Wilderness Act

West Virginia's conservationists regarded the 1975 act as one step forward in a long process of protecting more of the MNF through wilderness designation. The WVHC continued to support the Cranberry Backcountry as wilderness, but it also began to promote Laurel Fork, an area in the north-central part of the MNF that boasted pristine streams with teeming populations of trout, myriad wetlands, and ample opportunities to observe beaver dams and other aquatic wildlife.

Meanwhile, the very conditions under which the Forest Service managed the national forests were undergoing substantial change, partly because of the lack of public involvement during RARE. Since 1897, the Organic Act had guided management of the national forests. Congress, though, realized that it needed to update the management of the national forests, and in 1974, it passed the Resources Planning Act and, in 1976, the National Forest Management Act (NFMA).[21] The NFMA required each national forest to issue a forest plan that examined alternatives for managing the resources, recreation, and protection of the forest. Most important is that the public had to have opportunities to review forest plans before they were implemented. In addition, the managers of each national forest had to issue an environmental impact statement that assessed each of the alternative strategies proposed in a forest plan. After planning was completed, the national forest issued a Final Statement of Policy and Program, which guided the management of the forest and the budget requests for recreation, timber harvesting, and other activities. In addition, the law required the assessment of the condition of forests and rangelands every ten years. The NFMA reflected the increasing complexity of national forest management, requiring that timber production, recreation, environmental impact, wildlife protection, watershed management, and species diversity all be considered in the formal planning process.

After winning election as president in 1976, Jimmy Carter directed the secretary of agriculture to conduct a new, more complete study of roadless areas, known as RARE II. Commencing in 1977, RARE II included extensive surveys of some three thousand roadless areas encompassing sixty-two million acres, including millions of acres in the eastern national forests.[22] Numerous public meetings took place, and activists on both the left and the right participated in protests that, at times, became rowdy. Political conservatives asserted that the process was creating too much wilderness and excluding too much forestland from logging, grazing, and mining. In contrast, some environmentalists claimed that RARE II violated the Eastern Wilderness Act of 1975, which had placed the responsibility for designating wilderness in the hands of Congress.[23]

Under RARE II, the Forest Service studied nineteen roadless areas in West Virginia, developing the Wilderness Attributes Rating System (WARS) to rate potential wildernesses. Among the fifty-five study areas in the northern Appalachians, the Cranberry Backcountry received the second highest score, and in 1978, the Forest Service announced that it supported wilderness designation for the Cranberry area. The WVHC, meanwhile, added several other sites to its recommended list, including Laurel Fork (which was divided into a North and a South unit), Roaring Plains, and Seneca Creek. In three public hearings, West Virginians from around the state voiced strong support for these new wildernesses.

In 1980, Representative Cleve Benedict, a Republican from West Virginia, submitted a wilderness bill that included the Cranberry Backcountry and Laurel Fork North and South. The Seneca Creek Backcountry was on the original Forest Service list, but when local landowners opposed that wilderness designation, Laurel Fork was substituted. With support from Senators Byrd and Jennings Randolph, the Cranberry Wilderness Bill, as it was known, passed Congress in early 1983, and President Ronald Reagan signed it on January 13, 1983. It established the Cranberry Wilderness, which at 36,500 acres was the largest wilderness area in the East, and the Laurel Fork North and South Wildernesses, which together comprised twelve thousand acres. The bill, however, included language that "released" other parts of the MNF from being considered for future wilderness during the NFMA-guided forest planning process that culminated in the 1986 Monongahela National Forest Management Plan.[24] West Virginia's conservationists wanted more wilderness, but they would have to be patient.

The Wild Monongahela Act of 2009

After the passage of the 1983 Wilderness Act, a period of quiet followed during which the Forest Service managed the new wildernesses. Because conservationists could not propose any new wilderness for several years, they placed a priority on ensuring that the Forest Service preserved the wild characteristics of *potential* wilderness in the upcoming forest plan.

Several areas of the forest, such as Seneca Creek, had received high WARS ratings during RARE II, and the activists wanted to be sure that the Forest Service did not permit logging, mining, or road building in those areas.

Dr. Mary Wimmer, a professor of biochemistry at West Virginia University, had become involved in environmental issues in the state and, in 1984, helped found the state chapter of the Sierra Club. Wimmer explained:

> I wrote a separate addendum to our comments on the 1985 Draft Forest Plan on these areas and the justification for keeping them wild. To accomplish this, we helped the Forest Service design a new management prescription designated "MP 6.2" for back country protection. No timbering or road building would be done, with minimal human impact allowed, with emphasis on natural forces, semi-primitive non-motorized recreation, and remote wildlife habitat.[25]

The term *MP*, which is also called a Management Area, refers to areas identified in a forest plan and managed for specific ecological, economic, or social conditions. The Forest Service eventually designated seventeen areas in the MNF as Management Area 6.2, totaling 125,000 acres.[26] Another period of quiet followed as the Forest Service managed these new areas with no major controversies.

In 2001, the Forest Service started another required NFMA cycle of MNF plan review and revision, which would be completed by 2006. No longer bound by the release language of the 1983 Wilderness Bill, West Virginia's conservationists began to strategize for a new West Virginia wilderness law. The WVHC, the West Virginia chapter of the Sierra Club, and the Wilderness Society (later joined by the Pew Charitable Trust for America's Wilderness) formed a new organization—the West Virginia Wilderness Coalition (WVWC)—to spearhead the campaign for new wilderness. Mike Costello, who became involved in the campaign early on as a West Virginia University student and later served as one of the WVWC's coordinators, said, "It was around the issue of permanently protecting some of the Monongahela's best remaining wild lands by congressional wilderness designation that the West Virginia Wilderness Coalition got started."[27]

Several new leaders emerged. They included Dave Saville, who had earned his degree in forestry from West Virginia University, worked part-time for the WVHC for eight years, and then became an employee of the WVWC. Saville organized the first meeting that reignited the citizens' wilderness effort. Involved again was Helen McGinnis, the WVHC member of early West Virginia wilderness days, who had moved back to West Virginia after being gone for twenty-five years. Another leader was Beth Little, who lived in Pocahontas County and was, along with Wimmer, a member of the West Virginia Sierra Club. In addition, the WVWC hired Matt Keller as the wilderness campaign coordinator. He had graduated from Ohio University and had a degree in geography, with an emphasis on wilderness. Keller knew about mapping and geographic information systems, reflecting the professionalization and technological sophistication of the wilderness movement.[28]

The WVWC found that the Forest Service and state agencies, particularly the West Virginia Department of Natural Resources, were more resistant to wilderness protection than they had been during the 1980s. The mood was different, reflecting the reluctance of President George W. Bush's administration to take public forestlands out of circulation for logging, mining, grazing, and drilling for natural gas, even though publicly owned recreational lands were so scarce in the East. In addition, game managers contended that logging was needed for habitat improvement. The conflict reflected a growing division over the purposes of the public forests. Environmentalists increasingly asserted that the national forests must exist for the benefit of all the public and emphasized the importance of ecosystem protection and recreation. Most political conservatives subscribed to the view of the national forests as commodity producers. Indeed, energy production had gained greater prominence as new technologies unlocked the potential for extracting coal, oil, and natural gas from public lands. (Chapter 12 will examine the issue of natural-gas extraction in detail.) If forest supervisors seemed circumspect about supporting wilderness, it was often because the Forest Service was caught in the position of serving two very different masters.

The contrasting perceptions of the purpose of the public forests showed up in the differing proposals for wilderness in West Virginia in the first decade of the twenty-first century. After more than a year of detailed evaluation, on-the-ground inspection, and in-depth mapping discussions with various groups, the WVWC proposed a significant expansion of wilderness, encompassing fifteen areas that totaled 143,000 acres. During the same period, the Forest Service prepared several wilderness alternatives for its 2005 Draft Forest Plan. The Forest Service favored Alternative 2, which recommended only 27,700 acres of new wilderness and included timber harvesting and road construction in several areas designated MP 6.2, which had been previously protected.

Among the alternatives that appeared in the Draft Forest Plan, the WVWC decided to support Alternative 3, which included an expansion of Dolly Sods and several other new wildernesses, totaling 99,400 acres.[29] Wimmer said:

> Despite overwhelming public support for Alternative 3, the Forest Service did not move from their original wilderness preference in the 2006 Final Plan—only 27,684 acres, leaving out the spectacular 25,000-acre Seneca Creek Backcountry, the 10,000-acre East Fork Greenbrier, the 7,200-acre Dolly Sods North, and the rest of Roaring Plains. Furthermore, 20 percent of the original 6.2 acreage was lost to management prescriptions that allowed logging and road building, although some 6.2 areas were expanded and three new areas added.[30]

According to Costello of the WVWC, "The Forest Service opposed the Wilderness Coalition's preferred alternative, but there were 17,000 comments in favor of it, and we leveraged those numbers to persuade the legislators to support more wilderness."[31]

The WVWC began to execute a statewide campaign to build public opinion in favor of its expansive proposal for wilderness and to win the support of West Virginia's congressional delegation, which included Senators Robert Byrd and Jay Rockefeller, both Democrats; Representative Shelley Moore Capito, a Republican; and Representatives Alan Mollohan and Nick Rahall II, both Democrats. At that point, the conservationists needed a champion for their preferred alternative in Congress. Beth Little said, "Congressman Rahall [who had become chair of the House Resources

Committee in 2007] was that champion, which is what you need to pass wilderness bills."[32]

The WVWC's communication strategy focused on the economic benefits of wilderness to the state. Costello explained, "It was important to gain support from the business community, and more than 150 businesses supported more wilderness. Local mayors, including the mayor of Lewisburg, also spoke out in favor of wilderness."[33] Supporters pounded home the message that West Virginia boasted some of the finest wild forestlands in the East, which attracted thousands of outdoor recreationists—and their dollars—from the East Coast megalopolis that stretches from Richmond to Boston.

Costello added, "There was also support from the religious community. An organization called Christians for the Mountains was important in gaining support for more wilderness. In fact, the churches launched their own subcampaign, and these included the West Virginia Council of Churches and the United Appalachian Ministry."[34] Wimmer added, "Support also came from hunters and anglers, with West Virginia's Mountaineer Chapter of Trout Unlimited taking the lead. More than 130 health care professionals from around the state signed a wilderness support proclamation."[35] When the WVWC testified before the House Resources Committee, people from several different communities spoke, reflecting a broad base of support for wilderness in the state.

The most prominent opponent to the WVWC's preferred proposal was the game division of the West Virginia Department of Natural Resources, which is responsible for managing lands for hunting and fishing. The issue was not whether hunting would be allowed in wilderness; it would be. The issue was whether the Department of Natural Resources would continue to manage the forests as it deemed fit for game, especially deer, turkey, ruffed grouse, and trout. For instance, it wanted to be able to harvest timber in areas to create the edge environments that attracted deer and other wildlife. Wilderness supporters countered that only a small portion of the state was involved in game management and that boundary adjustments in proposed wilderness had been made to avoid limiting the department's management activities.[36]

Figure 9.2 *Seneca Rocks, Monongahela National Forest, West Virginia. Climbers throughout the East come to Seneca Rocks to scale these cliffs. The cliffs are part of the Spruce Knob– Seneca Rocks National Recreation Area but do not have wilderness protection. Photograph by Christopher Johnson.*

In 2007, a final bill took shape. Supported by Senators Byrd and Rockefeller and Representatives Rahall, Capito, and Mollohan, it included more than 37,000 acres. The legislation expanded the Dolly Sods, Cranberry, and Otter Creek wildernesses and added three new wilderness areas: Big Draft, an area of 5,144 acres that is in the southeastern part of the forest; Roaring Plains West, an area of 6,792 acres that lies about three miles southwest of the Dolly Sods Wilderness; and Spice Run, which consists of 6,030 acres toward the southeastern end of the forest. In March 2009, Congress approved the Wild Monongahela Act as part of the Omnibus Public Lands Act, and President Barack Obama signed it on March 30, 2009.[37]

Even amidst this triumph, though, some disappointment seeped in around the edges. Wimmer expressed frustration at not winning wilderness protection for the Seneca Backcountry's twenty-five thousand wild acres. The area's wilderness attributes had been highly rated as early as RARE II, and it has been protected from development since the 1986 Forest Plan. The ecosystem is adjacent to Spruce Knob, the highest point in West Virginia, and is a magnificent area that draws thousands of hikers. "Seneca Creek was the jewel that we wanted," Wimmer said, "but it was not included in the final bill."[38] The Forest Service currently manages the area as nonmotorized backcountry, with half being within the Spruce Knob–Seneca Rocks National Recreation Area (figure 9.2). Unlike permanent wilderness protection, backcountry management is subject to change with each new forest plan cycle.

Lessons Learned

West Virginia's conservationists used a variety of strategies to expand the amount of protected wilderness in the Monongahela National Forest, but it is useful to step back and look at these events from a broader perspective. Perhaps no one in recent years has done more to broaden that perspective than William Cronon, professor of history, geography, and environmental studies at the University of Wisconsin and the author of *Changes in the Land* and other landmark studies in environmental history. In 1995, Cronon published an essay titled "The Trouble with Wilderness; or, Getting Back to the Wrong Nature," which roiled the ranks of wilderness advocates. Cronon argued that the wilderness movement ran the danger of inadvertently promoting a division between the human and the natural by taking an approach to wilderness advocacy that was overly pristine and by creating a narrative of human history in which humanity's only effect on the environment has been negative. Cronon argued for "a middle ground in which responsible use and non-use might attain some kind of balanced, sustainable relationship."[39]

As West Virginia's wilderness advocates fought for more protected wilderness, they unquestioningly acted out of a deep commitment to the forests of their state, yet they also, as Cronon advised, took a responsible and sensitive accounting of the history and culture of their uniquely beautiful state. Their experiences tell us much about effective wilderness advocacy.

1. *Organization and economic arguments are essential for effective advocacy in favor of wilderness.* In their campaign for the Wild Monongahela Act, conservationists employed a realistic assessment of the economic impact of wilderness and leveraged that argument into an advocacy that effectively built public support.

2. *Respect local values and interests.* In the campaign for the 2009 law, the West Virginians successfully built a coalition of recreationists, business owners, church members, hunters, and anglers and worked within the context of the real-world needs of their legislators to demonstrate economic benefits for their state. In addition, West Virginia is a rural state, and the wilderness advocates successfully appealed to rural values, balancing preservation with economic development.

3. *"Compromise" was not a dirty word.* In the West Virginia campaign, the wilderness advocates engaged in realistic negotiations with legislators, the Forest Service, and other agencies, and these negotiations inevitably led to compromise. In the end, they successfully preserved more than thirty-five thousand acres of extraordinary forestland.

As a result of these and similar efforts, West Virginia has created an extraordinary network of national forests, state forests, and state parks that are managed for a variety of purposes, from backcountry recreation to timber production. As Wimmer noted, "West Virginia still has wild lands, unlike the developed areas along the eastern corridor, to get away to. Their rarity gives them added value."[40] Since the 1960s, the wilderness movement has had a wide-ranging effect on how the eastern national forests are managed. The notion of multiple use has expanded to include backcountry recreation, ecosystem protection, preservation of all species of native wildlife, and a return of significant amounts of forestland to their natural cycles of succession. The intact wilderness is also central to protecting watersheds, which was a major goal of the Weeks Act. Wildernesses have become natural laboratories in protecting some of America's most essential ecosystems.

CHAPTER 10

Boundary Waters Canoe Area Wilderness: Preservation versus Multiple Use

The sky hung like a gray blanket over the waters of Sawbill Lake, an island-dotted lake in the eastern part of the Boundary Waters Canoe Area Wilderness (BWCAW) in Minnesota's Superior National Forest. Chel Anderson, a plant ecologist for the Minnesota Department of Natural Resources who often guided people on ecological tours of the BWCAW, sat in the stern of a three-person Kevlar canoe and paddled it through the waters of the lake and toward a portage. "This is the southern border of the boreal forest," she explained to her two visitors, "which has white pines, red pines, jack pines, spruce, balsam fir, birch, and aspen. The aspen are more dominant than they once were. That's because timber harvesting replaced fire as the main management tool, and after timber harvesting, the aspen sprouts remained and regenerated."[1]

Boundary Waters, whose rivers and lakes straddle both sides of the U.S.-Canadian border (hence this ecosystem's name), has been protected as wilderness since the passage of the Wilderness Act in 1964, and it gained further protection from the Boundary Waters Canoe Area Wilderness Act, which Congress passed in 1978. The BWCAW is the only wilderness that is primarily aquatic, adjoining Canada's Quetico Provincial Park with its 1,180,000 wilderness acres. Boundary Waters followed an exceptional path—a path marked by conflict—to protection. The tensions surrounding the designation of this wilderness involved a fundamental conflict between preservationists, who favored keeping the area as pristine as possible, and multiple-use advocates, who favored continuing such uses as motorboat recreation, timber harvesting, and mining.

The conflict echoes tensions on numerous other eastern national forests. For example, in the Allegheny National Forest in Pennsylvania, which will be examined in chapter 12, wilderness advocates have been able to gain very little wilderness protection because industry wants to keep the national forest open to oil and natural-gas extraction. Similarly, in the early 1980s in New Hampshire, wilderness advocates disagreed over wilderness boundaries with representatives of logging and mining companies. There, however, the Society for the Protection of New Hampshire Forest and other organizations led negotiations that culminated in compromises over boundaries and the creation of the Sandwich and Pemigewasset wildernesses in 1984.

Wilderness protection has had a visible effect on Boundary Waters, just as it has had on parts of other eastern national forests. Anderson pointed out some of them as she led her visitors on the popular Kelso loop, which starts on Sawbill Lake, crosses the Kelso River, continues on to Alton Lake, and completes the circuit at the national forest campground on Sawbill Lake. The three canoeists reached their first portage, and Anderson pointed out faint traces of wood that were half-buried in the soil. "There used to be wooden canoe stands that paddlers could prop their canoes against," she explained. "In addition, the trail was much wider because the Forest Service cut back vegetation along the portage."[2] In 1978, the BWCA Wilderness Act limited human intervention in the natural environment, and in response, the Forest Service removed the canoe rests and reduced the cutting of vegetation along the portage. Now, the trail is only six feet wide, and the trio brushed against bushes as they traversed the sixty yards that separated Sawbill Lake from the Kelso River.

As they paddled into the river, they entered a completely different aquatic world. The waterway wound gracefully through lily pads and long, sweeping grass. Anderson explained that they were floating through a poor fen, or a wetland resting on a foundation of acidic peat and dominated by sedges. She pointed out pink and purple orchids, with their delicate-looking petals, and the carnivorous pitcher plants, which had bulbous pitcher-shaped leaves with red lines that resembled the arteries of a human body. She explained that insects enter the pitcher and gradually descend, but when they try to escape, tiny hair-like structures on the inner surface of the pitcher prevent them from climbing back out. Eventually, the insect drops into the acid liquid at the bottom of the leaf and dissolves, providing nitrogen for the plant.

The three paddled north on the Kelso River and came to a fork, where they turned left. Reaching the rocky entrance to the second portage, they disembarked, and Anderson pointed out another set of old wood beams partially buried in the soil. "There used to be rails that carried small carts," she said, "and people put their canoes on the carts to transport them from one body of water to another. Now only the wooden pilings are the remnants of the rail system."[3]

Anderson put the canoe back into Alton Lake, which, in contrast to Sawbill Lake, resembled a small sea, so choppy and restless were the waters. They pushed off, and the power of their three paddles pushed them steadily through the waves. They portaged to Sawbill Lake and returned to placid surface. Three bodies of water, three different moods, and yet it was all part of one interconnected system of water. As they paddled across the lake, Anderson grew reflective. "Today," she explained, "the critical issues facing Boundary Waters are nearby copper/nickel mining and the impact of climate change. Behind these issues is recognizing how important it is that these remain public forests, which means that they are to be managed for the benefit of everyone."[4]

The history of the BWCAW and the issues facing the region today reflect many of the questions that surround wilderness in the Monongahela National Forest and America's other eastern national forests. At the heart of the past and future of Boundary Waters is the old but still raging debate over preservation versus multiple use. What voice should local citizens have in decisions by the federal government regarding land-use policies? What has been the effect of wilderness protection on local communities? These essential questions blend ecology, politics, economics, and sociology.

A Circuitous Route to Wilderness Protection

Wilderness protection did not just happen in Boundary Waters. It took an enormous effort by a great many people over a great number of years. The first individual to press for environmental protection of the region was General Christopher C. Andrews, a Civil War general who developed a keen interest in forestry after the war. Andrews became Minnesota's first state fire warden and also served as a Minnesota forestry commissioner. His advocacy helped spur the creation of the Superior National Forest and Quetico Provincial Park in 1909. Andrews also persuaded the General Land Office not to sell 641,000 acres that had been scheduled for sale, and this land formed the nucleus of the Superior National Forest.[5]

That same year, a key player stepped onto the Boundary Waters stage, Ernest Oberholtzer (figure 10.1). Born in 1884 in Davenport, Iowa, Ober, as his friends called him, attended Harvard, where he contracted rheumatic fever. His doctor declared that Ober had only one year to live. Desperate to squeeze every inch of experience he could out of his remaining time, he traveled to Ely, Minnesota, and embarked on a summer-long exploration of the entire region. The experience left the young man feeling strengthened and renewed. Confident that he would outlive the doctor's verdict, he immersed himself in Boundary Waters, building a residence on Mallard Island, taking a heroic canoe expedition from Winnipeg to Hudson's Bay, studying the lifeways of the Ojibwa people, and penning exquisite essays about his beloved rivers and lakes.[6]

The other central figure in the early recognition of the unique beauty of Boundary Waters was Arthur H. Carhart, who, as the first landscape architect hired by the U.S. Forest Service, had been instrumental in protecting Colorado's Trappers Lake as a primitive area, as chapter 9 explained. In the summer of 1919, Carhart visited Boundary Waters for the first time and noted with alarm that cabins were sprouting up along the shores of several lakes. Having found an ally for preservation in Carl J. Stahl, assistant district forester, Carhart wrote a report in which he argued against construction of roads and cabins in the Superior National Forest. Stahl agreed and even removed funds for a road-building project. The Forest Service

Figure 10.1 *Ernest Oberholtzer (right) with Tay-tah-pa-sway-wi-tong (Billy Magee) and dog Skippy. Oberholtzer studied the lifeways of the Ojibwa people, wrote lyrically about Minnesota's north woods, and worked ceaselessly to protect the Boundary Waters Canoe Area from development. From the archives of the Ernest C. Oberholtzer Foundation, Minnesota Historical Society.*

then asked Carhart to conduct a survey of the recreational potential of Boundary Waters, and in 1921, he submitted that report, titled *Preliminary Prospectus: An Outline Plan for the Recreational Development of the Superior National Forest*, which laid out a comprehensive plan to create hundreds of miles of canoe routes.[7] Carhart saw that the Weeks Act could be used to justify wilderness preservation. In 1920, he argued, "Watershed protection depends directly on good forest cover, and the better the cover is maintained, the better will be the watershed. Both of these do not in the least interfere with timber growth when properly directed."[8]

In addition, he advanced a vision of wilderness recreation that was startlingly progressive for its day. "The best field in which to seek recreation," he wrote, "is in the great free fields of God's work. The plains, streams, hills, mountains, lakes, forests and valleys offer a form of recreation that surpasses any to be found where play is corralled within narrow walls and sold at so much per unit."[9] In 1926, in a major triumph for Carhart's vision, the Forest Service set aside part of the Superior National Forest as a primitive area—the Superior Roadless Area—in which there would be no timber harvesting, vacation resorts, or road construction.

The Backus Proposal

Even as the Forest Service was taking steps to preserve the primitive characteristics of Boundary Waters, an industrialist was planning to harness the power of the region's waters. He was Edward Backus, a lumber mogul who proposed to build seven dams to generate water power for lumber and paper mills.[10] The results would have been catastrophic, raising water levels on 14,500 square miles of waterways, sometimes by as much as eighty feet. Backus's proposal outraged conservationists. Carhart's writings and the *Preliminary Prospectus* had done much to raise awareness of the beauty and fragility of the Boundary Waters ecosystem, but it was Oberholtzer who emerged as a key leader of the movement to defeat the dam proposal.

In September 1925, Ober attended a three-day conference of the International Joint Commission (IJC), which had been created by a 1909 treaty between the United States and Great Britain on behalf of Canada. The treaty established processes for resolving disputes over the waterways along the U.S.-Canadian border. The purpose of the conference was to discuss whether the region's waters should be harnessed for industrial development in northeastern Minnesota. Backus made a forceful presentation of the economic benefits of his proposal, leading Ober to comment, "Mr. Backus had been so wholly successful whenever he had undertaken to get something from [either of the two] governments, that the prospects for stopping his program—even if it was a bad program—seemed very poor."[11]

To combat Backus's proposal, Ober formed an alliance with several lawyers who were strong conservationists, including Sewell Tyng, a junior partner in a New York law firm. Ober and Tyng penned a brief for the IJC that built a legal case against the Backus proposal, citing the financial harm that rising water levels would do to local tourist businesses and residences.

Ober then met with about twenty conservationists in the Twin Cities area who were alarmed by Backus's proposal. In 1927, Ober distilled the legal brief he had written with Tyng into a five-thousand-word analysis titled *Conservation or Confiscation: An Analysis of the Water Storage Project Proposed by Mr. E.W. Backus as Affecting International Boundary Waters, Particularly in Quetico Park and the Superior National Forest*. The Twin Cities group printed twenty-five thousand copies of the document and distributed it to libraries and other agencies throughout Minnesota.[12]

In addition to opposing the Backus proposal, though, the Twin Cities group believed that they needed a positive vision for the future of Boundary Waters. Ober sat down and generated another document—"the Program"—in which he proposed that the U.S. and Canadian governments manage the watershed as a bioregion. Ober recognized that the forests, rivers, lakes, and wildlife formed an ecologically integrated system in which the harm done to one part of Boundary Waters would inevitably ripple out and degrade other parts. In his program, he recommended zones that took into account geographical and historical factors. For example, Rainy Lake, which already had access to roads and railways, might be permitted some sustainable development, but areas at the heart of Boundary Waters would be kept completely wild.[13]

Once Ober had written his program, Minnesota's conservationists realized that they needed an organization to coordinate opposition to the Backus proposal. In 1927, they formed the Quetico-Superior Council in association with the Izaak Walton League. With some reluctance, Ober agreed to serve as president of the fledgling organization. Although the Quetico-Superior Council advocated for the international treaty first proposed by Ober, it also pressed for U.S. legislation that would permanently protect Boundary Waters. Fortunately, Senator Henrik Shipstead of Minnesota had come to believe that legislative protection of the region was essential. In 1928, he introduced a bill that would require congressional approval before water levels in Boundary Waters could be altered in any way. Representative Walter Newton of Minneapolis cosponsored the bill, which was introduced into the House as the Shipstead-Newton Act.[14]

Backus lobbied furiously against the bill. Then, in 1929, Newton resigned his seat in the House to become President Herbert Hoover's secretary, and fellow Republican William Nolan took Newton's place in the House. Because Nolan had been a close friend of Backus's, Ober and the Quetico-Superior Council feared that he would not support the bill in the House. Fred Winston, a highly respected public defender who came from a wealthy Minneapolis family, visited Nolan to explain why the bill was so essential to the future of Boundary Waters. Nolan listened closely and then told Winston that although he was a friend of Backus's, he didn't agree with him on the dam construction project.[15] Nolan even agreed to replace Newton as the bill's cosponsor.

The Senate approved the bill in 1930, but it faced difficulties in the House, where the Rules Committee would not let it come to the floor for a vote. Winston successfully lobbied the Minnesota congressional delegation and won its support, and on July 3, 1930, the Shipstead–Nolan Act passed the House, and President Hoover signed it into law. It was a landmark in wilderness protection, for it prohibited logging within four hundred feet of shorelines and prevented changes in lake levels on federally owned land. It was the first time in American conservation history that Congress had taken legislative action to protect wilderness.[16]

Growing Wilderness Protection of Boundary Waters

Throughout the 1930s and 1940s, Boundary Waters faced a fresh set of challenges because of continued timber harvesting, growing recreational use, and road construction. The region drew hunters, anglers, canoeists, and campers. Bruce Kerfoot, who owns and operates the Gunflint Lodge, some thirty miles north of Grand Marais, recalled:

> My grandparents came up here in 1928, when Minnesota was a frontier. They bought property and traded with the Chippewa Indians. In fact, their first guides were Indians. Hunters and anglers came in to do subsistence hunting and fishing. Gradually our family moved into operating a year-round trading post. During the post–World War II era, the lodge changed a lot. We put in amenities like indoor plumbing and built new cabins.[17]

The Kerfoots' experience was typical. Entrepreneurs established out-fitting companies, built campgrounds, hired out as guides, built resorts. By 1950, fifty resorts with upward of 300 buildings lined the shores of sixteen lakes, whereas another 150 private cabins were located near thirty lakes.[18]

The use of motorized vehicles for recreation also grew dramatically, driven in part by tourists' search for convenient ways to reach remote but desirable fishing and hunting locales. Motorboats, airplanes, and amphibi-ous craft carried visitors into Boundary Waters, jeeps and trucks trans-ported canoes over portages, and motorboats towed canoes to faraway lakes. The Forest Service built roads like the Gunflint Trail, which led from Grand Marais to a number of resorts, and the Sawbill Trail transported tourists to Sawbill Lake. At the same time, the Forest Service tried to strike a balance between timber harvesting and recreation. By 1941, the Forest Service had set aside 362,000 acres in which timber was not to be har-vested. This area was in "prime canoe country." On the periphery of Bound-ary Waters, however, logging for sawtimber and pulp expanded quickly.

After World War II, wilderness advocates began to pressure the For-est Service to reduce the commercial footprint in northeastern Minnesota. The conflict came into sharp focus over the issue of small airplanes. Flights carrying hunters and anglers into Boundary Waters had been a lucrative business since the 1920s, and by 1949, more than two dozen small compa-nies operated airplanes out of Ely, which had become the largest base in the United States for seaplanes.[19] The Izaak Walton League took the lead in objecting to the flights, which caused noise and air pollution. Conserva-tionists lobbied the White House to do something about the situation, and in 1949, President Harry Truman signed an executive order that banned both commercial flights and private flights that were lower than four thou-sand feet. The aircraft owners filed lawsuits, but by the mid-1950s, the courts had upheld the ban.

Two Wilderness Acts

During the 1950s, technical improvements made outboard motorboats more reliable, easing access to remote rivers and lakes and boosting tour-ism even more. Even canoeists got into the motorized act; according to one estimate, as many as one-fourth of canoes had square sterns, to which small motors could be attached. Similarly, snowmobile use proliferated during the 1950s and 1960s, attracting more people to the region but also negatively affecting wildlife and interfering with cross-country skiers. As a result, by 1980, the number of annual visitors to Boundary Waters had risen to more than one hundred thousand from a mere two thousand in 1940. In the same period, the length of the average stay had increased from long weekends (two to four-and-a-half days) to weeklong stays (six to seven days).[20] Logging also exploded, from 9.1 million board feet in 1940 to 97.5 million board feet in 1952. Moreover, the Forest Service was gradually removing land from the Superior Roadless Area and building logging roads.

In the face of the pressures for development, the wilderness constituency that had begun to coalesce nationally found strong voices in Minnesota. William Magie, for example, founded the organization Friends of the Wilderness, which advocated for the ban on low flights and for passage of the Wilderness Act in 1964. Another leading advocate was Ely resident Sigurd F. Olson (figure 10.2), who wrote a series of books, including *The Singing Wilderness* and *Wilderness Days*, which celebrated the wildness of Boundary Waters in lyrical prose. Olson was also a teacher, an environmental activist, and president of both the National Parks Association and the Wilderness Society.

By the late 1950s, wilderness advocates were pressuring Congress to pass legislation protecting pristine natural areas of extraordinary scenic, recreational, and ecological value. Minnesota benefited from having a strong conservation ally in Senator Hubert H. Humphrey, who cosponsored the first wilderness bill in 1957 with Representative John P. Saylor of Pennsylvania. In the years of debate leading up to passage of the Wilderness Act, Humphrey held town meetings throughout Minnesota and encountered sometimes intense criticism from people who feared the continued loss of jobs, a trend that had started with the decline of iron mining in the nearby Mesabi Range as steel companies overseas challenged the U.S. steel industry.

To respond to the criticisms, Humphrey wrote into his and Saylor's

Figure 10.2 *Sigurd F. Olson, ca. 1941. Olson brought Boundary Waters to national attention with books like* The Singing Wilderness, *advocated for wilderness protection of Boundary Waters, and served as president of the Wilderness Society and the National Parks Association. Wisconsin Historical Society (WHI-74112).*

bill several exceptions for Boundary Waters that were intended to mitigate the economic impact of wilderness designation. The law permitted the continuation of logging and the use of motorboats on rivers and lakes where they had customarily been used. The exceptions rankled wilderness advocates, and for the next fourteen years, the conflict would grow more heated between those who wanted full wilderness protection for Boundary

Waters and those who wanted to preserve the status quo. Indeed, the Wilderness Act left so much ambiguity regarding the BWCAW that the Forest Service was constantly writing rules regarding motorboats, snowmobiles, mining, and logging. Those rules often faced administrative challenges by conservationists and developers. As Kevin Proescholdt, Rip Rapson, and Miron L. "Bud" Heinselman noted in their book *Troubled Waters: The Fight for the Boundary Waters Canoe Area Wilderness*, "With each new challenge, it became increasingly clear that a long-term, permanent solution could only come from statutory changes that would take matters out of the realm of the inherently unstable Forest Service rule-making apparatus."[21]

The inadequacy of legislative and regulatory protections for Boundary Waters came into sharp relief in 1966, when the International Copper Company (INCO) signed leases to mine copper and nickel on land that bordered the Boundary Waters Canoe Area (BWCA). Chemicals from the proposed mining operations would almost certainly have leached into the surrounding rivers and lakes. Three years later, industrialist George W. St. Clair conducted feasibility studies for mining copper and nickel in the wilderness itself. He justified the prospecting by claiming that government regulation of Boundary Waters applied only to surface rights, not to the extraction of resources below the surface. In response to these threats, the Minnesota legislature in 1976 passed the BWCA Protection Act, which prevented the "development, exploitation, removal or adulteration of a natural resource" that was located in the Boundary Waters wilderness.[22]

Despite the state law, wilderness advocates feared that the BWCA protections remained weak, and their fears were confirmed in the aftermath of the Little Sioux forest fire in 1971, which blackened fifteen thousand acres near the Nina-Bouse and Ramshead Lakes. A member of the Minnesota Public Interest Research Group (MPIRG), which strongly supported wilderness protection, happened to be canoeing in the area soon after the fire, and she observed that a logging company was salvaging timber from the entrance into Boundary Waters at Moose River, land within the wilderness boundaries. Alarmed, MPIRG investigated further and called for a halt to the operations in November 1971. At the same time, snowmobile and motorboat use in the Boundary Waters Canoe Area was also increasing. According to Proescholdt, Rapson, and Heinselman in *Troubled Waters*, motorboats were plying their way over 62 percent of the water surface.[23]

Consequently, on July 27 and 28, 1974, several Minnesota conservationists met at the home of Heinselman, who had worked as an ecologist for the Forest Service for many years and, during the debates over the Wilderness Act in the early 1960s, had courageously spoken out in favor of wilderness protection. The group resolved to fight for a new federal law that would expand the wilderness protection of Boundary Waters and end mining, logging, and motorboat use entirely.

In the meantime, in 1974, Democrat Jim Oberstar had won election to Congress representing Minnesota's Eighth District, which included Boundary Waters. Oberstar recognized the need for economic development in his district, yet he also wanted to end the constant struggle over the future of the Boundary Waters Canoe Area. On October 14, 1975, he attempted to forge a middle way by recommending that Boundary Waters be divided into two zones: a wilderness area of 625,000 acres in which logging, mining, and motorboat use would be prohibited; and 527,000 acres that would be managed for multiple uses, including logging and motorized travel.[24]

To wilderness advocates, though, Oberstar's proposal was deficient. In 1976, Heinselman and other conservationists worked with U.S. Representative Don Fraser, who represented Minneapolis, to draft legislation that would expand wilderness protection to the entire Boundary Waters Canoe Area. National organizations, including the Sierra Club and the Izaak Walton League, become involved in the wilderness campaign. In 1976, several Minnesota-based conservationists formed a new organization, the Friends of the Boundary Waters Wilderness, which worked to build public support in Minnesota for the expansive wilderness bill that Fraser had written and submitted to the House.[25]

The next year, an opposing organization, the Boundary Waters Conservation Alliance, formed to support Oberstar's bill for dividing Boundary Waters into two zones. The alliance drew support from canoe outfitting operators, resort owners, and local unions, which were concerned about the disappearance of logging and mining jobs from northeastern Minnesota.[26] In 1977, a poll by the *Minneapolis Tribune* showed that the state was evenly divided, with 46 percent preferring the Oberstar bill and 49 percent backing Fraser's more expansive wilderness protection. Both bills languished in Congress, however.

In the fall of 1977, Representative Philip Burton, a California Democrat who served as the powerful chair of the House Subcommittee on National Parks and Insular Affairs, summoned Heinselman to his office in Washington, D.C., to help draft a bill that would blend the best features of the previously written bills. By November, Heinselman and Burton had drafted a bill that created a wilderness of approximately 1,065,000 acres. The bill prohibited all mining, logging, and snowmobiling but permitted ten-horsepower and twenty-five-horsepower motorboats on a handful of lakes and channels.

Burton added Representative Bruce Vento, a Democrat who represented St. Paul in Congress, as a cosponsor of the bill. On March 16, 1978, Burton and Vento submitted their wilderness bill. Oberstar, though, would not withdraw his own bill, the one that divided Boundary Waters into a wilderness area and a multiple-use area. Oberstar attacked the Burton-Vento bill, warning that it would "literally scare the hell out of the people of northern Minnesota—they have had it with federal regulations. You are putting a yoke on their necks."[27]

The Boundary Waters Conservation Alliance also fought against the Burton-Vento bill. The alliance cited a Forest Service study that banning logging would cost Minnesota $30 million and nearly twelve hundred jobs. The Minnesota Department of Natural Resources, however, produced a report showing that Minnesota could make up the difference in lumber and pulp by harvesting trees outside the BWCA.[28]

The Burton-Vento bill won the approval of the House Interior Committee in April 1978 and went to the floor of the House. There, Oberstar tried to dilute the bill by adding amendments that would expand motorboat and snowmobile use. Conservationists, though, lobbied hard in support of the expansive wilderness bill, which was now known as the Fraser-Vento-Nolan bill. On June 5, 1978, the House approved the bill and sent it to the Senate. In the Senate, though, Minnesota Senator Wendell Anderson opposed the Fraser-Vento-Nolan bill, claiming that it created too much wilderness.[29]

At that point, Senator James Abourezk of South Dakota tried to break the logjam by suggesting that the Senate Energy and Natural Resources Committee mediate a compromise between the two sides. Through grueling weeks of negotiation, the committee carved out a compromise that removed motorboats from several small lakes but preserved them on a few large lakes that had resorts and houses. The Friends of Boundary Waters Wilderness favored the compromise bill and persuaded Anderson to incorporate the compromise provisions into his bill. Anxious to bring the issue to closure, Anderson agreed to do so, revised his bill, and took it to the floor of the Senate, where it won a majority vote on October 9, 1978. Because of the compromises, the bill had to go back to the House, which passed it on October 14, 1978, and on October 21, President Jimmy Carter signed it into law.[30] The Boundary Waters Canoe Area Wilderness Act enacted the following provisions:

- It created 1,098,057 acres of wilderness.

- It changed the name of the area from the Boundary Waters Canoe Area to the Boundary Waters Canoe Area Wilderness.

- It prohibited logging, mining, and prospecting for minerals.

- It banned snowmobiles except on a few trails until 1984 and for grooming cross-country ski trails.

- It limited the use of motorboats to about twenty-five lakes, most of which were entry points into Boundary Waters.

- It reduced the number of lakes on which motorboats were allowed to sixteen in 1984 and fourteen in 1999, which represented approximately 24 percent of the surface water of Boundary Waters.

- It set maximum limits of ten-horsepower motors on smaller lakes and twenty-five-horsepower on larger lakes.

- It eliminated the use of motors on portages, with a handful of exceptions.[31]

Even this law did not settle the controversies over motor vehicles in the Boundary Waters Canoe Area, however. In the late 1990s, controversy once again arose over three motorized portages. Outfitters argued that their customers wanted trucks to be able to transport canoes over the lengthy portages, including the Trout and the Prairie portages, and in 1997, Oberstar and Minnesota Republican Senator Rod Grams introduced legislation to allow motor vehicles to return to the three portages. According to Chel Anderson:

> This dispute brought out all the old wounds and hostilities; it was like a psychotherapy session between the prodevelopment people and the environmentalists. The conflict was particularly prominent around Ely, which had been hurt economically by the decline in mining and in motorized sports. People also resented the loss of local control and the increase in regulation by the federal government.[32]

In 1998, a rider to a transportation bill permitted motorized vehicles to transport canoes on the Trout and the Prairie portages. The resolution of the conflict brought an era of relative peace to the region.

New Controversies over Mining

In recent years, though, the struggle between preservation and multiple use has sharpened once again over proposals by corporations to mine for copper and nickel in mineral-rich lands just outside the boundaries of the BWCAW. The situation shows that what happens outside wilderness boundaries can still deeply affect a wilderness itself. New technologies allow mining operations to extract copper and nickel from low-grade sulfide ore, which may contain as little as 1 percent of usable metal. As a result, mining corporations have filed more than one hundred applications since 2008 to conduct exploratory drilling in the Superior National Forest. One proposal by Twin Metals, a consortium formed by Duluth Metals and a company in Chile, would locate mines near the South Kawishiwi River and Birch Lake, less than three miles from the wilderness boundary.[33]

The proposal that had advanced farthest by 2012 was the NorthMet Project, developed by PolyMet Mining Corporation, a company based in Canada that proposed to build and operate an open-pit mine and a factory to process sulfide ore into several finished metals, including copper, nickel, and cobalt. The mine and processing plants were to be sited on eight hundred acres south of Ely and near Hoyt Lakes, an environmentally sensitive wetland connected to the Lake Superior watershed.[34] According to PolyMet, the operation would generate about 360 long-term jobs. The land was in the Superior National Forest, for which PolyMet claimed that it had subsurface mineral rights. The Forest Service disagreed, however, asserting that no company could operate an open-pit mine on national forest land. PolyMet and the Forest Service started discussions on a land exchange that would move part of the wilderness boundary to exclude the mine.[35]

In October 2009, the Minnesota Department of Natural Resources and the U.S. Army Corps of Engineers filed a draft environmental impact statement (DEIS) for the PolyMet project. The DEIS identified a number of serious environmental concerns. One risk was that nickel, manganese, and other metals would contaminate groundwater at the mine site. Metals would seep into groundwater, with significant contamination of the surrounding wetlands. The Lower Partridge River contains wild rice that local Native Americans harvest, and the DEIS projected that sulfide would contaminate the rice. In addition, as sulfide leached into the nearby rivers and lakes, the waters would be at high risk for mercury contamination. Even more alarming was that this water flows into Lake Superior, potentially adding to the mercury contamination in the lake. Wildlife would also be affected because the construction of the mines and accompanying contamination would reduce habitat for the Canada lynx and the gray wolf. Fish were also at risk because of the release of mercury into wetlands and rivers.[36]

In February 2010, the U.S. Environmental Protection Agency (EPA) issued a report that was far more critical of the NorthMet Project than was the DEIS. The agency cited numerous serious contamination problems from the leaching of mercury, copper, nickel, and other metals into the surrounding network of waters. The executive summary stated, "The project's proposed operation and post-closure management plan for acid-generating waste rock and wastewater is inadequate and needs to be improved."[37] Of particular concern was the formation of sulfuric acid when rain falls on the exposed sulfide ore, presenting the threat of serious harm to wetlands and surrounding waterways. The report noted as well the severe effect that heightened mercury levels would have on the Fond du Lac and the Grand Portage bands of the Minnesota Chippewa tribe because the people's diets include large amounts of fish. Finally, the agency took PolyMet to task for not including funds for the inevitable cleanup of the site and nearby waterways after the closing of the mine, which was projected to operate for twenty years. The EPA gave the PolyMet DEIS the lowest possible rating—Inadequate (3)—forcing PolyMet to revise its proposal and issue a revised draft environmental impact statement.[38] In September 2011, PolyMet reduced somewhat the scope of its proposal by eliminating the plan to produce copper metal on site. State and federal agencies and an outside contractor started preparing a revised DEIS.[39] In May 2012, though, PolyMet announced that the revised DEIS would be delayed until the first quarter of 2013.[40]

The Friends of the Boundary Waters Wilderness, WaterLegacy, and other Minnesota conservation organizations firmly opposed the mining proposals. According to Paul Dancic, executive director of the Friends of the Boundary Waters Wilderness:

> We are spending 70 to 80 percent of our time fighting against the mining proposals. The mining will create waste that will leach into Boundary Waters, which feeds water into three out of the four major watersheds in North America: the Mississippi, the St. Lawrence, and the Arctic. When the companies mine, they use only 1 percent of the ore. The rest is waste, tons of waste. It will cause more pollution than anything that was done here in the past.[41]

To support the warnings, the Friends of the Boundary Waters Wilderness in 2010 reported that acid was leaching from the exploratory drillings that INCO had done for copper and nickel in 1974.[42]

The NorthMet Project has added a further dimension to debates over wilderness by raising the issue of whether industrial activities such as mining should be permitted near wilderness boundaries. Boundary Waters has received wilderness protection because it is an extremely valuable ecosystem in which the water, air, vegetation, fish, and wildlife form an integrated whole. A mining operation may lie outside the wilderness bound-ary, but the mine is still part of that ecosystem and can have a negative effect on the ecological health of all life within the system. For these rea-sons, Friends of the Boundary Waters Wilderness has opposed the copper and nickel mining proposals, and industrial activities near wildernesses on other eastern national forests are receiving closer scrutiny.

Lessons Learned

In Boundary Waters, the wilderness idea found some of its most eloquent voices in Arthur Carhart, Ernest Oberholtzer, and Sigurd Olson. At the same time, however, the impact of wilderness on the local economy has been a constant concern to the people who make northeastern Minne-sota their home and to the legislators who represent them in Minnesota's legislature and the U.S. Congress. Consequently, the region has reflected the conflicting priorities, values, and perceptions that have driven the decades-old battle over wilderness. For conservationists, the struggle to preserve the pristine quality of Boundary Waters points to several themes.

1. *The changes in how the Boundary Waters Canoe Area was managed resulted largely from grassroots democratic action.* Local conservationists were, without question, the driving force behind the preservation of Bound-ary Waters as a wilderness. Concerned people organized to express their principles and values and learned to find their way through the maze of political action. The Friends of the Boundary Waters Wilder-ness emerged from the crucible of local citizens who dedicated them-selves to preserving the magnificent beauty of Boundary Waters. On the opposing side, the Boundary Waters Conservation Alliance grew out of the concerns of outfitters and other business owners who feared losing their markets and their control over how they managed their businesses.

2. *Compromise was a positive step for change.* Despite the passions that have sometimes boiled over in the debate over preservation versus multi-ple use, the history of Boundary Waters demonstrates that productive compromise is possible. The original Wilderness Act of 1964 contained

compromises that distressed many of Minnesota's conservationists when the law permitted motorboats and motorized portages, yet the idea of wilderness proved powerful enough that wilderness advocates accepted the compromises and then continued to persuade the public to support incremental changes that led, over time, to more extensive wilderness protection.

3. *Wilderness created unforeseen recreational and business opportunities.* Despite the concerns about the economic impact of wilderness protection, a constellation of businesses has sprouted up in Ely, Grand Marais, and other towns in northeastern Minnesota, businesses that could not even have been imagined forty years ago. People travel to Boundary Waters today because of the wilderness, and while they are there, they buy clothing and equipment, hire guides, and employ outfitters. According to Bill Hansen, who owns and operates Sawbill Lake Canoe Outfitters, "Several outfitters started an advertising campaign to promote Boundary Waters as a family experience and to introduce children to the area." The initiative was a response to Richard Louv's book *Last Child in the Woods*, which addresses the lack of outdoor activity among young people in contemporary American society. Hansen added, "The Gun Flint Trail Outfitter Association led the initiative and published a book, *Becoming a Boundary Waters Family*. They have also worked with the Forest Service."[43] Hansen's comments reflect that wilderness has opened up opportunities for entrepreneurs to serve the needs of people drawn to the region's unparalleled opportunities for adventure, solitude, and harmony with nature.

Perhaps the greatest lesson learned is that the vast natural system we call Boundary Waters has benefited tremendously from wilderness protection. Biologist Chel Anderson made this point about the restoration of the ecosystem as she led her two visitors across Sawbill Lake and back to their starting point at the Sawbill Lake campground. She stopped paddling for a moment and pointed to the trees lining the shore of the lake. "Since the days of logging, the forests have regenerated nicely," she said, in a tone of deep appreciation for the renewed vigor of the north woods. She continued, "As a botanist for the Minnesota Department of Natural Resources, I've surveyed the forests, and I have to say that I am pleased with their condition."[44] They were simple words, but they said much.

Suddenly the three of them spied a loon about fifty yards from the canoe. The bird dipped its neck and buried its head for a few moments under the surface of the water, searched for fish, and then drew its head back up. With its elegant neck, the bird held its head high against the background of the emerald forest. The sighting of the loon was a tiny event in the endless processes of nature, yet the three human visitors accepted it as a small gift as they glided toward the Sawbill Lake landing. It was a small gift from an extraordinary world of water that, against all odds, has retained its pristine power to reinvigorate the human spirit.

CHAPTER 11

Ottawa and Hiawatha National Forests: The Return of the Wolf

The gray wolf appears like a ghost at the edge of the wood, about one hundred yards away, and lopes down a hill and around a small pool, striding on loose-jointed legs. As it moves, its haunches ebb and flow, and its paws leave a five-inch footprint in the soil. Its fur is luxurious enough to withstand temperatures of forty degrees Fahrenheit below zero. The animal surveys its surroundings with eyes that have adapted to pierce the night, when it is most active. Its jaws are hinged by prodigious muscles that exert pressure of fifteen hundred pounds per square inch, twice the pressure of a German shepherd.[1] L. David Mech, a wildlife biologist at the University of Minnesota and world-class expert on wolf behavior and conservation, remembers seeing a wolf clamp onto the snout of a moose and continue to bite down even as the moose swung it wildly from side to side.[2] Rolf O. Peterson, professor of wildlife ecology at Michigan Technological University and another renowned wolf researcher, tells of seeing a wolf grab a moose's rear leg and refuse to let go even as the beast dragged it over yards of rugged terrain.[3]

In addition to these astonishing capabilities, the wolf (figure 11.1) has a remarkable ability to adapt. According to the International Wolf Center—which Mech and other wolf specialists founded in Ely, Minnesota, in 1985—two species of wolves exist, the gray wolf and the red wolf, but the number of subspecies is often listed as thirty-two. The wolf has thrived in

Figure 11.1 *Gray wolf. By the 1970s, the gray wolf had virtually disappeared from Michigan's Upper Peninsula. In 2009, the U.S. Fish and Wildlife Service reported that the wolf population in the Upper Peninsula had recovered to 577. Jupiterimages/Photos/com.*

an amazing variety of regions and habitats, including Europe, northern India, southwest Asia, Russia, and the Arctic. In North America, wolves range from the vicinity of Mexico City all the way north to Greenland.[4]

One of the most astounding aspects of the gray wolf, or *Canis lupus*, has been its resurgence in eastern national forests, including the Ottawa and Hiawatha National Forests in Michigan's Upper Peninsula (UP), the Superior and Chippewa National Forests in Minnesota, and the Nicolet and Chequamegon National Forests in Wisconsin. The recovery of wolves in Michigan's two national forests in the UP is particularly compelling for several reasons. First, wolves had nearly disappeared from Michigan, and their reappearance is a remarkable story in itself. Second, state and federal agencies, including the U.S. Forest Service, have engaged in an impressive collaboration to encourage repopulation by wolves. Finally, the wolf is a

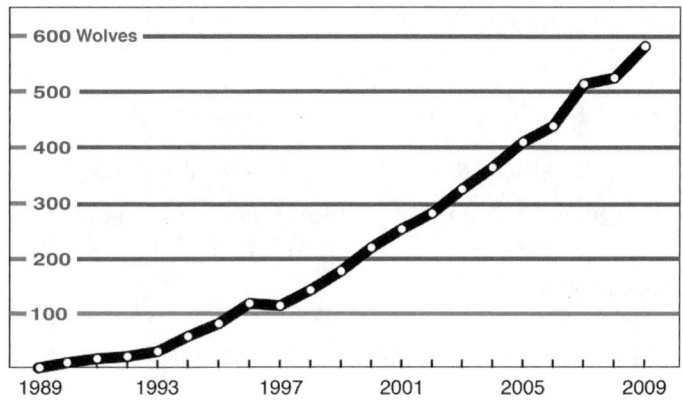

Figure 11.2 *Number of wolves in Michigan. Source: Michigan Department of Natural Resources and Environment. Graph by Christopher Clark.*

charismatic symbol of the trend toward restoring wilderness characteristics in the eastern national forests.

Soon after the Endangered Species Act passed Congress in 1973, the gray wolf was one of the first species listed as endangered, and even fifteen years later, in 1988, the Michigan Department of Natural Resources (DNR) estimated that the UP had only a tiny wolf population of three.[5] (The Michigan island of Isle Royale, covered later in this chapter, has had a population of wolves since the late 1940s.) By 2009, the population of gray wolves in the UP stood at 577 (figure 11.2). They had returned in force to the Hiawatha National Forest, which covers 880,000 acres in the eastern and middle sections of the UP, and the Ottawa National Forest, which covers almost one million acres in the western part of the peninsula.[6] Even in Michigan's Lower Peninsula (LP), with its relatively dense human population, residents have reported occasional wolf sightings. In 2003, a lone gray wolf set foot on the icy surface of the Straits of Mackinac and trotted south into Presque Isle County.[7]

The successful recovery of wolf populations points to a larger trend: the protection of habitat for species with declining populations in eastern national forests. The U.S. government places these species into three categories:

1. Endangered species, which face the danger of extinction
2. Threatened species, which are likely to become endangered
3. Sensitive species, which regional foresters manage to maintain populations and avoid having the animals listed as threatened or endangered[8]

Protection of habitat for these species has become an increasingly important part of the management of eastern national forests. Chapter 8, for example, explained how the Forest Service in Florida has used prescribed burning to preserve habitat for the red-cockaded woodpecker. On the Huron-Manistee National Forest in the LP, the U.S. Fish and Wildlife Service (USFWS) and the Forest Service protect habitat for Kirtland's warbler, an endangered neotropical migratory bird that inhabits Michigan and, more recently, Wisconsin during the warm months before migrating to the Bahamas for the winter. The federal government has formed an eight-county wildlife management area in which it restricts public access during nesting season and plants stands of the jack pine that the bird prefers for nesting.[9] Similarly, in the White Mountain National Forest, the dwarf cinquefoil, a rare perennial that belongs to the rose family, nearly became extinct. One problem was that hikers on the Appalachian Trail were trampling the flower. The Forest Service worked with the Appalachian Mountain Club to reroute the trail, setting the stage for a resurgence of the flower.[10]

The recovery of wolves in Hiawatha and Ottawa National Forests reflects similar changes in habitat management as well as in attitude. In 1900—even in the 1960s—the wolf was feared and vilified. Since then, human views of and behavior toward wolves have undergone tremendous change, reflecting emerging ideas about wildlife protection and the preservation of biodiversity. The story of wolf recovery also shows how the Forest Service has developed closer working relationships with other federal agencies, particularly the USFWS, state conservation agencies, and private wildlife protection organizations.

Decline of the Wolf

To understand how policies and attitudes toward wolves have changed, it helps to take a brief look back into American history. Beginning during the colonial era, settlers aimed to subjugate the forest and regarded wolves as fierce antagonists that killed livestock and people. Reflecting typical attitudes among pioneers, E. Billings wrote an article in 1856 for the *Canadian Naturalist and Geologist* in which he characterized the wolf as "a cruel, savage, cowardly animal, with such a disposition that he will kill a whole flock of sheep merely for the sake of gratifying his thirst for blood."[11]

Wolves had existed in Michigan since the last glaciers slowly retreated approximately ten thousand years ago. Their prey was large mammals, including bison, woodland caribou, moose, elk, and white-tailed deer, the last two of which were particularly common in Michigan. Scientists have estimated that Michigan supported a population of three thousand to six thousand wolves in Michigan before European-American settlement. Wolves formed an integral part of the culture of the Ojibwa people, who inhabited the region and regarded the wolf as sacred.

When settlers began to splay west across the lands of the Northwest Territories in the early nineteenth century, they targeted wolves with a relentless vehemence (figure 11.3). When Michigan joined the Union in 1837, the ninth law that the newly formed state legislature passed was a wolf bounty: "An act for the destruction of wolves." In 1840, the state paid bounties on 279 wolves. The legislature kept raising the amount of bounties, and from 1838 to 1921, the state paid out an astounding $154,000. By 1910, the wolf population in the LP had been wiped out, although wolves survived somewhat longer in the UP. In 1959, the state paid a bounty on only one dead wolf, indicating that the population was near zero.[12]

Aldo Leopold's "Fierce Green Fire"

Even as states were extirpating wolves, tiny bits of evidence were beginning to emerge in the early twentieth century about the essential role that wolves and other top-of-the-food-chain predators played in healthy forest ecosystems. In 1906, the Kaibab Plateau, which abuts the north rim of the

Figure 11.3 *Wolf furs and trappers. Hunters in the United States engaged in a centuries-long campaign to extirpate wolves. Since the 1970s, American attitudes about wolves have changed dramatically, and for many, the wolf is now a potent symbol of wilderness. Courtesy of the Arizona Historical Society/Tucson. Photo number 5929 from Photo Collection 10, folder 417.*

Grand Canyon, became a game preserve, and the U.S. Bureau of Biological Survey (USBBS)—forerunner of the USFWS—set out to eliminate predators, including wolves, coyotes, bobcats, and mountain lions. What's more, the agency banned deer hunting. Not surprisingly, the population of deer exploded, and Bambi and his myriad cousins foraged on tree bark, young trees, and any other vegetation they could find. According to Curt Meine, Aldo Leopold's biographer, "Among professional resource managers and the public alike, the Kaibab episode became a starting point for reconsideration of the role of predators."[13]

In the 1920s, another stage in the evolution of attitudes toward wolves unfolded in the Southwest, where Leopold worked as a young forester in New Mexico's Gila National Forest. An avid hunter, he believed in the need to control populations of wolves and other predators. In "The Varmint Ques-

tion," an article published in 1915, he wrote, "It is well known that predatory animals are continuing to eat the cream of the stock grower's profits,
and it hardly needs to be argued that, with our game supply as low as it is,
a reduction in the predatory animal population is bound to help the situation." He went on to identify "wolves, lions, coyotes, bob-cats, foxes, skunks,
and other varmints" as worthy of reduction by hunting and trapping.[14]

In 1936, Leopold went on a hunting expedition to Mexico's Sierra
Madre Mountains, which had remained the domain of the Apache Indians
and had never been settled by the Spanish or the Mexicans. Leopold was
astonished by what he saw. Here was an ecosystem in which the human
footprint was exceedingly light because of the sparseness of the Apache
population. Wolves and mountain lions prowled the mountains, and Leopold observed a fine equilibrium between predators and prey. "Deer irup-
tions [sic]," he wrote, "are unknown. . . . Mountain lions and wolves are still
common." Leopold realized that wolves had controlled the deer population, preventing the disastrous ecological effects that had spread across
the Kaibab Plateau earlier in the century. He made a radical proposal:
"Would not our rougher mountains be better off and might we not have
more normalcy in our deer herds, if we let the wolves and lions come back
in reasonable numbers?"[15]

In 1944, Leopold wrote an essay titled "Thinking Like a Mountain,"
which later appeared in *A Sand County Almanac*. He remembered an episode from his days as a youthful forest ranger in New Mexico when he had
shot a wolf:

> We reached the old wolf in time to watch a fierce green fire dying in
> her eyes. I realized then, and have known ever since, that there was
> something new to me in those eyes—something known only to her and
> to the mountain. I was young then, and full of trigger-itch; I thought
> that because fewer wolves meant more deer, that no wolves would
> mean hunters' paradise. But after seeing the green fire die, I sensed
> that neither the wolf nor the mountain agreed with such a view.[16]

In the thirty years that had passed since he had shot that wolf, Leopold
had come to a much greater understanding of the role of wolves and other

predators in a variety of ecosystems, from the boreal forests of northern Minnesota to the desert mountainscape of the Sierra Madres.

Olaus Murie was another conservationist who added to the knowledge of wolves and questioned the predominant attitude toward them. A biologist who worked for the USBBS, Murie began a study in 1927 on predator behavior in Jackson Hole, Wyoming, just south of Yellowstone National Park. Murie found that the effects of predators on prey populations were far more complex than had traditionally been realized, and he started to criticize the USBBS's policies of extirpating predators. He wrote to Leopold, "I have felt that too much attention has been given to the predatory animal factor. . . . I do not find the coyote a bad fellow at all."[17]

The Wolves of Isle Royale

The years after World War II saw a steady expansion in scientific knowledge about wolves, particularly their importance in controlling the populations of natural prey such as deer. The growing insights of biologists and ecologists were creating a picture of a healthy ecosystem as a dance in which the different performers and movements blended to form an intricate and beautiful choreography. Removing one player, such as the wolf, could upset the equilibrium of this dance of nature.

A turning point in wolf research occurred in the late 1940s on Isle Royale, an island in Lake Superior that is some forty-five miles long and nine miles wide and lies approximately twenty miles off the coast of Ontario, Canada. The island, a part of Michigan, has been a national park since 1931. The experiences here contributed to scientists' understanding of the role of wolves in forest ecosystems and helped set the stage for the recovery of wolf populations on the Hiawatha and Ottawa National Forests. During the 1920s and 1930s, no wolves existed on Isle Royale. Because the moose on the island encountered no natural predators, their population exploded, numbering as many as three thousand. During this period, wildlife biologist Adolph Murie, the brother of Olaus Murie, observed more than thirty moose at a time browsing at one of the lakes that dot the island. In the mid-1930s, hundreds of moose died of starvation during several brutal winters. The island was an ecosystem that was out of equilibrium.[18]

Then, in 1948, observers reported seeing what they thought were several wolf tracks. By 1951, biologists confirmed that wolves had migrated to the island, crossing the ice bridge that connected the island to the mainland during the winter. By the mid-1950s, wildlife biologist Durward Allen of the USFWS had established a full-time research project to study wolves, moose, and their interactions. Participating in this research was David Mech, who was completing his doctorate in wildlife ecology from Purdue University and would later publish the book *The Wolves of Isle Royale*. In 1963, Allen and Mech coauthored a *National Geographic* article titled "Wolves versus Moose on Isle Royale," in which they explained that after migrating to the island, the wolves had "found a ready food supply in the moose, which had overbrowsed the overpopulated island."[19] One photograph showed a moose foraging on aspen twigs, and the caption noted that a moose might consume twenty-five pounds of forage a day, reducing habitat for birds and small mammals.

Allen and Mech emphasized the positive ecological impact of the return of the wolves. "Today," they wrote, "wolves limit the herd to about six hundred, and young trees and brush are coming back."[20] The wolves were also culling the older and diseased moose, helping strengthen the remaining herd. The two biologists concluded with an anecdote intended to counteract the general public's fear of wolves. While walking through the woods, they came upon a kill scene, with a pack of wolves surrounding the bloody carcass of a moose. As the men approached, the wolves retreated into the nearby woods. Mech circled the carcass and then stepped onto it. The wolves returned to the edge of the woods, about sixty feet from where Mech was. They looked at him, sniffed, and then returned back into the woods, bringing an end to the episode and countering the stereotype of the wolf as an aggressive attacker of humans.[21]

In the early 1970s, Allen retired as the director of the research project on Isle Royale and selected as his successor a young biologist who was finishing his PhD, Rolf Peterson. Ever since, Peterson has studied the wolves on Isle Royale, attaching radio collars to the animals and tracking their movements in detail. During the 1990s, for example, Peterson was able to observe the adult life cycle of an alpha female known as Wolf 450. He traced her rise to dominance in what was known as the East Pack, her leadership in hunting expeditions, and the growth of the pack as she gave birth to ten pups. According to Peterson, "Her ten surviving pups between 1991 and 1994 helped stay the threat of extinction."[22] In January 1995, Wolf 450 and the East Pack happened upon a moose that had been killed by one of her sons and an accompanying female. The pack chased the two wolves and engaged in a bloody fight with them. Peterson continued, "The old alpha female was carefully eating flesh from the dead wolf's rib cage. . . . Never before had I seen a wolf eat another, even after killing it. She had emerged the victor in this battle for dominance."[23] Peterson informed readers about wolf behavior, but he also countered any temptation to romanticize the animal.

Although Isle Royale was the site of this natural-world experiment in wolf recovery, Michigan, Minnesota, and Wisconsin continued their policies of wolf extirpation. Even as the early as the mid-1940s, however, ecologists and conservationists began to call for an end to the bounties. In 1957, Wisconsin stopped the bounties and established legal protection of the wolf.[24] Michigan ceased its wolf bounty in 1960 and, five years later, granted legal protection to wolves. Minnesota, which still had a wolf population that numbered between 350 and 700, continued its bounty until 1965 and then instituted a directed predator control program from 1965 to 1974. In this program, registered trappers were authorized to remove wolves, coyotes, bobcats, lynxes, and foxes that had attacked livestock or other wildlife. The program was aimed primarily at coyotes that were attacking sheep in northwestern Minnesota.[25] In 1966, the U.S. Congress passed the Endangered Species Preservation Act, and in 1967, the USFWS listed the gray wolf as an endangered species. The protection applied mainly to wolves on federal properties, such as national forests. The Federal Endangered Species Act (ESA) passed Congress in 1973 and further strengthened the protection of species at risk of extinction, and in 1974, the wolf was listed as an endangered species, making it illegal to hunt or trap wolves. In 1976, Michigan further strengthened wolf protection with the passage of the state's Endangered Species Protection Act.[26]

Beginnings of Wolf Recovery in Michigan

The passage of the ESA and state laws set the stage for the recovery of the wolf population on the Hiawatha and Ottawa National Forests and catalyzed changes in how the national forests were managed. According to the ESA, federal agencies, including the Forest Service, had to "implement a program to conserve fish, wildlife and plants . . . to ensure their actions do not jeopardize the continued existence of any threatened or endangered species or result in the destruction of adverse modification of critical habitat." The National Forest Management Act also required national forest managers to protect habitat for "native and desired nonnative vertebrate species . . . well distributed in the planning area."[27] The Forest Service was primarily responsible for maintaining forest habitat to support viable populations of endangered and threatened species, whereas the USFWS and state wildlife agencies developed the strategies for protecting the species. As a result, a high degree of cooperation was required among the Forest Service, the USFWS, and state conservation agencies. This cross-agency cooperation reflected a trend toward interdisciplinary approaches to forest management that encompassed foresters, wildlife biologists, botanists, hydrologists, and other scientists.

Consequently, the Hiawatha and Ottawa National Forests became the setting for a grand initiative in wolf recovery, and the Forest Service began to manage both forests to protect habitat for wolves and to work with other agencies during the process of wolf recovery. According to Dr. Dean Beyer, a wildlife biologist who works for the Michigan DNR, "A species is considered recovered when it is no longer at risk and long-term survival is reasonably assured."[28] In promoting wolf recovery, wildlife biologists faced a serious dilemma: Should they actively reintroduce wolves, or should they wait to see whether wolves would return by migrating from Minnesota and Canada? In March 1974, wildlife biologists at Northern Michigan University undertook a bold experiment. Working with the USFWS and the Huron Mountain Club, they released four wolves that had been captured in Minnesota into the forests of the UP. Tom Weise, who spent many years as a wildlife biologist for the Michigan DNR until his retirement in 2005, recalled, "I was working on my master's degree at Northern Michigan and became involved in the effort. We chose the land of a private club in northern Marquette County and released the wolves, but by November, all four had died. They were shot, trapped, or killed by automobiles."[29] They had, however, survived long enough to demonstrate that the UP provided enough favorable forest habitat for wolves.

Then, in 1988, researchers were stunned to discover territorial markings by three wolves, followed by the birth of cubs in the spring of 1989, confirming that the wolves had begun to reproduce. This exciting discovery occurred in Dickinson County in the central part of the UP, and the pack was nicknamed Nordic ("north" plus "Dickinson"). Researchers captured one of the wolves and attached a radio collar so that they could track its travels. The discovery of this pack marked the beginning of the recovery of the wolf population, and from 1989 to 1996, the number of wolves in the UP grew to more than one hundred. The size of the average pack also increased, from 3.0 wolves per pack to 4.6 wolves per pack.[30]

It seemed a long bet that wolves would return to the more heavily populated LP, but in November 2003, researchers in the UP captured and put a radio collar on a female wolf, known as Wolf 4918. The researchers followed the animal's journeys over the next four months until they lost the signal. Then, in October 2004, a coyote trapper mistakenly captured and killed Wolf 4918 in a trap in the northern part of the LP. It was the first time that a wolf had been positively identified in the LP since 1910. One month later, researchers found the tracks of two more wolves in the LP.[31] Since then, signs of wolves in the LP have been sporadic, but even those occasional signs reflect the growing presence of wolves in Michigan.

According to Tom Weise, "The Endangered Species Act has played a major role in wolf recovery."[32] The passage of the ESA created conditions in which the natural process of population recovery could take over. As researchers studied the resurgent population, they learned more about the adaptability of the animal and dispelled certain myths about the habitat required to sustain a wolf population. At one time, biologists had believed that wolves required extensive wilderness to thrive. In 2010, however,

David J. Mladenoff, the Beers-Bascom Professor of Conservation at the University of Wisconsin, reported that wolves were reentering areas with some human population, although that population was relatively sparse. If enough prey existed, then wolves could thrive even in nonwilderness habitat. "In our new model," Mladenoff wrote, "the best predictors of wolf habitat are lack of agricultural land and low road density."[33] Those factors explained why wolves had returned in force to the UP but not to the LP, which has far more roads and farms and less contiguous forest.

Managing Eastern National Forests for Wolves

The recovery of wolves in Michigan's UP reflected several trends that have drastically changed the management of the eastern national forests since the passage of the Weeks Act. Wildlife biologists and other scientists were the drivers of field-based research because they increasingly recognized the role of wolves in the forest ecosystem and wolves' importance in controlling populations of deer and other prey. They also encountered far less resistance from a public whose attitudes toward wolves had softened considerably.

These changes also evidenced themselves in the Forest Service's response to the Endangered Species Act and the repopulation by wolves. That law and the National Forest Management Act required the Forest Service to maintain habitat for viable populations of endangered and threatened species. In addition, a new generation of Forest Service managers and staff had been trained in wildlife biology, ecology, and related fields, and their training affected the direction of wildlife policy on the eastern national forests. A culture developed in which forest managers did not want to lose species on their watch, reflecting their professional pride and actual concern about the species.

Consequently, as the gray wolf population recovered, the managers of the Hiawatha and Ottawa National Forests wrote forest plans that described how they would maintain forest habitats to encourage wolf recovery. Those plans reflected the strategies that the Michigan DNR expressed in a key document, *Michigan Gray Wolf Recovery and Management Plan*, issued in 1997. The Michigan Gray Wolf Recovery Team included representatives from the Michigan DNR, the U.S. Forest Service, the National

Park Service, the USFWS, and the Great Lakes Indian Fish and Wildlife Commission. Weise served as team leader.

The document spelled out specific strategies that the state and federal governments would follow to ensure that the wolf population remained stable. The recovery team stated that wolves were better able to sustain themselves when road densities were less than one mile of road per square mile. Fewer roads reduced collisions between autos and wolves and—just as important—reduced forays into wolf habitat by poachers. The report noted, however, that wolves could sustain themselves in an area with more roads "if people are generally tolerant of wolves or if it is adjacent to an extensive roadless area."[34] The recovery team also recommended that roads be closed when they were no longer needed for timber harvesting.

A major habitat issue was the protection of wolf dens. According to the recovery team, "Several studies . . . suggest that human disturbance can cause den abandonment or movements to new dens."[35] Moreover, researchers found that wolves had abandoned dens because of adjacent logging or road construction. They realized that for a territory of one hundred square miles inhabited by a wolf pack, a home site of eight acres would need protection. The recovery team proposed that state and federal agencies create regional corridors for wolves, known as habitat linkage zones. The corridors would provide areas for wolves to move through regions that contained human populations.[36]

The Ottawa and Hiawatha National Forests incorporated many of these strategies into their forest plans. For example, the Ottawa National Forest created a 265,000-acre Remote Habitat Area, which, according to the forest's *Fiscal Year 2008 Monitoring and Evaluation Report*, was "to be managed to provide habitat for species that require some degree of remoteness from human activity, including the gray wolf."[37] The forest plan placed special emphasis on protecting wolf dens and rendezvous sites and on maintaining populations of wolf prey, including white-tailed deer. The report also pointed out that although the forest had experienced 10 percent annual growth in wolf populations in the 1980s and 1990s, the rate of growth had slowed in recent years because wolves had populated much of the suitable habitat.

The Hiawatha National Forest also identified a Remote Habitat Area, which consisted of 64,000 acres. According to the forest managers' Environmental Impact Statement (EIS) filed in 2006, the population of wolves on the forest had stabilized, but the statement established the goal of providing "sufficient amount of young forest and conifer cover for prey species"; in addition, increasing trails for snowmobile use and off-road vehicles would have a negative effect on wolves' prey.[38] In another report issued the same year, the Forest Service traced the causes of eight wolf mortalities that had occurred between March and December of the year. Of those deaths, poaching was responsible for four.[39] The data pointed to the need for continued efforts to safeguard wolves and manage human-wolf interactions as effectively as possible.

Forest Service policies have extended to the protection of habitat for a wide range of other species. The secretary of agriculture has directed the Forest Service to manage "habitats for all existing native and desired non-native plants, fish, and wildlife species in order to maintain at least viable population of such species."[40] Reflecting this priority, the forest plan for Hiawatha National Forest includes strategies for protecting not only the gray wolf but also the bald eagle, the Canada lynx, Kirtland's warbler, Hine's emerald dragonfly, and the piping plover.[41] The Ottawa National Forest has been managing habitat to protect the trumpeter swan, the Canada lynx, Kirtland's warbler, and other species.[42] In addition, the Forest Service has in recent years increased its monitoring of sensitive species, for which a reduction in population could mean that they become endangered or threatened. For each eastern national forest, a regional forester's sensitive-species list includes mammals, birds, mollusks, insects, and plants for which the agency is trying to maintain the viability of current populations.

Managing Wolf-Human Interactions

Sometimes, though, the recovery of an endangered species can create new forest management challenges for the Forest Service and other agencies, and that is certainly true of the wolf. Pat Hallfrisch, who worked as a unit manager for the Michigan DNR for many years and still makes his home in the UP, noted, "There is still a faction that says that the only good wolf is a dead wolf." He added, "Another faction says to leave wolves alone and let them live in the forest."[43]

A key issue has been increased wolf depredations against livestock. According to Tom Weise, "There have been a few depredations, but these get sensationalized."[44] Data collected by the Michigan DNR supported Weise's point. In its annual report for 2009, the Michigan DNR reported twelve instances of wolf attacks on livestock and two on domestic dogs. These numbers were lower than the previous year, when the report observed that "wolves killed a relatively large number of small animals."[45] According to the report, the UP has nine hundred livestock farms, the majority of which raise beef cattle. In 2009, wolves attacked eight farms, or fewer than 1 percent of the total. In the eleven years from 1998 to 2009, fifty-four farms—or 6 percent—experienced wolf depredations. Thirteen of the farms were the victims of wolf attacks more than once, whereas coyotes were involved in twelve other attacks on farms. The Michigan DNR estimated that for every increase in the wolf population of one hundred, approximately three more wolf attacks would occur.[46] In addition, the Michigan DNR verified that wolves attacked and killed thirty-three dogs and injured nine others between the years 1996 and 2008.

To help farmers manage their losses—and to counteract negative public opinion—Michigan, Wisconsin, and Minnesota have all instituted compensation programs for farmers. The Michigan Department of Agriculture paid up to $4,000 per animal, but the farmers had to prove that the loss was due to wolves or coyotes. Defenders of Wildlife also established a compensation fund in the amount of $10,000, which the International Wolf Center administers. In 2009, all Michigan farmers who lost livestock to wolves received a total of $4,686.50 in compensation.[47] The Forest Service and the Michigan DNR used nonlethal methods to reduce wolf depredations against livestock, including strobe lights, flashing lights, sirens, rubber ammunition, and the use of animals to guard livestock.

As the number of wolves increased, disagreements have also emerged between those who want to continue federal protections of the wolf under the ESA and those who want to end that protection, a process called delisting. The states moved first. By 2004, the wolf count in Michigan had been higher than two hundred for the fifth year in a row, prompting the state to move the wolf from its endangered list to the threatened list. In 2007, the state started the process of removing wolves from the list of threatened species.[48]

Delisting the wolf at the federal level proved to be more contentious. In 2007, the USFWS removed wolves from the list of endangered species in the western Great Lakes region, but a court decision in 2008 overturned this decision and sent the ruling back to the USFWS, which once again delisted the wolf in 2009. The agency did not allow the required time for public comment, however, and withdrew its ruling in July 2010.[49] In December 2011, the USFWS delisted the wolf in the western Great Lakes region, and the delisting took effect in January 2012. In addition, the managers of the Ottawa National Forest indicated that they would shift their focus "from recovery to management" by protecting habitat, educating the public about wolves, and helping the Michigan DNR monitor wolf populations.[50]

Lessons Learned

The changes in attitudes and policies toward wolves has been nothing less than extraordinary, reflecting an embrace of wildlife by many Americans and a celebration of wilderness values, of which the wolf is a potent symbol. In looking at these new attitudes and how they have affected the management of the eastern national forests, several important themes emerge.

1. *Wolves and other large predators are important in preserving and enhancing the biodiversity of forest ecosystems.* Wolves and other predators have a wide-ranging effect on other species in a forest ecosystem. Zoologists have coined the term *trophic cascade*, which means that the removal of a predator like the wolf has a cascading effect on prey populations that can lead to dramatic changes in vegetation and smaller animals. In examining this effect, researchers on Isle Royale have discovered through dendrological studies that balsam fir grew in greater annual increments during periods when the wolf population was high. By reducing the moose population, wolves permitted a resurgence of the

firs. Researchers have also observed increased growth of willows, aspens, and cottonwoods within wolf territories.[51]

2. *The Forest Service has adapted to new priorities in managing endangered, threatened, and sensitive species.* As has been noted, the Ottawa and Hiawatha National Forests have created large remote habitat areas that have a minimum of roads, large contiguous forest, low density of human populations, and sizable amounts of prey. Moreover, the two national forests have written the preservation of these areas into their forest plans, reflecting the shift away from commodity production and toward the promotion of biodiversity on the eastern national forests.

3. *Managing human-wolf interactions is important to the continued viability of wolves and other large predators such as bears in eastern national forests.* A major reason for the success of wolf recovery in Michigan has been changes in people's attitudes. According to Kevin Schanning, who teaches at Northland College in Ashland, Wisconsin, the change in attitudes toward wolves has occurred gradually, starting in the 1930s.[52] In 2004, the Sigurd Olson Institute at Northland College sent out five thousand surveys to Michiganders regarding their attitudes toward wolves and found the following:

   - 68 percent of the respondents were in agreement with the statement that "the wolf is a symbol of the beauty and wonder of nature."

   - 51 percent of the respondents expressed agreement that "wolves are a part of our vanishing wilderness and should be protected."

   - 57 percent were in agreement that "wolves are essential to maintain the balance of nature."

   The data point to an overall increase in acceptance of wolves, especially among urban dwellers who use the national forests for recreation.

4. *Successful wolf reintroduction depends on education and public acceptance.* Education of young people to understand wolves, their behavior, and their place in forest ecosystems has played an enormous part in public acceptance of wolf recovery. As Weise explained, "Attitudes toward

wolves have changed because of a number of factors. There is more education about them, and young people learn about them from documentaries."[53] Michigan, for example, has undertaken an extensive program to educate residents about wolves. In 2009, Michigan's wolf coordinator made twenty-two presentations about wolves to live audiences and another thirty-nine presentations to media.[54]

One of the major centers of education and information about wolves is the International Wolf Center. Jess Edberg, information services director for the center, said:

> Our primary purpose is educating the public about the science of wolves. We also provide a means for funding the compensation to farmers in Michigan for losses because of wolves. People come to the forest because of wolves. In our programs, we emphasize their role in the ecosystem. We also offer field trips, which is a way of connecting people to wolves. On the trips, participants look for scat, listen to wolf howls, and learn about wolf-tracking equipment. The feedback we get is that people gain a greater appreciation of wolves.[55]

Edberg described a project cosponsored by the International Wolf Center in which children in Minnesota, Canada, and Mexico City wrote about wolves in journals. Edberg recalled that one student wrote, "Before the project, I didn't like wolves. Now I do."[56]

The child's words speak not only to the recovery of wolves and other endangered single species but also to the stewardship of the forest ecosystems that support those species. In a fundamental way, the return of the wolf highlights that many Americans have come to perceive and value the eastern national forests as complex ecosystems that foster habitat for ecologically valuable flora and fauna. This evolution in the perception of the national forests has yielded numerous environmental benefits, including the preservation of biodiversity, as the Forest Service and other government agencies seek a balance between commodity demands and the protection of species for future generations.

CHAPTER 12

Allegheny National Forest: The Challenges of Shale Oil Drilling

In 2009, George Zimmermann, who owned a winery and tomato farm in southwestern Pennsylvania, noticed something peculiar: his water did not taste right. Some of his neighbors had noticed the same thing and also complained about discoloration. Zimmermann then had his water and soil tested. Much to his chagrin, he learned that the water had high levels of such chemicals as arsenic, benzene, and naphthalene, and the soil contained dangerous levels of mercury, selenium, and other chemicals. "There are substances that can't be made by nature and that's what's in the ground," he said to a reporter from the Reuters News Agency.[1]

Zimmermann blamed the contamination on Atlas Energy, Inc., which had recently started hydraulic fracturing for natural gas in the area. During the fracturing process, companies inject millions of gallons of water and chemicals under enormous pressure into the ground to fracture shale and capture oil and natural gas—a process called "shale oil drilling," "oil shale drilling," or "shale gas drilling." In Pennsylvania, the gas is embedded in the Marcellus Shale, which lies some five thousand to eight thousand feet below the surface of the earth and extends from the Finger Lakes in New York through Pennsylvania, Ohio, Virginia, West Virginia, and parts of Tennessee and Kentucky. Zimmermann claimed that his water had been clean and pure before the fracturing—often called "fracking"—had started the year before. Later in 2009, Zimmermann sued Atlas Energy, claiming that the operation had made his 480 acres of land in Washington County nearly worthless.[2] By 2012, the outcome of the suit had not been determined.

Some 170 miles north of Washington County, hydraulic fracturing has also come to the Allegheny National Forest (ANF), and more conventional drilling for oil and natural gas has skyrocketed within the forest. The increased drilling activity has catalyzed opposition by environmental groups and some local citizens, who claim that drilling in the forest reduces habitat for wildlife; pollutes the water, soil, and air; contaminates underground sources of drinking water; reduces the number of trees and other vegetation in the forest; and decreases land available for recreational uses.

Proponents of drilling respond that companies have the right to extract oil and natural gas because they control the subsurface mineral rights in the forest and that the revenues are critical to the rural economy of this part of the Keystone State. According to the Pennsylvania Independent Oil and Gas Association (PIOGA), the ANF produces six hundred thousand to eight hundred thousand barrels of oil a year, using traditional methods of extraction. At $100 a barrel, the value of that production is approximately $70 million. In addition, companies capture approximately fifteen billion cubic feet of natural gas from the ANF; revenues in 2008 were estimated to be $52 million.[3] Furthermore, the companies are important employers. The PIOGA estimates that in Pennsylvania, the oil and gas industry produces more than $7 billion in annual revenues and employs twenty-seven thousand people.[4]

The prospect of hydraulic fracturing in the East has raised the economic stakes even higher. In a 2012 report, the Energy Institute at the University of Texas at Austin declared, "Shale gas is widely considered a 'game changer' in the energy picture for US."[5] The U.S. Geological Survey estimated in 2002 that the Marcellus Shale contained some 1.9 trillion cubic feet of natural gas. By way of comparison, the United States uses about 24 trillion cubic feet of natural gas a year. By early 2008, two geology professors at Penn State and the State University of New York at Fredonia upped the ante by estimating that the Marcellus Shale might contain as much as 500 trillion cubic feet of natural gas, of which 50 trillion cubic feet were recoverable with current technologies. This amount could supply the entire United States with natural gas for two years or more. The value of the gas was estimated to exceed $1 trillion.[6]

Since the Arab oil embargo of 1973, the United States has sought energy independence, and in recent years, natural gas has emerged as a strategic resource in attaining that goal. In February 2012, *Bloomberg News* reported, "The U.S. is producing so much natural gas that, where the government warned four years ago of a critical need to boost imports, it now may approve an export terminal."[7] The same article speculated that the United States could become the number one producer of energy in the world by 2020, creating jobs, increasing incomes, and growing the revenues of business and government alike. The *Oil and Gas Investments Bulletin*, which reports on investment opportunities, called for the United States to expand rapidly its plants for liquefying natural gas for export. According to an article in the bulletin, the market for liquefied natural gas will balloon by 40 percent between 2010 and 2015.[8]

Consequently, the discovery of such vast reserves of natural gas in the Marcellus Shale has led to an intense conflict over resource extraction and its effect on water quality and supply, wildlife, and vegetation in national forests. As we have seen, the issues of timber harvesting, prescribed burning, wilderness protection, and wolf recovery all point to substantial management reforms driven largely by the desire to restore the original natural characteristics of the eastern national forests. Despite this trend, though, the forests remain valuable resources for timber, oil, and natural gas, and stark disagreements have emerged over the amount of resources that should be extracted from the forests. Like most resource conflicts, this one comes down to how much: how much wilderness, how much timber harvesting, how much oil extraction, how many trails. Those favoring resource extraction say that 95 percent of the forest was a brush patch at one time, that oil and gas companies now are affecting only 20 percent of the area, and that the forest will grow back. What is the greater good, providing natural gas to the nation or importing it from a foreign country? Can we conduct resource extraction for the national welfare in a way that protects the environment? These are difficult questions that will profoundly affect the ANF and eastern national forests. The questions have become even more relevant as natural gas has, in recent years, supplied a greater portion of the country's energy needs. In August 2012, the U.S. Energy Information Agency reported that carbon dioxide emissions had fallen to their lowest levels in twenty years. The government agency attributed this important decline to three factors: (1) greater use of natural gas, which emits significantly less carbon than coal; (2) a mild winter of 2011–2012 that reduced demand for energy use; and (3) lessened demand for gasoline.[9] The advantages of natural gas will undoubtedly lead to greater use of hydraulic fracturing in the near future. As a result, it is imperative that we examine closely the effects that this relatively new method of drilling has on eastern national forests.

The use of hydraulic fracturing in the East started in southwestern Pennsylvania in 2003, and since then, the potential for fracking has become an issue on several eastern national forests. On West Virginia's Monongahela National Forest, for example, private companies own nearly 40 percent of the mineral rights, and by law, the companies have the right to access those resources. Expecting increased drilling, the U.S. Forest Service has awarded a research grant to West Virginia University to study levels of methane in water wells before and after hydraulic fracturing.[10] In Virginia in 2010, a company announced plans to start hydraulic fracturing operations near Virginia's George Washington National Forest, but the company abandoned its plans after local citizens protested.[11] In Alabama in 2012, the Bureau of Land Management, which manages mining and drilling operations on federal lands, announced that it would lease thousands of acres of land for drilling in the Talladega and Conecuh National Forests.[12]

Nowhere, though, have the disagreements over oil and natural-gas extraction been as sharp as they have been on the ANF, which is located about 110 miles north of Pittsburgh. This forest of 513,000 acres sits on vast oil and natural-gas reserves, including natural gas embedded in the Marcellus Shale. Private companies or individuals own subsurface mineral rights on 93 percent of the forest. The situation is known as a split estate and is not unusual on eastern national forests. On national forest lands, the U.S. Forest Service analyzes proposals to drill or mine for subsurface minerals, including oil and natural gas. The Bureau of Land Management, which has jurisdiction for the management of oil and natural gas on federally owned lands, then offers the lands for lease and ovesees the drilling or mining.

Since the ANF became a national forest in 1923 under the auspices of the Weeks Act, companies have exercised their rights to drill for oil and natural gas, but as energy prices have risen, the number of conventional wells, also known as shallow wells, has risen dramatically. In 2007, the Forest Service stated that the ANF had eight thousand conventional wells. In 2009, the PIOGA placed the number of active wells at eleven thousand.[13] Some estimates, however, have placed the number of wells as high as fifteen thousand.[14] An article in the *Pittsburgh Post-Gazette* estimated that companies would drill two thousand new wells in 2012 alone.[15] In addition, since 2009, operators have drilled a minimum of six hydraulic fracturing wells on land within the ANF's proclamation boundaries, with two sites lying on Forest Service land.[16]

Conventional Oil and Gas Drilling on the ANF

Regional history and culture are inescapable factors in the dispute. The ANF sits at the heart of the region that pioneered the oil industry in the United States. Only forty miles south lies the town of Titusville, where, in 1859, "Colonel" Edwin L. Drake drilled the first commercially viable oil well in the country. Oil and natural-gas industries, along with timber, brought good jobs and prosperity to the region, and Pennsylvania became famous for its high-quality motor oils and other lubricants. This economic legacy has had a major effect on the ANF. Before the federal government purchased it, the forest had been cut over so much that locals derided it as "the Allegheny Brush-Patch." The hemlocks and beech that once dominated were gone except in a few areas. After the national forest was created, hardwoods like maple and black cherry, which are highly desirable for furniture, grew in. As the second growth established itself, songbirds, deer, and grouse returned.

In the decades since, the ANF has become the site for thousands of conventional oil wells. It is a rare trail that does not take a hiker past a shallow-well drill slowly pumping oil with the steady motion of a mechanical weightlifter. Nor is it unusual to walk past fenced-in batteries of oil tanks or to witness heavy trucks hauling crude on the roads that wind through the forest. The ANF still boasts beautiful places like the Allegheny River Valley, which slices through the heart of the forest like a ribbon of gold, but it does not take long to see the widespread evidence of oil and gas drilling on the forest.

To gain access to minerals, companies clear away the vegetation, build well pads, and construct roads. The role of the Forest Service is to ensure that the clearing and drilling are carried out in ways that minimize the effect of surface disturbances on the forest ecosystem. One of the major current disputes about the ANF is how far the Forest Service can go in regulating private companies' drilling activities. According to the Forest Service, each well site removes 1.3 acres of wildlife habitat.[17] The clearings for the hydraulic fracturing wells are considerably larger.[18] In 2009, Leanne M. Marten, who was then the forest supervisor, stated that approximately 9,790 acres of habitat had been cleared for wells. According to the ANF Forest Plan, 191,000 to 241,000 acres of the forest are subject to future oil and gas development. The plan assumes that one well can be drilled every five acres, so the potential exists for as many as 48,200 wells.[19] In 2011, oil and gas companies had built 2,083 miles of roads within the ANF for the trucks that transported equipment and crude oil. These roads existed in addition to 1,243 miles of other Forest Service roads.[20] The result is forest fragmentation, which is visible even to casual visitors, because hiking trails intersect frequently with well pads, roads, gravel pits, and water-containment pits (figure 12.1). In addition, heavy trucks pound the roads, leaving mud and deep tracks along the sides and contributing to erosion. The ANF's last final environmental impact statement (FEIS), which was issued in 2007, stated, "There are concerns that management activities, such as harvesting, fertilizer use, road construction, and ATV/OHM [all-terrain vehicle/

Figure 12.1 *Roads in the Allegheny National Forest in Pennsylvania. The U.S. Forest Service estimated in 2011 that oil and gas companies had built 2,083 miles of roads within this forest. These roads exist in addition to another 1,243 roads used for timber harvesting and administration. The roads cause habitat fragmentation for wildlife and vegetation. © Carl Heilman II/Wild Visions, Inc. with flight support from LightHawk.*

off-highway motorcycle] and equestrian use could contribute to adverse effects on soil resources."[21]

Furthermore, the FEIS stated that fragmentation has had a negative effect on several animal species in the forest. For example, the Forest Service expects lower populations of northern goshawks. According to the FEIS, "The change in outcome for the northern goshawk results from anticipated future oil and gas development. Although there will continue to be large areas on the ANF where oil and gas activity is not expected to occur, there are expected to be some large gaps in remaining suitable habitat."[22] The Forest Service expects similar outcomes for the wood turtle and the Eastern box turtle. According to the FEIS, "The short-term and long-term outcome for these two turtles decreases from the present condition primarily due to increased oil and gas activity that will adversely impact suitable habitat."[23] The FEIS states as well that continued construction of roads will cause sedimentation in streams, affecting populations of fish, mussels, and other aquatic creatures.

Conventional oil and gas extraction have also had an effect on recreational activities and the aesthetic values of the forest. In 2007, local snowmobilers complained that oil- and gas-related road construction was disrupting the Allegheny Snowmobile Loop, a 114-mile track that is a magnet for snowmobilers from all over the region. During winter, companies clear snow from roads to allow access by their trucks, preventing use of part of the snowmobile track. In addition, at one trailhead, snowmobilers observed workers felling trees and clearing away slash with bulldozers to make way for a well pad. In an interview, snowmobile enthusiast Karen Atwood said, "This is supposed to be a forest, not an oilfield."[24] Residents Jan Burkness and Bruce Burkness recounted how they had recently purchased a home with a beautiful view of the Allegheny River Valley, only to have bulldozers and steam shovels invade the area and clear the forest to make way for well pads, roads, and gravel pits. Jan Burkness stated, "It used to be a beautiful, beautiful, drop-dead beautiful forest. Not anymore. That road goes for miles; I could take you down there, and it'd be more of the same. People used to camp there—they don't anymore."[25]

The Marcellus Shale and the Technology of Hydraulic Fracturing

Although conventional oil and gas drilling have been responsible for most of these effects to date, hydraulic fracturing threatens to exacerbate the impact, leading environmental groups to oppose both expanded conventional drilling and fracturing. Leading the opposition have been the Allegheny Defense Project (ADP), an environmental group with approximately three thousand members in the region; Forest Service Employees for Environmental Ethics (FSEEE), which was formed in 1989 with the mission of protecting national forests; and the Pennsylvania chapter of the Sierra Club. The FSEEE has initiated a campaign to prevent land in national forests from being leased to companies intending to conduct fracturing operations.[26] The environmental groups claim that because the Marcellus Shale wells use an average of 5.6 million gallons of water per operation, hydraulic fracturing will divert water from rivers and streams.[27] They also allege that the operation worsens forest fragmentation, pollutes waterways and aquifers, creates air pollution, and causes dangerous migration of methane gases. Oil and natural-gas companies respond that hydraulic fracturing within the forest will open up major reserves of natural gas, helping the United States achieve energy independence and bringing significant economic benefits to the region.

To understand both sides of this dispute and examine the potential effects of hydraulic fracturing, an understanding of the technology is essential. Hydraulic fracturing has been commonly used in the West since the 1950s, but only recently have oil and gas companies started to apply the technology in the eastern United States. Hydraulic fracturing is now being conducted throughout the United States (figure 12.2). Extracting natural gas from the shale is difficult for two reasons. First, reaching the shale requires deeper drilling than is true of conventional gas drilling. Second, capturing the natural gas from the shale is difficult to do in an efficient and profitable manner. The gas molecules reside in three places in the shale: pore spaces, naturally occurring fractures, and organic material. The pore spaces are poorly connected, however, making it difficult to open up pathways that allow the gas to flow to a well, where it is brought to the surface. If an operator can solve these problems, then the process can be very efficient and profitable.[28]

In 2003, Range Resources–Appalachia LLC started to drill in Washington County in southwestern Pennsylvania and applied three relatively new technologies that had been used to capture natural gas from the Barnett Shale in Texas: horizontal drilling, hydraulic fracturing, and multiple wells extending from one well pad.[29] By 2005, the company had produced a profitable flow of natural gas from the well. Two years later, Pennsylvania had granted more than 375 permits to drill for gas embedded in the Marcellus Shale.

To access natural gas embedded in the shale, the operator must drill vertically to the level of the shale and then deviate the drill in a horizontal direction to travel through the shale (figure 12.3). The horizontal drilling allows the operator to access as much as possible of the shale and the gas.

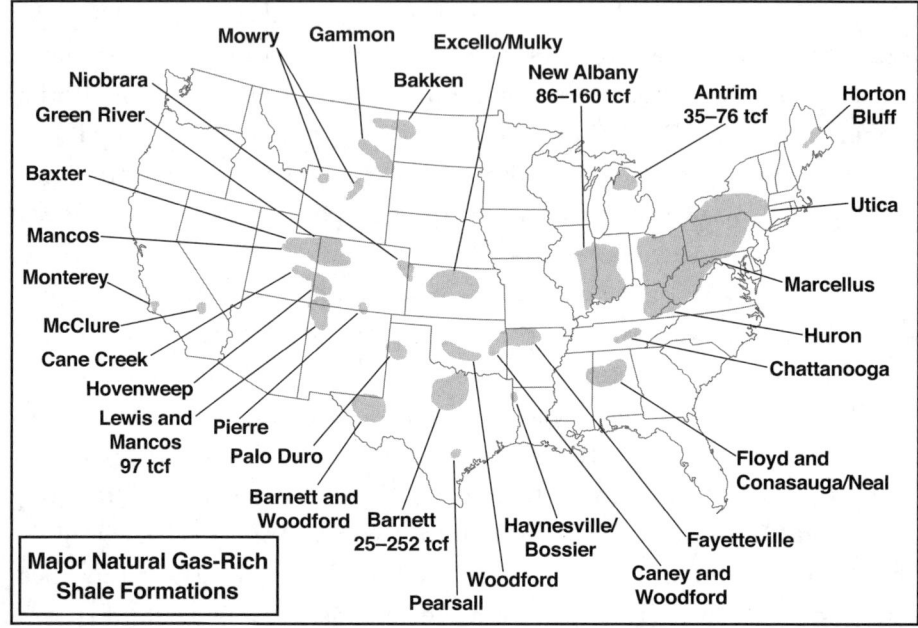

Figure 12.2 *Major shale formations rich in natural gas. Map by Christopher Robinson.*

After drilling the well, the operator pours cement to serve as a casing to protect the pipe and insulate the well from the surrounding earth. The well bore may travel through an underground source of drinking water, which typically lies at a much shallower depth than the shale. Consequently, the cement must maintain its integrity to protect the groundwater from being contaminated by methane or the chemicals involved in the fracturing process.[30]

Once the pipe is in place and the cement poured, the goal is to multiply the number of fractures in the shale to stimulate as much natural gas as possible to move to the pipe, which has tiny holes in it. The operator uses explosives to loosen the shale and then injects several millions of gallons of water laced with chemicals at extremely high pressure to fracture the shale, giving the gas a path to the well. One well pad usually has numerous wells, allowing the operator to capture the natural gas from many parts of the shale. While injecting the water and chemicals, the operator iso-

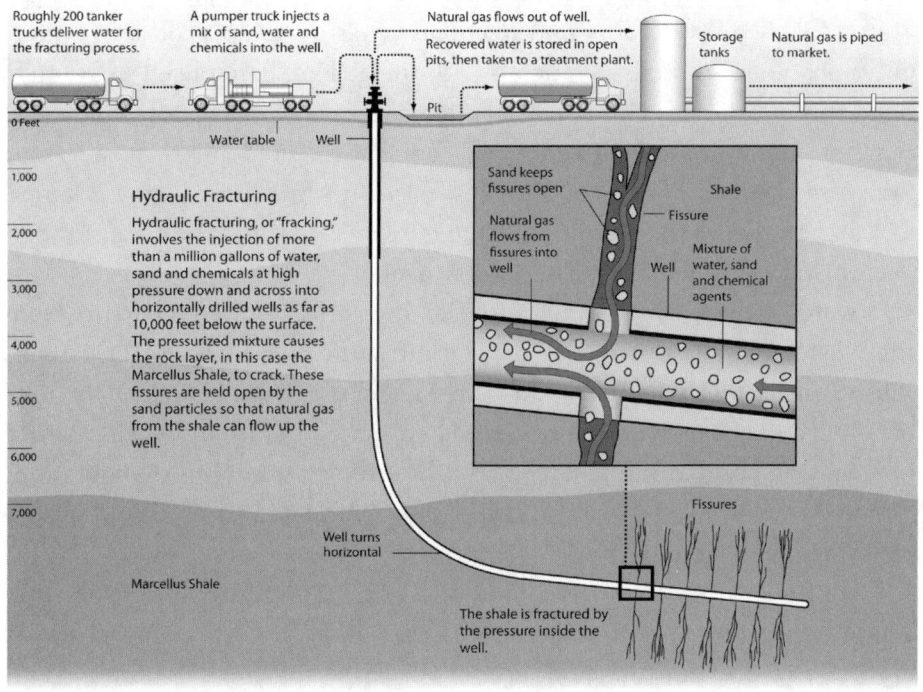

Figure 12.3 *Hydraulic fracturing. Al Granberg/ProPublica.*

lates part of the well and seals off the rest, allowing the creation of the high pressure required. Along with the water, the operator pumps tons of sand or ceramic beads through the pipe and into the shale, propping open the fractures so that natural gas can continue to flow into the well even after the pressure is turned off. The sand and ceramic beads are known as proppants.[31]

After the fractures are opened, the gas, water, and some of the chemicals are brought back to the surface. Researchers estimate, however, that between 10 percent and 90 percent of the water and chemicals remain in the targeted shale area.[32] The mixture that returns to the surface, called flowback, typically includes chemicals used in the process, such as benzene, toluene, ethylbenzene, and xylene. The mixture enters a heater treater, where a glycol solution separates and absorbs the water, and the heater boils off the water from the gas. The boiled-off water enters a separate tank, where it cools and becomes what is called produced water. Oily substances that also come up with the gas are condensed in another tank; this substance is known as condensate water. This water, which still remains contaminated with chemicals, may be injected into the ground for long-term storage. In late 2011, however, the injection process was blamed for earthquakes in the vicinity of Youngstown, Ohio, and the state stopped permitting the injections. The operator may also transport the condensate water off site for treatment as wastewater.[33] Removing toxins from the wastewater, however, has proven difficult. According to the FSEEE, "Disposing of this heavily contaminated water is a major headache for miners—and a bigger one for the environment. The water is either left to settle in a holding pond, often resulting in toxic leaks that kill forest vegetation; dumped into old wells, where it can pollute connected aquifers; or trucked to sewage treatment plants."[34]

Another by-product is the mud and brine that come out of the well. These substances are often deposited into nearby containment pools that are lined to prevent the fluids from contaminating the surrounding soil and groundwater. The operator may also build permanent tanks that hold the fluids until trucks carry them off for treatment. This alternative for storage is considered safer than on-site pits because fluids cannot spill onto surrounding soil or evaporate into the atmosphere and add to air pollution.[35]

Potential Effects of Hydraulic Fracturing

Environmental groups and other critics have identified a number of environmental effects from hydraulic fracturing. Four that have garnered the most attention are the potential for contamination of underground sources of drinking water, the migration of methane gases, the emission of air pollutants, and increases in forest fragmentation.

To ascertain the possible effect of fracturing on groundwater, researchers have been identifying the chemicals used in the process. In 2011, Theo Colborn, Carol Kwiatkowski, Kim Schultz, and Mary Bachran of the Endocrine Disruption Exchange, a nonprofit organization that collects and distributes data about the health and environmental impacts of chemicals, published a study of the chemicals used in the hydraulic fracturing process.[36] To identify the chemicals, they had to use material safety data sheets, which list components for the workers who are handling the chemicals. The researchers could not learn the composition of the fracturing fluids directly from the oil and gas companies because the 2005 Energy Policy Act exempted hydraulic fracturing from the Safe Drinking Water Act, which would have required that the companies list the chemicals used. The exemption has come to be known as "the Halliburton loophole" because it allowed Halliburton and other companies involved in fracturing to protect the chemical composition of the fracturing fluids as proprietary information. Dick Cheney, the former chief executive officer of Halliburton, supported the exemption when he was U.S. vice president under George W. Bush.[37]

Colborn and her associates identified 944 products containing 632 chemicals.[38] The chemicals serve a variety of functions: to improve the injection ability of the fracturing fluids, to make water slick to reduce friction and decrease the horsepower needed to generate the enormous pressure needed for fracturing, to prevent clogging of soil in the well, to improve fluid viscosity, to reduce corrosion of pipes and fittings, to prevent the plugging of fractures in the shale, and to inhibit the buildup of fluids in the pipes.

Of the 632 chemicals that they identified, 70 were chemicals that have had ten or more adverse health effects. The authors wrote that "75% of the chemicals on the list can affect the skin, eyes, and other sensory organs, the respiratory system, the gastrointestinal system, and the liver. More than half the chemicals show effects on the brain and nervous system."[39] Many of these chemicals are toxic and are known cancer-causing agents. Kerosene and diesel fuel, methanol, formaldehyde, ethylene glycol, hydrochloric acid, and sodium hydroxide are toxic. Petroleum distillates like kerosene contain benzene, which is a known carcinogen.[40] The chemicals can cause numerous other symptoms, such as stomach cramps, vomiting, dizziness, headaches, asthma, numbness, fainting spells, and convulsions.[41]

Oil and gas companies maintain, however, that they have taken steps necessary to ensure that the water laced with these toxic chemicals does not leak into surrounding aquifers. Chevron, for example, explains:

> Our wells have a combination of up to eight layers of steel casing and cement, which forms a continuous barrier between the well and the surrounding formations. We run pressure tests to ensure the well's integrity. We also conduct a combination of tests over the life of the well to verify long-term integrity. Our wells are designed to protect groundwater for the life of the well.[42]

In addition, the company claims that it stores the flowback in lined pits or steel tanks and "is working to capture and reuse 100 percent of the fracturing fluids and water produced with the natural gas. This reduces our freshwater consumption as well as our need for water trucking, transfer and disposal."[43] The companies also point out that the fracturing process takes place several thousand feet below the aquifers. The PIOGA has minimized the dangers of fracturing fluids intermingling with underground sources of drinking water:

> The well stimulation process has not been identified as the source of groundwater contamination, as it takes place at depths between 5,000–8,000 feet below ground surface. Most groundwater aquifers are found between 100–200 feet below the surface, typically more than a mile above the shale being fractured.[44]

In February 2012, the Energy Institute at the University of Texas at Austin issued a report, *Fact-Based Regulation for Environmental Protection in Shale Gas Development*, that supports the industry's assertions that fracturing is not responsible for groundwater contamination. In addition to the Energy Institute's researchers, the survey of fracturing wells included staff from the Environmental Defense Fund, which often works with corporations to find solutions to environmental problems. According to the study, "No evidence of chemicals from hydraulic fracturing fluid has been found in aquifers as a result of fracturing operations."[45]

Dr. Anthony Ingraffea, a professor of engineering at Cornell University, disputes the claims by the industry that chemicals cannot escape from well casings nor pollute underground sources of drinking water. Ingraffea, who has called for a moratorium on fracturing on the Marcellus Shale until further research confirms the technology's safety, made a presentation on March 17, 2012, to participants in a two-day conference in Bethlehem, Pennsylvania, titled *Marcellus Shale Exposed*. In his presentation, Ingraffea demonstrated the results of an experiment that he and colleagues had performed at Cornell. They had inserted a pipe encased in cement into a rock foundation. He showed photographs demonstrating that the cement had failed and explained, "No matter how many layers of cement there are, what matters is the outermost layer, which we can't observe. We can't tell when there is a break in the cement." He added, "If cracks occur and joints fail, then the chemicals can leak through the casing."[46]

The U.S. Environmental Protection Agency (EPA) has gathered some evidence that such leakage may occur. In December 2011, the agency issued a draft report, *Investigation of Ground Water Contamination Near Pavillion, Wyoming*. The study was a follow-up to reports by residents of smelly and ill-tasting water near the town of Pavillion, Wyoming. The researchers examined shallow sources, such as pits and septic systems, and deep sources, which were the gas production wells, and concluded, "Detection of high concentrations of benzene, xylenes, gasoline range organics, diesel range organics, and total purgeable hydrocarbons in ground water samples . . . indicates that pits are a source of ground water contamination in the area of investigation."[47]

Another potential danger that critics have cited is the migration of methane gas, which was the apparent cause of two house explosions in Bradford, Pennsylvania, near the ANF, in 2011. When natural gas comes out of the earth, it contains methane, propane, butane, pentane, ethane, and small amounts of heptanes and hexane. Natural-gas companies separate out the other compounds to leave methane, which burns cleanly. Critics have been concerned that both shallow-well drilling and fracturing can cause the migration of methane gas into drinking water and surface outlets, where it presents dangers of asphyxiation, fire, and explosions.

In addition, if methane escapes into the atmosphere, it could exacerbate global warming. According to the EPA, the gas is more than twenty times as effective as carbon dioxide at trapping heat in the atmosphere.[48]

In 2011, Stephen G. Osborn, Avner Vengosh, Nathaniel R. Warner, and Robert B. Jackson of Duke University published a report on research on methane migration in Pennsylvania and New York. They wrote, "Our results show evidence for methane contamination of shallow drinking-water systems in at least three areas of the region and suggest important environmental risks accompanying shale-gas exploration worldwide. . . . Methane concentrations were 17-times higher on average."[49] The researchers also tested water supplies to see whether fracturing had caused the upward migration of the chemical-infused fluids used in fracturing. They wrote, "Based on our data, . . . we found no evidence for contamination of the shallow wells near active drilling sites from deep brines and/or fracturing fluids."[50] The sample was small, however, and the primary purpose of the researchers was to measure methane, not fracturing fluids.

The researchers cited two likely mechanisms by which methane could migrate to the surface. One was that the gas could pass through leaks in gas-well casings. The other was that methane could travel to the surface through fractures in the earth created by the fracturing process. In addition, the researchers stated that gas could travel through "the many older, uncased wells drilled and abandoned over the last century and a half in Pennsylvania and New York."[51] Pennsylvania probably contains more than one hundred thousand abandoned and uncapped wells.

The third concern about hydraulic fracturing is its contribution to air pollution. A potential source of air pollution is volatile organic compounds (VOCs), gases that are emitted from solids and liquids and that may negatively affect people's health. The University of Pittsburgh's Center for Healthy Environments and Communities found that these organic compounds enter the atmosphere as gases as they sit in the containment pits where the flowback is often stored. According to researchers, the wastewater "will offgas its organic compounds into the air. This becomes an air pollution problem, and the organic compounds are now termed Hazardous Air Pollutants."[52] One of these gases is methanol, which can produce dizziness, headaches, nausea and other stomach problems, conjunctivitis, blurred vision, and sleeplessness. Such noxious fumes affected a Pennsylvania woman named June Chappel, who had a wastewater pit just beyond her property line. She smelled an odor that was a combination of gasoline and kerosene, forcing her to stay indoors and leaving a film on her windows. Finally, Chappel and her neighbors hired an attorney, and the owner of the tank, Range Resources, agreed to move the pit.[53]

In *Fact-Based Regulation for Environmental Protection in Shale Gas Development*, the Energy Institute at the University of Texas examined whether VOCs from fracturing wells near the Dallas–Fort Worth International Airport are contributing to high levels of ozone in the region. They concluded, "Investigations in the Fort Worth area have found that most VOCs are not associated with natural-gas production or transport."[54] In contrast, in DISH, Texas, a town north of Fort Worth that also has fracturing wells, researchers have found high levels of xylene, naphthalene, and benzene in the air. The authors of the study state, "Modeling studies indicate that 70 to 80% of benzene is from fugitive emissions of natural gas, but that other VOC constituents are from motor vehicle emissions."[55] Consequently, the fracturing may be worsening air pollution problems that already exist.

Finally, fracturing has the potential to accelerate forest fragmentation, with harmful effects for wildlife, trees and other vegetation, water quality, and soil. Figure 12.4 shows an aerial view of a typical hydraulic fracturing site. At another site on the ANF that was being prepared in March 2012, the operator had constructed three permanent tanks on concrete pads that were about fifty yards away from the well location, which was being readied for drilling. Piled up nearby were lengths of plastic pipe, which will carry the natural gas from the forest. The cleared site appeared to comprise approximately five acres. Half a mile away was a recently dug pit, filled with water and protected from leakage with a liner. It was a containment pool, where the operators planned to capture water from nearby streams to use in the fracturing process. The operators had cleared

Figure 12.4 *Marcellus Shale well drilling in Elk State Forest, Pennsylvania. Shale oil drilling uses heavy equipment, requiring the clearing of up to five acres of forestland. In addition, each fracturing operation draws between five million and six million gallons of water from forest rivers and streams. © Carl Heilman II/Wild Visions, Inc. with flight support from LightHawk.*

about five more acres for the containment pit. As hydraulic fracturing is expanded on the ANF, such sites will multiply, fragmenting the forest even further and degrading habitat for wildlife and vegetation.

Disagreements over the Future of the Allegheny National Forest

The prospect of hydraulic fracturing and the rapid increase in conventional drilling have dramatically heightened tensions surrounding the ANF. Without question, the economic stakes are high. Ted Howard, vice-president of nearby Howard Drilling, told the *Pittsburgh Post-Gazette* in 2009, "We've owned these [mineral] rights for 100 years."[56] The alternative, he continued, was for environmental groups or the federal government to purchase the mineral rights from the companies that own them.[57] Mineral-rights owners have the right to extract oil and natural gas, but the Forest

Service also has the responsibility to protect the forest's water, soil, wildlife, and vegetation. The key issues involve two central questions:

1. At what point does the exercise of property rights impinge upon the environmental health of the forest and the rights that visitors have to enjoy and benefit from the publicly owned surface of the land?

2. What are the rights and responsibilities of the Forest Service, the oil and gas companies, and the general public under the National Environmental Policy Act (NEPA) of 1969, the law that established the country's basic environmental policy and established the Council on Environmental Quality?

The specific vehicle through which conflict over drilling in the ANF has played out is the forest plan. In February 2007, the Forest Service approved a new Allegheny National Forest Plan, which replaced the plan of 1986. During the intervening years, oil and gas extraction had expanded rapidly, but environmental groups did not think that the revised plan had taken adequate account of the resulting environmental impacts. On November 20, 2008, the Sierra Club, the FSEEE, and the ADP filed a lawsuit, known as *Forest Service Employees for Environmental Ethics, Allegheny Defense Project, Sierra Club vs. U.S. Forest Service*. The environmental groups claimed that in formulating the 2007 forest plan, the Forest Service had not complied with NEPA by conducting environmental analyses or allowing adequate time for public comment.[58] According to Cathy Pedler, forest watch coordinator for the ADP, "The public's right to administratively appeal a project is part of the NEPA process, which attempts to ensure transparency, sufficient environmental analysis, and public participation in the federal decision making process."[59]

In April 2009, the three environmental groups announced an out-of-court settlement agreement with the Forest Service that affected the procedure by which oil and gas companies applied for permits to drill on the forest. Typically, a rights holder files a sixty-day notice of intent to drill, and the Forest Service evaluates the drilling proposal and issues a notice to proceed (NTP). In the settlement, the Forest Service announced that it would allow fifty-four drilling projects currently under consideration to proceed, involving 588 wells. The agency would not, however, grant further NTPs while it studied the environmental impact of drilling and prepared a new forest-wide environmental impact statement (EIS).[60]

Oil and gas companies disagreed with the suspension of NTPs during the preparation of the EIS. Consequently, on June 1, 2009, the Minard Run Oil Company, the Pennsylvania Independent Oil and Gas Association, and the Allegheny Forest Alliance—a coalition of local school districts and businesses—filed a lawsuit against the Forest Service, the FSEEE, the Sierra Club, and the ADP, claiming that the Minard Run Company was prevented from exercising its mineral rights because of the Forest Service's decision to suspend the NTPs. The district court granted an injunction against the Forest Service, which appealed the decision to the Third Circuit Court in Erie, Pennsylvania. The court upheld the injunction against the Forest Service, ruling that the agency must resume considering NTPs during the customary sixty-day review periods.[61]

In its decision, the court cited the Organic Act and the Weeks Act. The Organic Act, the court held, granted the federal government authority to establish rules and regulations regarding the national forests. According to the court, though, the Weeks Act did not grant additional rights to the Forest Service to regulate actions by the owners of mineral rights. Consequently, the court upheld the rights of property owners to exercise their mineral rights as long as they followed Pennsylvania state law governing split estates.[62]

In response, the FSEEE, the ADP, and the Sierra Club filed a motion to vacate the injunction granted by the circuit court. The Forest Service filed a separate appeal of the injunction. At issue was whether Congress intended for the Forest Service to have the same power to regulate Weeks Act forests that it did to regulate national forests that had come into the system by other means. In a brief submitted on March 6, 2012, the three environmental organizations stated:

> This motion is about whether this Court, in its preliminary ruling, unnecessarily and erroneously ruled that the USFS does not have the

legal right to place reasonable conditions on access to the surface of the national forests in order to protect that federally owned property. Other than national forests acquired under the "Weeks Act" about a hundred years ago, the courts have held that the USFS does have that right.[63]

Bill Belitskus, president of the board of directors of the ADP, explained, "There have been accusations that we have a two-tier national forest system. One forest can't be managed differently from another. Other forests apply NEPA to national forests, so NEPA should be applied to the Allegheny National Forest."[64] For other eastern national forests, the Forest Service has conducted environmental analyses of proposed drilling actions by private companies. In 2007, for example, the Ottawa National Forest conducted an environmental assessment of a proposal by Trans Superior Resources to conduct exploratory drilling for approximately forty minerals in the forest. Trans Superior had leased the mineral estates in one area of the forest.[65] In its environmental assessment, the Forest Service wrote:

> Federal actions such as permitting and/or authorizing access and surface occupancy for the exercise of private mineral rights must be analyzed to determine potential environmental consequences pursuant to the National Environmental Policy Act of 1969 (NEPA).[66]

On September 6, 2012, however, U.S. District Judge Sean McLaughlin denied the request by the U.S. Forest Service and the three environmental groups for an injunction to prevent the resumption of drilling on the forest. In addition, he overturned the 2009 settlement issued by the Forest Service, which had effectively created a ban on drilling during the process of creating a new environmental impact statement.[67] The ruling upheld the previous process by which the Forest Service evaluates oil and gas drilling applications during sixty-day review periods.

This lawsuit is only one of many that have been filed regarding hydraulic fracturing and, to a lesser extent, conventional drilling. For example, in Susquehanna County, Pennsylvania, residents have filed lawsuits alleging that fracturing fluids contaminated underground sources of drinking water. In Bradford County, Pennsylvania, residents have filed suit for similar causes. By late 2011, plaintiffs had filed more than three dozen lawsuits in Pennsylvania, New York, Texas, Arkansas, and other states on the basis of groundwater contamination, air pollution, soil contamination, and loss of property value.[68]

Lessons Learned

For the ANF, the expansion of conventional drilling and the prospect of extensive hydraulic fracturing have laid bare deep divisions over the proper uses of the eastern national forests and the role that the forests play in our national life, pitting interests that favor increased resource extraction against those who emphasize forest sustainability, habitat preservation, recreation, and aesthetics. One does not have to oppose sustainable uses of national forest resources, such as timber harvesting, to be concerned about the long-term effect that hydraulic fracturing and expanded conventional drilling may have on the ecological health of the forest and the human health of those who live near the forest, however. The FSEEE has stated, "National Forests are far more valuable for their enduring and sustainable supply of water than for a one-time exploitation of underground natural gas."[69] What makes the issue of oil and gas extraction on the ANF so difficult, though, is that western Pennsylvania's economy has depended so much on these industries. Several lessons learned from other eastern national forests can place the ongoing dispute in the ANF into a broader perspective.

1. *The trend in managing eastern national forests has been to understand forests as large ecosystems.* Decisions about the national forests have increasingly taken into account water quality, air quality, wildlife protection, and the protection of trees and other vegetation. These factors all come into play because we know so much more about the interrelationships in a forest ecosystem than we did not-so-many years ago. Consequently, it is appropriate that the Forest Service examine in scientific terms the forest-wide environmental effects that hydraulic

fracturing and conventional oil and gas drilling may have on the ANF. Moreover, because the public has a vested interest in the health of our public forests, it must be made aware of the forest-wide effects of these industrial activities. It is worth remembering that the driving purpose of the Weeks Act was "the protection of the forests and the water supply of the States."[70]

2. *A collaborative approach has been most effective in managing disputes over decisions affecting a forest.* For example, in the Holly Springs National Forest, a cooperative and inclusive approach by forest managers yielded workable compromises among different interest groups over timber harvesting. In resolving disputes, effective forest managers have listened, disseminated information, and created transparent procedures for decision making. Such approaches seem essential in finding solutions that balance the goal of U.S. energy self-sufficency with the good of the ANF and the interests of mineral-rights holders.

3. *The economic arguments for protecting the ANF from environmental degradation are highly persuasive.* The ANF remains the only national forest in Pennsylvania, and it is accessible to millions of people in Pittsburgh, Cleveland, Cincinnati, and other cities of the East and Midwest. The forest affords ample opportunities for hiking, camping, hunting, fishing, and canoeing. In fact, the Forest Service stated that its preferred management alternative "would provide increased opportunities for both semi-primitive remote types of recreation and ATV/OHM trail riding areas."[71] If hydraulic fracturing and conventional oil and gas drilling come to dominate on the forest, the result could be a serious opportunity cost: lost dollars due to reduced tourism.

Of all the eastern national forests studied for this book, the ANF reflects most acutely the deep divisions between those who take a utilitarian view of nature and those who emphasize its inherent value. One reason for the creation of national forests was to find a balance between these two views. As the experiences of the Superior, Holly Springs, Monongahela, and other eastern national forests show, the nurturance of biodiversity, provision of recreational opportunities, and protection of scenic values have practical economic value as well as value for the species that inhabit the national forests. The words of Gifford Pinchot come to mind: "Where conflicting interests must be reconciled, the question will always be decided from the standpoint of the greatest good of the greatest number in the long run."[72] Without question, the ANF faces the potential of major disruptions to its ecology: to its water quality, air quality, soil integrity, and protection of high-quality habitat to ensure biological diversity. As the future of this forest hangs in the balance, Pinchot's counsel to consider the best interests in the long run is as appropriate and timely as it was when he penned it.

CHAPTER 13

Michigan's National Forests: The Invasion of the Emerald Ash Borer

It looks like a fat green cigar with wings. It sports an iridescent emerald-colored body and is about three-fourths of an inch long and one-sixteenth of an inch wide. Its wings run almost the length of its body, and it has a thick, squat head. It looks harmless enough, somewhat like a grasshopper (figure 13.1).

It's not harmless, though. Far from it.

It is the emerald ash borer (EAB), an insect native to China. Since 2002, when the EAB was first detected in Detroit's suburbs, it has spread like an epidemic to forests and woodlands, covering forty thousand square miles in twenty states across the Midwest and the East. Therese Poland, a research entomologist for the U.S. Forest Service who studied wood-boring insects for her doctorate degree and has been heavily involved in research on the EAB, said, "It's difficult to determine just how many trees have been destroyed by the insect so far, but a good guesstimate is fifty million."[1] Moreover, this beetle threatens to destroy the eight *billion* ash trees that populate the United States.[2] By 2005 and 2006, the EAB had spread north from southeastern Michigan all the way to the state's national forests—the Huron-Manistee National Forests in the Lower Peninsula and the Ottawa and Hiawatha National Forests in the Upper Peninsula—but the EAB was not just a threat to national forests. Indeed, the beetle refused to recognize such human constructs as national forests and invaded state forests, state

Figure 13.1 *Emerald ash borer. This insect, a native of Asia, has infected ash trees throughout the Lake States and the East and poses a major threat to millions of trees. David Cappaert.*

and city parks, lawns, parkways, and wherever else there were ash trees. According to the Forest Service, "The current situation has been described by pest specialists as the largest known outbreak of a flatheaded wood borer ever recorded, anywhere in the world."[3] Unlike other destructive insects, which prey on weakened hosts, the EAB has been able to lodge itself in healthy trees and spread from there to uninfected areas. In addition, at the time of the beetle's first detection, it had no known natural enemies, vastly complicating the task of slowing the spread of the insect.

Invasive Species: Causes and Effects

Since the passage of the Weeks Act and the creation of the eastern national forests, biologists, ecologists, and foresters have learned an extraordinary amount about managing forests in sustainable ways. They have learned to harvest timber in self-sustaining quantities, restore wildlife, harness fire, and protect wilderness. In many ways, America's forests are among the healthiest and the best-protected in the world.

Invasive species like the EAB, however, pose a serious threat to this extraordinary progress in restoring the health of America's eastern forests. According to Roger Mech, a program leader for the Michigan Department of Natural Resources, "Many insect and disease problems occur in cycles. Some are tied to drought cycles that only surface every decade or so, but some are tied to the ever-changing conditions of the forest."[4] Two conditions—globalization and climate change—have hastened the spread of invasive species, though. Globalization has radically increased the movement of plants, animals, and diseases around the world, often with ruinous effects. Harold A. Mooney, professor of environmental biology at Stanford University, has written, "In aggregate, these invasions are global in extent and are having consequences that are generally unappreciated but quite threatening to many human activities."[5]

Since the European age of exploration started in the late 1400s, thousands of species of plants, animals, and diseases have spread across the globe. Hawaii, for example, has become the home to some thirty-five hundred alien species of insects and plants. Similarly, California has been the recipient of more than one thousand invasive plant species in addition to the sixty-three hundred native species found in the state.[6] Scientists estimate that throughout its history, the United States has become the host for some fifty thousand alien species. Some of these species, such as rice, wheat, and cattle, have become staples of the American diet. A majority of invasive species do no harm, but a minority do, and the damage they cause is extensive and growing worse. Dr. David Pimentel, who heads a team of researchers on invasive species at Cornell University, estimated that the amount of damage caused by invasive species amounts to some $120 billion a year in the United States.[7]

Alien species spread throughout the world in a variety of way. Ships transport organisms in water used as ballast. Ornamental plants are transported for decoration, and because of a lack of biological enemies, they can quickly overtake their new ecosystems. For example, U.S. nurseries in the nineteenth century imported buckthorn from Europe as a decorative hedge, and now it has become a major headache in the Midwest, crowding out grasses and wildflowers in woodlands, prairies, and savannas. On weekends, saw-wielding volunteers hike into forests, cut buckthorn, and apply herbicide to keep the noxious weed from resprouting. Another highly publicized example of an alien species is the Burmese python, which irresponsible pet owners have released into the Florida Everglades, upsetting the fragile ecosystem of that extraordinary network of wetlands.

Biologists are also beginning to understand the effect that climate change is having on invasive species, creating climate conditions in which they can thrive. If forests have higher mean temperatures, alien species are spared the colder temperatures that would once have killed them. The National Wildlife Federation reports, for example, that garlic mustard, kudzu, and purple loosestrife—invasive species all—have expanded their habitats because of warmer temperatures and resulting changes in rain and snow patterns. In the western United States, the mountain pine beetle, which survives winters in great numbers because the region has higher mean temperatures than it once had, has killed millions of trees that carpet the majestic sides of the Rockies.[8]

The damage caused by alien species can be pervasive, devastating the health of an ecosystem. They can crowd out native species, spread disease, deplete water supplies, stimulate fire, disrupt fisheries, clog water works, and destroy gardens.[9] They may alter the food web so that nutritional sources decline, harming native species and reducing biodiversity. Kudzu has taken over many ecosystems in the South, replacing diverse vegetation with unbroken walls of this monocultural plant. Invasive species can also modify the conditions of ecosystems so that they are more vulnerable to natural events like fire. A prime example is the Floridian plant titi, which has waxy leaves that burn hot during forest fires. In Florida, as was explained in chapter 8, the Forest Service is trying to push the titi back to its native swamplands as part of its program to manage wildfires. Invasive species are a particular threat to endangered and threatened species. The federal government has listed 958 such species, and invasives have had an adverse effect on 400, or 42 percent, of these endangered and threatened species, preying on them, outcompeting them for food and habitat, carrying diseases, and killing young growth.[10]

Forests are as vulnerable as any other ecosystem to invasive species. Plant pathogens, which are organisms such as fungi and viruses that cause plant disease, have harmed American forests since the nineteenth century.

During the 1800s, the chestnut blight fungus decimated the magnificent chestnut trees that populated forests up and down the East coast. Dutch elm disease, which migrated to the United States from Europe during the 1930s in diseased logs, has killed millions of trees throughout the United States, which spends about $100 million a year to remove diseased elms.[11]

Invasive insects represent another severe threat to forests. The Asian long-horned beetle is a destructive insect that first made its appearance in Brooklyn, New York, in 1996 and has spread quickly to woodlands in the East and the Lake states. Another pernicious insect is the hemlock wooly adelgid, a small native of Asia that was accidentally transported to Virginia in the early 1950s. Since then, it has migrated to forests from Georgia to Maine. Once the insect infests a hemlock, the tree typically dies in four to ten years.[12]

As trade, migration, and travel forge ever more connections among different parts of the world, invasive species will continue to pose a serious threat to forests and other ecosystems. A close examination of the EAB provides insights into the threat posed by a specific pest, but it also serves a more general purpose, revealing how scientists formulate an overall strategy for fighting an invasive species through early identification, public cooperation, and the use of chemical and biological measures. The case of the EAB also underscores the collaborative efforts that the Forest Service and other government agencies have undertaken to slow the spread of the insect and mitigate its damage. Early on in the EAB crisis, the Forest Service noted that it "is well positioned to be a leader nationwide and worldwide in the battle against invasive species. Our challenge is to learn to lead collaboratively."[13] As will become apparent, the Forest Service has taken on this leadership role in conducting research and implementing strategies to combat the spread of EAB and other invasive species in forests.

Detecting the Emerald Ash Borer

Biologists, social scientists, and other specialists who have grappled with the EAB have followed a process that is typical in identifying and combating alien species:

1. Identify the species and gain an understanding of its biology.

2. Learn about the mechanism or mechanisms by which the species spreads and the extent of the migration.

3. Develop strategies to slow or halt the spread of the species.

4. Formulate ways of minimizing the harm that the species causes and begin to restore the health of host ecosystems.

The EAB threatens one of America's premier trees, the ash tree, which numbers 850 million in Michigan alone. Ash provides a strong but flexible wood that makes it ideal for manufacturing baseball bats, furniture, tool handles, and guitars. The loss in lumber, pulp, furniture, and other products manufactured with ash would be immense and is estimated at some $1.7 billion. Across the United States, ash trees constitute 40 percent of trees with large diameters and 7.5 percent of the hardwood timber used for wood products. The value of these products has been estimated to be at least $25 billion. The EAB also threatens the landscapes of America's cities and suburbs, where ash trees have been planted in quantity since the 1940s. In nine cities in southeastern Michigan, for example, ash trees comprise 12 percent of all trees on city streets. In Atlanta, Baltimore, Boston, Chicago, New York, Oakland, Syracuse, and Philadelphia, ash trees contribute approximately 14 percent of the leaf areas.[14]

In addition, ash trees supply numerous ecological services for America's forests. They offer browsing areas and protect wildlife such as rabbits, porcupines, and beavers. The plentiful seeds of the tree feed songbirds, game birds, ducks, insects, and small mammals. Green ash is found in more parts of the United States than any other ash species and forms a dominant canopy, keeping forest soils moist and providing habitat for amphibians and reptiles. Black ash grows in swamps and bogs, and the deaths of black ash and green ash would have major effects on the ecology of those ecosystems.

The first detection of the insect in the United States occurred when homeowners in Detroit and Windsor, Ontario, noticed that ash trees were dying in unusually large numbers. The situation was particularly alarming because communities had planted millions of ash trees to replace elms destroyed by the Dutch elm disease. The ash grew fast, and it was immune to disease, or so botanists had thought.

When Michigan residents reported the diseased trees, local officials called in entomologists, who carved into the wood and extricated samples of an iridescent green beetle. They transported the specimens to Michigan State University's Department of Entomology, and in July 2002, after a month of painstaking examination, Dr. Eduard Jendek, a world-class expert on Asian beetles, identified the specimens as *Agrilus planipennis*, or the emerald ash borer. The beetle was a native of Japan, Korea, Mongolia, Taiwan, eastern Russia, and the northeastern corner of China. Until 2002, it had never been found outside its Asian home, nor had it been a threat in Asia because ash trees there had evolved to resist the beetle.[15]

The next mystery to be solved was how the EAB had migrated to the United States. One mode of entry was in trees that nurseries had imported from Asia; another was in the wood of packing materials, such as crates and pallets, which were used to protect goods entering the country. According to Therese Poland and Deborah G. McCullough, an assistant professor of entomology and forestry at Michigan State, "In addition to the EAB, at least 10 nonindigenous forest insects associated with solid wood packing material have been discovered in the United States or Canada since 1990."[16]

When Jendek identified the EAB, entomologists in the United States undertook a crash course to discover how the EAB reproduced, attacked ash trees, and spread from one area to another. In addition, they had to identify herbicides and natural enemies—if any—that would destroy the insect. Finally, public officials had to develop effective ways to enlist the public's aid in identifying new infestations and slowing the spread of the insect.

The researchers were able to glean essential facts about the EAB's biology and life cycle from Chinese sources. In spring, the adult beetles eat their way through the wood of the ash tree until they emerge through the bark, leaving D-shaped holes that are three to four millimeters wide. They then consume leaves, sometimes eating enough of the vegetation to mar the appearance of a tree but not do any real harm. The true danger to the tree takes place as the result of the mating process. After feeding for five to seven days, the adult male mates with a female, which lays her eggs in fissures in the bark. In two weeks, the larvae hatch and then dig galleries through the bark and feed on the phloem, or inner bark, and the cambium, which is the thin tissue between the bark and the wood of a tree. By feeding on these tissues, the larvae disrupt the distribution of nutrients. Poland explained, "The beetle usually is found first in the canopy of a tree. It feeds on the leaves, and the larvae appear first in the branches. As the branches die, the insect works its way down to the trunk."[17] The larvae finish feeding in October or November and then enter the prepupal stage, the inactive period of development before later emergence into adulthood. The prepupae spend the winter in cells in the bark or wood of the tree and emerge as adults in the spring. The mating cycle begins anew, creating a new generation of larvae that feed on the phloem and cambium. As this cycle repeats itself, the ash tree succumbs in one to three years.

Researchers, however, have found that some of the insects follow a longer life cycle of two years. When autumn comes, some EABs spend the winter as larvae rather than prepupae, and these insects require another year to reach maturity. Entomologists do not yet understand why some of the insects require a two-year development cycle, but they believe that the phenomenon may be related to low-density populations or resistance by healthy trees. The discovery of a two-year cycle has complicated mitigation efforts because researchers run the likelihood of underestimating EAB populations. The two-year cycle also means that herbicide applications might not kill all larvae embedded in the bark of a tree.[18]

Once researchers formed a basic understanding of the biology and life cycle of the insect, they examined the extent to which the EAB had spread, the mechanism by which it had migrated, and the amount of damage to woodlands. In 2002, the United States and Canada convened a panel of respected scientists to conduct risk assessments of the EAB. At its first meeting, the panel agreed that the EAB posed a major risk to woodlands, threatening major losses of ash trees and a projected economic loss in the range of $20 billion to $60 billion.[19]

On September 30 and October 1, 2003, a group of scientists gathered in

Port Huron, Michigan, for several key purposes: (1) to communicate their early research on the EAB, (2) to find areas of cooperation on future investigations, (3) to identify potentially duplicative research initiatives, and (4) to decide upon future areas of research. In a reflection of the emerging collaborative approach, the group brought together experts from the Forest Service, state departments of agriculture and natural resources, nursery research centers, and research universities.[20] The EAB Advisory Board, as this group came to be known, met under the auspices of the Forest Service's Forest Health Technology Enterprise Team, which the agency had formed in 1995 to provide forest management services to managers of privately owned and publicly owned forests.[21]

Determining how widely the EAB had spread from southeastern Michigan proved difficult because of the invisibility of the larvae. To solve the problem, entomologists devised an ingenious solution. In the summer of 2004, the Michigan Department of Natural Resources created "trap trees" at more than one hundred sites, including national forest campgrounds, state parks, and state forests.[22] The researchers girdled the trees by removing the bark from part of the trunk. They then placed sticky traps on the trees and monitored them every two weeks during the period of the season when the beetles took flight. Poland explained, "The EAB was attracted to stressed ash trees, such as trees that were near campgrounds and along roads. We developed lures to detect the beetle and have worked on developing optimal traps. Purple and a shade of bright green seem to be the most attractive colors to the beetle, but there isn't one magic trap that works all the time."[23] Researchers also inspected piles of firewood throughout Michigan.

Researchers studied the rings of diseased trees and concluded that the EAB had probably been in the United States for at least ten years before its detection in 2002. According to Poland, "This meant that the beetles were freely moving around. When people had an ash tree that was dying, they would cut it into firewood. That's one way in which the beetle spread. So, the spread of the EAB was human-assisted. This was probably how it spread to Michigan's national forests."[24] Expanding the area of investigation, the Michigan Department of Agriculture established ten thousand trap trees throughout the Lower Peninsula and the Upper Peninsula, again placing them near campgrounds and along roads, where stressed ash trees were likely to be found. Researchers also spearheaded investigations of nurseries, sawmills, and campgrounds and found more of the insects in Michigan as well as in Ohio and Indiana. The beetle was spreading, and it was spreading quickly.

The dispersal of the EAB resembled a military campaign by an aggressive organism that encountered no natural enemies. The insect would establish a satellite colony in a previously uninfested area. Rodrigo J. Mercader and colleagues in the Department of Entomology at Michigan State University explained how the insect conquered territory from this beachhead:

> Following establishment, these satellite colonies typically grow, coalesce and ultimately greatly increase the speed of the invasion front. Consequently, any action to reduce their formation or growth can significantly decrease the spread of an invasive species. The small population sizes typical of newly formed isolated populations also make successful management more practical.[25]

Researchers also discovered that the EAB could fly, an important mechanism by which the infestation was spreading. In 2004, for example, the Forest Service and Ohio State University reported on experiments with twenty-eight EABs that were tethered. They found that half of the insects covered fewer than fifty meters a day, but one energetic insect covered 5.2 kilometers in 40 hours, flying for 70 seconds, resting for 130 seconds, and reaching speeds of up to 3.5 kilometers per hour. Researchers also discovered that mated females flew faster, longer, and farther than the males, dispersing their eggs over a wider area than had previously been suspected. These data helped explain how the EAB had been able to spread so far and so fast.[26]

In the mid-2000s, the Michigan Department of Agriculture kept looking for more ways to measure the insect's population size accurately. After girdling trap trees, the researchers discovered outlier sites that were more far-flung than had previously been suspected, and by 2005, they realized that the EAB had scattered to twenty counties in Michigan, seven counties in Ohio, and two counties in Indiana. It had also started to infest trees on the borders of the Huron-Manistee National Forests on Michigan's Lower Peninsula. Poland explained, "The appearance of the EAB in the Huron-Manistee National Forests is still a little patchy. The EAB prefers open-growth trees, like those along roads or riparian areas. We have found the beetle on the edges of woodland areas, where they can attack stressed trees, but they are just starting to spread into the interiors of the forests."[27]

On August 5, 2008, researchers found the insect in Houghton County, in the Hiawatha National Forest in the Upper Peninsula, the northern-most point at which the EAB had been found in North America.[28] By 2009, reports of the EAB had surfaced in twelve additional states—Illinois, Iowa, Kentucky, Maryland, Minnesota, Missouri, New York, Pennsylvania, Tennessee, Virginia, West Virginia, Wisconsin—and in the province of Quebec.

Mitigating the Effects of the Emerald Ash Borer

The EAB's rapid spread and its destructive power brought into sharp relief that invasive species have reached near-epidemic proportions in the United States, threatening to degrade not only national forests but also national parks, state forests and parks, and the natural landscapes of cities and towns. In 2004, the Forest Service synthesized ongoing research and mitigation strategies and issued *National Strategy and Implementation Plan for Invasive Species Management*, a book intended to guide federal and state policies on the control and eradication of invasive species.

For the people charged with responsibility for protecting the country's natural areas, invasive species are daunting because they present the simultaneous challenges of researching the species, disseminating the findings, developing strategies to identify and prevent the spread of the species, and building public awareness and support. Meeting those challenges has required collaborative efforts, and the Forest Service, the National Park Service, state departments of natural resources, local authorities, and research departments of universities have improved their level of coordination. In *National Strategy and Implementation Plan for Invasive Species Management*, the Forest Service spelled out the following four-part mitigation strategy that emphasized interagency cooperation.

1. *Prevention*: Collaborate with the Animal and Plant Health Inspection Service and other agencies to assess the risks from pests, establish which invasive species should have the highest priority, work with countries of origin to stop the transfer of pests, and develop programs to prevent the spread of pests.

2. *Early detection and rapid response*: Establish an early detection and rapid response emergency fund to cover the costs of eradication efforts, translate research in other languages into English, and coordinate efforts with other agencies to identify infestations by invasive species.

3. *Control and management*: Coordinate efforts to conduct inventories and map the locations of invasive species; assess the risks posed by the species; identify top priorities for developing strategies for controlling noxious weeds and terrestrial and aquatic species in national forests and adjacent areas; do research on physical, chemical, and biological controls; and develop strategies to prevent the spread of these species.

4. *Rehabilitation and restoration*: Collect and disseminate information about successful efforts to rehabilitate and restore natural areas, issue scientific findings on ecosystem restoration, work with other agencies to distribute seeds and other planting materials, and develop stocks of native plants that resist the spread of noxious weeds.[29]

Fighting the Spread of the Emerald Ash Borer

The EAB Advisory Board used this general plan of attack as a starting point and then developed specific strategies based on the unique characteristics of the EAB. According to Poland and McCullough, "This task [was] especially difficult given the scale of the infestation and our lack of knowledge about EAB biology and ecology."[30] In 2002, federal and state agencies established quarantines against moving ash trees—including cut trees and firewood—that might be infested (figure 13.2). The agencies also quarantined wood chips and bark chips larger than one inch in diameter. Michigan banned nurseries from selling or transporting ash trees, inspected goods traveling over the Mackinac Bridge between the Lower

Figure 13.2 *Quarantine sign at a Wisconsin campground.
The EAB Advisory Board, a multiagency task force to spearhead strategies against the emerald ash borer, established a
quarantine on moving firewood that might contain the insect.
© Drake Fleege/Alamy.*

Peninsula and the Upper Peninsula, and confiscated any materials containing the EAB. The state also conducted "inspection blitzes" of firewood
along major highways.[31]

In addition to establishing quarantines, the Michigan Department of
Agriculture and other agencies disposed of wood that had already been

infested. When inspectors found infected trees, they cut all the ash trees within one-half mile of the infestation, ground the wood into chips, and burned the chips at an electricity plant, using the chips as biofuel. They also sprayed herbicide on the stumps so that the trees could not resprout. As of 2006, agencies had removed more than 290,000 trees from eight eradication sites in Michigan, six in Ohio, and three in Indiana. Researchers also devised additional uses for the wood from infected trees in an attempt to salvage some economic return from the destroyed trees. After the wood was treated, it was turned into railroad ties, lumber, pulp, handles for tools, packaging, and landscaping materials.[32]

After the EAB spread to Illinois, that state's Department of Agriculture took a somewhat different approach. Instead of cutting *all* the trees in a stand, they cut only the infested ones. About Illinois's efforts, Therese Poland wrote, "It is hoped that removal of the infested trees will help reduce populations and contain the insect's spread while allowing time for development of a scientific remedy that could control EAB without having to cut down all of the ash trees."[33]

Insecticides became a weapon in the arsenal against the EAB, but researchers were careful to minimize the use of insecticides so that chemicals did not cause unintended consequences, such as harming wildlife. The researchers experimented with a wide number of insecticides that reduced the number of larvae in a tree when sprayed on trunks or injected into them. In 2004, for example, researchers sprayed infested trees with three different insecticides: Merit (imidacloprid), Onxy (bifenthrin), and Astro (permithrin). They sprayed the bark of the logs in the middle of May to early June and discovered that the mortality rate of the larvae ranged from 66 percent to 94 percent.[34]

Another promising approach was to use insect-pathogenic fungi that were natural enemies of immature EAB. In 2004, researchers began to test BotaniGard, which is a biopesticide that controls insects that infest crops, trees, and forests. In the laboratory, the research staff found that the BotaniGard had an 80 percent effectiveness rate in destroying EAB larvae before maturation. In another study, researchers sprayed uninfected trees that had been transported from a nursery. To test the persistence of Botani-

Gard, they exposed leaves to EAB adults at different intervals after the application of the pesticide: zero days, four days, seven days, and eleven days. The pesticide had, respectively, 100, 96, 88, and 78 percent success rates in killing the EAB, depending on the number of days after treatment that the leaves were exposed to the insects. "This is good persistence for a biopesticide," the researchers noted.[35]

In 2009, Mercader and other entomologists at Michigan State University reported on a promising approach to impede the establishment of satellite colonies. The researchers girdled a small number of ash trees in the expectation that the stressed trees would attract female beetles. They then applied an insecticide before the larvae could metamorphose into adulthood. After applying the insecticide to girdled trees within seventeen hundred meters of the original infestation, they achieved a 90 percent reduction in the spread of the EAB. The researchers concluded that applying the insecticide "at the leading edge of the invasion wave would be ideal," although identifying the leading edge of the wave could be difficult.[36]

These approaches, though, were not practical for use in forests. Poland explained, "In national forests, it's harder to use insecticides because of environmental regulations, and it's also harder to cut infested trees because of regulations over timber harvesting."[37] The long-range solution, entomologists believed, was to identify biological enemies that would inhibit the spread of the insect in forests. Indeed, finding the natural enemy of an invasive species has become a desirable strategy because it minimizes the use of chemicals. In the case of the EAB, researchers looked in China for the insect's natural enemies, and in a stroke of luck, they identified three species of parasitic wasps that had caused 60 percent mortality of the EAB at an experimental site. According to Poland:

> Two of these wasps were new to us, and they had to be named, described, and classified. After identifying these wasps, which are highly specialized, we did studies of them for five years. We released them under tightly controlled conditions among stands of trees that had infestations and then compared the mortality rates to control stands of trees. The wasps were found to be highly effective in killing the larvae of the EAB.[38]

Although the wasps were natural enemies of the EAB, they did not harm other trees or vegetation.

Another promising natural enemy of the EAB has been the woodpecker. In 2004, David Cappaert, McCullough, and Poland reported that woodpeckers feasted on EAB larvae, suggesting that the birds might prove as effective in reducing EAB populations as pesticides were. "Woodpeckers work over the entire core region for free," the researchers wrote, "and are always more popular than spray trucks."[39] The researchers examined twenty-four sites in southeastern Michigan to determine the effect that woodpeckers had on EAB larvae and found a high rate of mortality. The woodpeckers ate the larvae after pecking holes in the infested tree. Researchers have even experimented with putting out suet to attract more woodpeckers, but they want to avoid overpopulating a forest with woodpeckers, which would then have difficulty surviving the winter.

As researchers identified strategies to prevent infestation and destroy the insect, they blended these approaches into an integrated plan of attack. For example, in 2009, Steven A. Katovich, who works for the Forest Service in St. Paul, Minnesota, reported on what was called the SLAM (SLow Ash Mortality) pilot project in Michigan's UP. The program has six components: (1) surveys of the EAB to determine its population density and distribution; (2) investigation of ash trees to identify the trees that had been infested; (3) use of multiple tools to destroy the EAB; (4) implementation of regulations, such as prohibitions against removal of firewood from infested areas; (5) collection and management of data regarding the EAB; and (6) communication and outreach to the public and to outdoor professionals in national forests, state parks, and other outdoor environments.[40] To reduce the EAB infestation, they girdled trees, applied insecticides, and developed programs to treat and salvage the wood from infested trees. Katovich and colleagues applied the SLAM strategy in UP forests. In 2012, Katovich reported that the SLAM research is continuing and shows some evidence of slowing the advance of the EAB. Katovich added that the integrated SLAM approach seems more effective in urban areas but less effective in the national forests, where there is a good possibility that the EAB will eventually kill most ash trees.[41]

The final stage in the process of recovery is ecosystem restoration, and states have been working to replace dead ash trees with trees that resist known invasive species. Michigan, for example, awarded fifty-four grants totaling more than $850,000 to plant ten thousand trees that could not host the EAB. "Although these efforts help," Poland and McCullough wrote, "only a fraction of the dead urban ash trees in southeastern Michigan has been replaced."[42] The Forest Service also sponsored rehabilitation and restoration programs, not only to replace infested ash trees but also to develop white pine stock that resists white pine blister rust, disease-resistant American chestnuts, and elms that can fend off Dutch elm disease.[43]

Lessons Learned

As the different corners of the globe are knitted ever more closely together through global trade and international travel and as climate change continues to lengthen growing seasons and precipitation patterns, the problem of invasive species can only be expected to grow in coming years, posing major threats to forests, savannas, woodlands, prairies, and other natural areas. The experience of the EAB demonstrates that the agencies charged with protecting natural areas must develop integrated, multipronged mitigation strategies. The campaign to control the EAB has pointed to a number of strategies that show promise in protecting natural areas from the assault of destructive species.

1. *Early detection and rapid response are essential.* In 2002, the first reports about the EAB surfaced in southeastern Michigan. In a remarkably short time, the Forest Service, the Michigan Department of Agriculture, and local officials collaborated to identify where the infestations were, communicate the threat to the public, and place a quarantine on the removal of infested ash trees from the core infested areas.

2. *The public needs to be informed and involved.* The agencies charged with protecting natural areas recognized that the public had to play a central role in preventing the spread of the EAB. The agencies developed posters and other communication materials that informed the public about the threat of the EAB and clearly explained the guidelines for

the quarantine on the movement of firewood and other materials that might contain the insect. Moreover, the public, by and large, cooperated with the efforts by government agencies because people wanted to save ash trees. To help control costs and improve mitigation efforts, the Forest Service and other agencies are increasingly relying on citizens for early detection and rapid response to control invasives before they become a bigger problem. For example, botanist Chris Mattrick of the White Mountain National Forest trains both employees and citizen volunteers in early detection and is closely involved in a cooperative invasive species management area with other agencies and organizations.

3. *Collaboration among public and private agencies is essential.* One hallmark in the fight against the EAB has been the cooperation among the Forest Service, the Michigan Department of Agriculture, the Michigan Department of Natural Resources, Michigan State University, other state universities, and privately owned nurseries. As the Forest Service emphasized in *National Strategy and Implementation Plan for Invasive Species Management*, the challenge to the Forest Service is "to learn to lead collaboratively." This cooperation improved the feedback loop for identifying new infestations, disseminating information about the EAB, and implementing strategies for destroying the insect. The EAB Advisory Board has been an outstanding example of this collaborative effort.

4. *Solutions to invasions by nonnative species are multifaceted.* The public agencies that have collaborated to mitigate the spread of the EAB have used a variety of strategies to detect the insect, slow its spread, and destroy it. For example, Michigan State and other research universities have reported on the use of biological controls *and* chemical controls to destroy the insect.

5. *Biological controls show great promise for the future.* Researchers have traveled to China to find out whether the EAB has natural enemies in the environment there. Why have they made this effort? They have done so because control by biological measures means that we are put-

ting fewer pesticides and herbicides into the environment. Such chemicals may be effective, but they add potentially harmful substances to ecosystems, causing unintended consequences. The use of biological controls against purple loosestrife is instructive. First introduced from Europe, this purple-headed weed spread rapidly into wetlands throughout the United States, crowding out native plants. In the mid-1990s, scientists discovered that *galerucella* leaf-eating beetles fed on the flowers, stems, and leaves of purple loosestrife but did not harm other plants in wetlands. Thus, the beetles helped reduce populations of purple loosestrife. Several states introduced the beetle into wetlands that had been invaded by the plant. Connecticut, for example, introduced the Beetle Farmer Program in 2004 to encourage schools, conservation groups, and scout troops to raise beetles for the state to introduce in more wetlands. By 2011, the state had released more than 1.5 million beetles into approximately one hundred wetlands in Connecticut.[44]

The fight against the EAB has been marked by the relatively rapid application of science, cooperation between the public and private sectors in natural resources, and a commitment by a good part of the public to help prevent the dispersal of this invasive beetle. As trade and travel continue to shrink the world and countries face the growing threat of invasive species, however, a global strategy becomes more imperative than ever. At the heart of this strategy must be cooperation to regulate the movement of flora and fauna, inspect shipments of organic materials, share biological knowledge, and discover biological controls that can minimize unintended harm to forests. "Free trade" is not free if insect, disease, or plant invaders hitch a ride on shipments coming into this country. Furthermore, reducing regulations and eliminating agencies like the Animal Plant and Health Inspection Service might produce savings in the short term, but in the long run, such moves could be catastrophic for the eastern national forests and other irreplaceable ecosystems.

CHAPTER 14

National Forests of Vermont and North Carolina: Loving the Forests to Death

When the coauthors of this book first met to start planning our collaboration, David asked a startling question: "Are the eastern national forests being loved to death?"

National forests did not exist east of the 100th meridian in the early twentieth century (see figure 4.2). Today, the United States is blessed with extraordinary eastern national forests that draw millions of people a year to hike, ski, camp, backpack—to "re-create" themselves in any number of ways—yet their widespread popularity has created the problem David articulated in his question. Because of deep-seated changes in our culture since the early 1900s, millions of Americans have decided that they want to live close to national forests. The trend has had a profound and insidious effect on the eastern national forests. Because these forests are close to large metropolitan areas, this cultural change has affected these easily accessible public lands in particular. People want the rural lifestyle, the proverbial peace and quiet, and relief from the stresses and strains of modern life.

This trend, however, is having huge unintended consequences: fragmenting forests, reducing wildlife habitat, causing pollution and erosion, spreading invasive species, and threatening fragile wildflowers and other flora that are integral to healthy forest ecosystems. Indeed, if growth near the eastern national forests is not managed properly, the inevitable result will be the gradual deterioration of the very forests Americans have worked so hard to restore.

Vermont and North Carolina have both experienced significant growth near national forests, and the demographic trends are similar in both states. In Vermont, people attracted to a bucolic landscape have been purchasing primary and secondary residences near the Green Mountain National Forest (GMNF). North Carolina is a magnet for retirees, second-home purchasers, and workers in the state's booming high-tech sector. Incoming migrants have purchased residences near North Carolina's four national forests: the Nantahala, the Pisgah, the Uwharrie, and the Croatan. The federal government, state governments, and conservation groups are trying to conserve these magnificent forests for future generations. In addition, private forestland owners are under pressure to sell lands for development, while conservation organizations and government agencies have been assisting owners in holding on to their lands.

Dimensions of Development near National Forests

Starting around 1990, the migration of people to rural areas began to accelerate. Between 1990 and 2000, rural counties in which national forest lands comprised at least 10 percent of the land experienced an 18 percent increase in population. To examine the scope and effect of increased population density near national forests, the U.S. Forest Service partnered with the Pacific Northwest Research Station to undertake a thorough study of demographic trends near national forests. In 2007, the agencies issued their report, titled *National Forests on the Edge*, which provided a comprehensive picture of growing development pressures. The report included a wealth of data measuring the increase in housing density near national forests and reached this sobering conclusion: "Counties with national forests and grasslands already are experiencing some of the highest population growth rates in the Nation as people move near public lands."[1] It furthermore stated that fully 25 percent of the U.S. population lives in a county that has land that is part of the national forest system.

The researchers measured the density of housing at three different distances from national forests and grasslands: one-half mile, three miles,

Table 14.1 Eastern National Forests with at Least 25 Percent of Adjacent Lands Expected to Experience Increased Housing Density by 2030

		Percent of Lands Lying within:			
Eastern National Forest	State	0–0.5 Mile	0.5–3 Miles	3–10 Miles	0–10 Miles
Chattahoochee-Oconee	Georgia	31	35	35	35
Cherokee	Tennessee	30	36	31	32
Croatan, Nantahala, Pisgah, and Uwharrie	North Carolina	26	29	30	30
Huron-Manistee	Michigan	31	32	26	28
Land Between the Lakes National Recreation Area	Kentucky, Tennessee	5	23	31	28
Green Mountain and Finger Lakes	Vermont, New York	28	31	25	27

Source: Susan M. Stein et al., *National Forests on the Edge* (Washington, DC: USDA Forest Service, 2007), 9.

and ten miles, distances at which increased housing density can affect forests. By examining current growth rates, the researchers projected that by 2030, 8 percent of the lands within ten miles of national forest boundaries will experience increases in housing density, affecting nearly twenty-two million acres of land. These trends are particularly pronounced in the East, the South, and the Lake states. As table 14.1 shows, six groups of eastern national forests are projected to experience greater housing density on at least 25 percent of the lands within ten miles of their boundaries.[2]

In addition, ten groups of eastern national forests are near five hundred thousand acres of rural lands that are expected to increase in population density. Eight of the forest groups are in the South, which has been experiencing explosive rates of urban and suburban development. Table 14.2 lists the ten groups of eastern national forests lying near land that is expected to undergo significant increases in housing density by 2030.[3]

Another major challenge for the eastern national forests is that public lands are interspersed with privately owned lands. For example, on the Huron-Manistee National Forests in Michigan's Lower Peninsula, privately

Table 14.2 National Forests with More Than 500,000 Acres of Adjacent Land Expected to Experience Increased Housing Density by 2030

Acres of Adjacent Private Lands with Projected Housing Density Increases

Eastern National Forest	State	Acres (in Thousands)
George Washington–Jefferson	Virginia, West Virginia	1,424
Mark Twain	Missouri	1,326
Chattahoochee-Oconee	Georgia	1,176
Croatan, Nantahala, Pisgah, and Uwharrie	North Carolina	1,073
Bienville, Chickasawhay, Delta, DeSoto, Holly Springs, Homochitto, Tombigbee	Mississippi	1,071
Bankhead, Conecuh, Talladega, Tuskegee	Alabama	834
Francis Marion–Sumter	South Carolina	720
Daniel Boone	Kentucky	650
Green Mountain and Finger Lakes	Vermont, New York	590
Cherokee	Tennessee	544

Source: Susan M. Stein et al, *National Forests on the Edge* (Washington DC: USDA Forest Service, 2007), 10.

owned lands are interspersed with a few large blocks and many smaller blocks of public land. Because of this patchwork pattern of land ownership, the Huron-Manistee has issues with access and other conflicts over uses of the public lands.[4] Such situations call for high levels of private-public cooperation.

A related challenge is the existence of numerous inholdings, in which private parties own land within national forests. In the United States as a whole, inholdings represent approximately 17 percent of lands within the proclamation boundaries of national forests and grasslands. Inholdings are much more prevalent in the East than in the West, representing approximately 46 percent of eastern national forest lands. These lands are under the ownership of a variety of parties, including individuals, corporations, Indian tribes, and state and local governments. In 1950, there were approximately 500,000 houses on inholdings. In 2000, the number of houses had tripled to 1.5 million, adding to the pressures of development adjacent to eastern national forests.[5]

Consequences of Growth for National Forests

This growing human population density is having a significant ecological impact on national forests. One effect is forest fragmentation, which the Sustainable Forests Partnership defines as "the process of a contiguous land base being divided into smaller pieces."[6] Fragmentation reduces connectivity, or the physical and ecological connections that characterize an ecosystem. Reduced connectivity affects wildlife, preventing animals from reaching habitats they need for food, shelter, and reproduction. Road construction and growing traffic increase animal mortality rates, and the construction of new houses and roads affects migration patterns. Another effect of higher population density is invasive species, which are often introduced into an ecosystem through foot traffic, motor vehicles, and nursery plants.

Higher population density can also reduce access to public forestlands. In 1999, national forest managers reported that approximately 14 percent of national forest lands were more difficult to reach because of construction on adjacent private lands. For example, when a new housing development is constructed, it may cause the closing of a road that formerly led into part of a national forest. The opposite problem can also occur. Greater population and more roads can result in increased usage of national forests, upping pressures on wildlife habitats and increasing noise and light pollution. Forest managers need to plan for decreased *and* increased access, a daunting management challenge.

In addition, the risk of wildfires increases dramatically when the human population is greater near national forests. The threat is particularly great in the East, where approximately 75 percent of the fires on national forests are caused by humans. The proximity of residential housing to national forest boundaries makes firefighting more complex, raising the costs of fire protection. Water quality is also vulnerable to the impact of human population. Housing and road construction damage the banks of rivers and streams, lower water quality, and interfere with hydrologic cycles. Construction also creates more hard surfaces, slows the absorption of precipitation into soils, and increases stream flow, causing floods and dispersal of pollution.

As people move closer to national forests, crime rates may also rise, as evidenced by the ubiquitous signs in West Virginia's Monongahela National Forest that warn visitors not to leave valuables in the car. Irresponsible people living near national forests dump trash illegally, use and deal drugs, and grow marijuana. Encroachment is a problem, with some private landowners using public lands as gardens, storage sites, and garbage dumps. People cut timber illegally, poach wildlife, and use off-road vehicles on unsanctioned trails, causing erosion and destroying fragile plants.

At the same time, Forest Service funding has been stretched thin in recent years, making it even more difficult for the agency to keep pace with the challenges posed by greater population density. The Forest Service faces higher costs in measuring the environmental impact of forest plans because forest managers must account for a wider constellation of threats to forest ecosystems. In summary, the report concludes, "Increased housing development in rural areas bordering America's National Forest System lands could alter the ecological, social, and economic resource and services provided by those public lands and increase their management costs."[7]

Rapid Development in Vermont

If ever there was a forest that was in danger of being loved to death, it is Vermont's Green Mountain National Forest (GMNF), which rises and falls across the Vermont landscape like a blanket of green and granite. Formed in 1932, the GMNF presents all the challenges facing eastern national forests: vacation homes, ski resorts that bump up against the sides of mountains, and a world-class system of hiking trails that attracts thousands of hikers every year. Since 1990, the rural culture of Vermont has attracted an influx of exurbanites seeking respite from the pressures of New York, Boston, Philadelphia, and points in between, causing inflation of land prices. The consequence has been to heighten pressures on forestland owners and farmers to sell lands near the GMNF, leading to parcelization and fragmentation. Parcelization results when large tracts of land are divided into smaller tracts. One consequence of parcelization is forest fragmentation, with far-reaching effects on plant and animal species, wildlife habitat, and water quality.

In August 2006, the Vermont Natural Resources Council (VNRC), which is dedicated to protecting Vermont's natural resources, convened a group of diverse individuals and groups to devise strategies for protecting the state's forests from the risks of overdevelopment, parcelization, and fragmentation. In 2007, the group, known as the Forest Roundtable, issued a report that painted a comprehensive portrait of Vermont's changing land-ownership patterns and their effect on the GMNF, state forests, and privately owned forests. According to Jamey Fidel, the forest and biodiversity program director of the VNRC, "The Roundtable was created in part to unite various stakeholders after significant debates on new wilderness designation in the mid-2000s, when the Forest Service was revising the forest plan. About thirty people first met as part of the Roundtable, including people from the forest products industry."[8] The Roundtable was created to focus attention on the vitality of private forestland and has since expanded to include about 180 different agencies and individuals, including the forest supervisor of the GMNF and representatives from the Vermont Department of Parks and Recreation, the Vermont Fish and Wildlife Department, and the state's Council on Rural Development.

The Roundtable's 2007 report presented a clear picture of the trends that are threatening the GMNF. The study found an unprecedented level of real estate activity. The forests in Vermont encompass some twenty-six million acres, and during the years from 1980 to 2005, almost twenty-four million acres were sold to new owners.[9] Fidel explained, "Lots of privately owned forestland started changing hands, including timber corporations that were starting to pull out of Vermont. One of the major sales was by the Champion International Paper Company. The company had heavily logged forestland in the Northeast Kingdom and sold approximately 132,000 acres, which fortunately ended up being conserved by a partnership of public and private entities."[10] A partnership of the Conservation Fund, the U.S. Fish and Wildlife Service, the Vermont Land Trust, the Vermont Housing and Conservation Board, the Vermont Agency of Natural Resources, and The Nature Conservancy of Vermont purchased and protected the land. Since the sale in 1999, twenty-six thousand acres of the property have become part of the Silvio O. Conte National Wildlife Refuge, and another twenty-two thousand acres formed the West Mountain Wildlife Management Area. The Plum Creek Timber Company purchased more than eighty thousand acres and employs sustainable forestry methods to manage the lands under a working-forest conservation easement with the Vermont Land Trust and the Vermont Housing and Conservation Board.[11]

During the same period, Vermont experienced a major appreciation in land values. From 2001 to 2006, the average price of residential houses and condominiums rose from $126,000 to $185,000, whereas vacation homes and condos jumped in value from $110,000 to $200,000. Between 2001 and 2005, land also became much more expensive, particularly tracts smaller than twenty-five acres, which were easier than large tracts to purchase, rezone if necessary, and subdivide. Tracts smaller than twenty-five acres shot up in price from $4,505 an acre in 2001 to $10,000 an acre in 2005, whereas during the same period, tracts larger than 25 acres rose in value from $974 an acre to $1,580 an acre. As a result, land speculators had considerable incentive to divide larger tracts into smaller tracts and resell them, reinforcing the trend toward parcelization.[12]

Between 1982 and 1992, the state's population grew by 10 percent, but the amount of developed land spiked by approximately 25 percent. As a consequence, the state experienced a virtual explosion in construction. In 1970, the state had about 165,000 housing units, but by 2000, that number had catapulted to about 294,000 units. The building trend continued after 2000, with more than 17,000 building permits being issued between 2000 and 2005.[13]

These trends will continue. According to Forest Service projections, the density of housing on privately owned forestlands in Vermont is expected to grow by 5 percent to 40 percent across the majority of watersheds by 2030, with much of this growth occurring along the Connecticut River. In addition, many longtime owners of rural properties are aging. According to the Roundtable's report, people older than 65 years of age own upwards of 25 percent of the private forestland in the United States. Unless these people make provisions in their wills, the land will be divided among the heirs, exacerbating the trend toward parcelization.[14]

The Forest Roundtable described how parcelization and fragmentation can affect Vermont's forest ecosystems. For example, clearing forestlands can reduce evergreens and conifers that provide habitat for deer during the winter. In addition, migratory corridors that black bears and other wildlife use may be diminished. Some residents engage in "woodscaping," or removing the understory to create more of a park-like appearance, which may result in the loss of habitat for certain small mammals and songbirds. As trees have been cut down to make way for development, Vermont's forests have lost some of their ability to regulate the flow of rivers and streams. The report explained, "This leads to changes in streamflows, increases in sediment, [and] reshaped stream bottoms and banks."[15] The consequences can be degradation of water quality and reduction in the diversity of aquatic life.

Parcelization and fragmentation have had a number of economic effects on the state, both positive and negative. Developers have invested in the state and made money, and the ski industry brings in some $750 million annually for local economies.[16] At the same time, as private timberland owners have sold their lands, the amount of timber harvesting has decreased. In addition, after subdivision, the remaining privately owned forestlands are often smaller, and owners are less likely to harvest timber that is within sight of their homes. To some extent, the wonders of nature that have traditionally drawn tourists to Vermont are diminishing. Today, hikers may climb to the top of a mountain, look down, and see a subdivision rather than uninterrupted forest (figure 14.1).

Protecting Vermont's Forestlands

The stakes of overdevelopment are enormous, and as befits those stakes, Vermont is taking innovative steps to try to slow down and reverse the trends toward parcelization and fragmentation. The state's initiatives fall into four broad categories: (1) provide financial incentives to slow the trend of selling forestlands for the purpose of subdividing them, (2) improve

Figure 14.1 *Forest fragmentation in northwestern Vermont. Residential development, agriculture, and road construction are all carving up Vermont's forests, creating fragmented habitats that have a wide-ranging effect on wildlife and vegetation. Eric Sorenson, Vermont Fish and Wildlife Department.*

regional and local planning and zoning to preserve forestlands and other green spaces, (3) strengthen statewide management of the forest ecosystem, and (4) promote the forest products industry.

The Roundtable made a number of recommendations to strengthen the financial incentives for keeping forests intact on private lands near the GMNF. Such efforts are critical because the majority of U.S. forests are in private hands (figure 14.2). In Vermont, the Roundtable argued that the state should strengthen its Use Value Appraisal (UVA) program, which ensures that if land is kept in use for farming or forestry, it will be taxed for the value of its actual or current use. The program is intended to counter the traditional approach to property appraisal, which requires that the

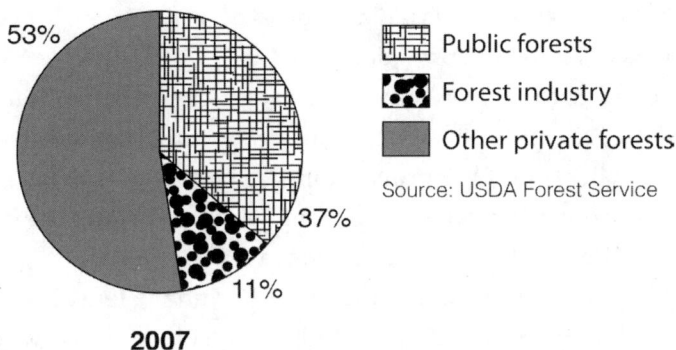

53%

Public forests

Forest industry

Other private forests

Source: USDA Forest Service

37%

11%

2007

Figure 14.2 *U.S. forestlands by ownership class. Graph by Christopher Clark.*

property be appraised for its "highest and best use," implying that the property will inevitably be developed for commercial or residential use. Under the UVA program, a privately owned forest continues to be appraised as undeveloped land. Under this valuation, towns lose property tax revenue, but the state reimburses them for the lost revenues. As Fidel wrote in the 2007 Roundtable report, "The UVA Program has been a very successful tool for reducing the effects of parcelization and forest fragmentation."[17]

A related recommendation is to ensure that, in the future, property owners who put conservation easements on their property will receive lower tax assessments. A conservation easement is a voluntary agreement by an owner not to allow certain types of development on a piece of property. According to the VNRC, however, because of inconsistent approaches to property valuations by government agencies, a property owner who currently receives an easement does not necessarily receive a lower property tax bill.[18]

In addition, the Roundtable urges conservation groups and state government to inform the public of the ecological and economic benefits of forests. For example, tree roots purify water without any cost to the people who use the water. According to the Forest Service, "Many of these goods and services are traditionally viewed as free benefits to society, or 'public goods'—wildlife habitat and diversity, watershed services, carbon storage, and scenic landscapes, for example."[19] It is difficult to put a dollar measure on these, but one estimate placed a value of $33 trillion every year on the global services that forests provide.[20]

The second category of recommendations involves better planning and zoning to protect green spaces. Such planning begins with accurate information collection. One problem in Vermont has been that information about parcelization rates is scattered across various state and local agencies, resulting in data collections that are organized in different ways. The Forest Roundtable recommended that Vermont's legislature pass legislation that would fund and standardize the collection of parcelization data, which, in turn, would serve as a more accurate baseline for assessing conservation efforts. As part of this effort, the state would develop maps showing the growing parcelization in different parts of the state.[21]

Vermont has also put into place improved tools for regional planning. Fidel explained, "There are thirteen regional planning commissioners, and the VNRC is cooperating with them to develop comprehensive policies that include the protection of forestlands. We want to strengthen zoning protections, promote clustered housing, and protect viable forestlands."[22] In clustered housing, homes are grouped closely together on a tract of land so that more land can be set aside as woodland, for recreation, or for farming. Regional planners see clustered housing as an important tool of sustainable development because it reduces the footprint of housing on a landscape.

In Vermont, many of the development pressures are evident in popular recreation towns like Stratton, Warren, Shrewsbury, and Killington. "Many towns are doing significant things to protect forestlands," Fidel pointed out. "Bennington has a forest reserve district, in which zoning has slowed development. Shrewsbury has a wildlife overlay district, which is specially designed to protect wildlife. The Mad River Valley, which includes Warren, Fayston, and Waitsfield, has developed a forest, wildlife, and communities project, which has the goal of reducing fragmentation near the GMNF."[23] These towns and regions have developed land-use planning models that encourage clustered housing and prevent certain kinds of development.

The third group of initiatives Vermont has undertaken is to strength the statewide management of its forest ecosystem, which includes the

GMNF but also encompasses state and privately owned forests. Parcelization and fragmentation have cut off corridors for migrating wildlife. In addition, because of climate change and global warming, winter temperatures in the Northeast have been increasing by 0.5 degree Fahrenheit per decade (winter temperatures by 1.3 degrees Fahrenheit per decade) since 1970. As a result, cold-weather species have migrated farther north to locate appropriate food and shelter, but they still move south during the spring. As they migrate south, they encounter more and busier roads.[24]

To address this problem, conservation organizations and state agencies have banded together to create several projects, such as the Staying Connected Initiative and Vermont's Critical Paths Project, to develop strategies to improve migratory corridors. One solution is to leave gaps in guardrails to allow animals to move more easily from one side of a road to the other. Other solutions include lowering speed limits, using less salt, and putting up road signs warning motorists about wildlife. Pipes carrying water underneath roads can also be made large enough to allow amphibians, small mammals, and snakes to travel beneath the roads. Overpasses and underpasses provide routes for wildlife to cross roads safely. As of 2010, Vermont had built approximately fifteen overpasses and underpasses.[25] In addition, the state can cut down trees and thick undercover from areas close to highways, allowing drivers better to see wildlife. It can remove salt licks near highways to prevent animals from being attracted to those areas and plant roadside vegetation to encourage animals to cross the highway in places with lower traffic volume. Finally, the state has conserved land near highways by creating nature sanctuaries and granting conservation easements, enabling animals to find safe places to cross highways. Land conservation also improves habitat connectivity, which is critical to maintaining wildlife populations.[26]

The fourth area of initiative by the Roundtable has been to promote Vermont's wood products industry. Fidel explained, "There is wide agreement that part of preserving open space and forestland is to have working forests, which should be managed using sustainable forestry. A working landscape helps keep the forest intact."[27] According to the Roundtable, Vermont's forests bring more than $1.4 billion into the state's economy and create nearly fourteen thousand jobs.[28] As a result, one of the Roundtable's goals has been to strengthen the viability of privately owned forests by encouraging greater cooperation among landowners, logging companies, sawmills, paper manufacturers, and lumberyards to promote the value of working forests. In addition, the Roundtable encourages efforts to promote and brand Vermont wood and urge local architects and contractors to use this local resource.[29]

One pioneering initiative is Vermont Family Forests (VFF), a nonprofit conservation organization that has worked since the mid-1990s to combine promotion of the state's wood products industry with informed stewardship of forestlands. The driving force is David Brynn, founder and executive director. Brynn was a forester for more than thirty years, many of them as a Vermont state forester. "Years ago," he explained, "I noticed more private forestland owners who were asking about values other than timber production, such as recreation and wildlife protection. We decided to create a different kind of organization that would emphasize forest health, and that became Vermont Family Forests."[30]

The organization espouses three guiding principles: (1) informed stewardship of forests, (2) sound economic returns gained from ecological forestry, and (3) a land ethic based on community values and strengths. Brynn said that he has been most influenced by the writings of Aldo Leopold and Wendell Berry, the renowned Kentucky writer and farmer who has written that "the two great ruiners of private land are ignorance and economic constraint."[31] Like Berry, Brynn has one foot in the practical need to make a living and the other in the visionary goal of protecting the land.

VFF has developed a forest health conservation checklist of more than forty practices that will ensure water quality, diversity, carbon storage, and the productivity of the forest for the long run.[32] Similarly, it has created a certification system for Vermont-grown timber to guarantee that the wood has been harvested and processed according to standards of sustainability. "Our strategy," according to Brynn, "has been to help private forestland owners pool their resources together and figure out ways of adding value to timber—to sell products while acting as good stewards of their forests."[33]

Brynn explained how VFF recently helped promote significant timber sales in the state. Middlebury College was building a new science center, Bicentennial Hall, and VFF worked with the college and its architects to supply the wood from local forests. Originally, the architects had planned only for white oak, but VFF's foresters showed how they could use nine different species, including oaks, beeches, and maples. Brynn added, "We also showed them how to use wood that traditionally has been considered defective because it has knots and other defects. We call this 'character wood.'"[34]

Another innovation has been to create a model for community-based timberland investment management organizations, which can buy and manage forestland collectively. "Such an arrangement," the Roundtable states, "could help individual landowners pool their resources and share the costs of ownership and management."[35] If forests are profitable and productive and if tax policies promote forest conservation, then owners will have more incentive to preserve forests and the critical habitat they provide for the mammals, amphibians, songbirds, wildflowers, ferns, and trees that constitute the bountiful natural life of Vermont.

Similar Threats in North Carolina

North Carolina is another state in which population growth and migration are placing enormous pressures on national, state, and privately owned forestlands. The state's four national forests, which encompass 1.25 million acres and attract more than 8.5 million visitors a year, stretch from the Atlantic to the Appalachian Mountains and boast an extraordinary range of terrain, from gently rolling hills to the rugged, granite-capped mountains of western North Carolina.

In 2010, the Forest Service's Southern Research Station Headquarters in Asheville, North Carolina, issued a report titled *WNC Report Card on Forest Sustainability*, which examined major demographic changes that are reshaping North Carolina and deeply affecting ecosystem connectivity, wildlife, water quality, air quality, and the wood products industry in the state's eighteen westernmost counties. The report described four categories of surface use: (1) developed, which has considerable amounts of pavement covering the soil; (2) undeveloped, which includes farms, forests, and grasslands; (3) protected lands, which are national forests, state forests, parks, and land conservancies; and (4) water.

In 1976, according to the report, western North Carolina still retained much of its rural character, with a mere 1 percent of its land being developed. During the 1980s, however, development in North Carolina exploded, and the uses of land underwent a revolution. Interstate 40 linked Raleigh, Durham, Chapel Hill, Greensboro, and Winston-Salem to the western forests, making weekend getaways to vacation homes in the mountains highly convenient. In addition, the state's robust economy generated high-paying jobs as well as the means to afford vacation homes. Furthermore, the state's mild climate and natural beauty enticed retirees.[36] Brent Martin, a transplanted Georgian who lives in one of the western counties and serves as the southern Appalachian regional director for the Wilderness Society, said, "We've had the development of large gated communities in western North Carolina, which attract wealthy retirees. People move here looking for a better quality of life and a preferable climate. But as a result, these developments raise multiple issues. There is much less forestland and farmland than there once was." He continued, "If you look at the Nantahala and Pisgah National Forests, cities surround them like a bathtub ring, and those cities are drawing water from the national forests."[37]

The statistics echo Martin's words. By 2006, 7.1 percent of the land in western North Carolina had been developed, meaning that an average of six acres had undergone development every day since 1976. In addition, individuals who had purchased property were occupying an average of 0.22 acre per person, compared with 0.06 acre per person in 1976, reflecting the growing footprint of people on the land. Many of these new residents bought old farms, but others built at ever higher levels on mountains, where residences supplanted forests. The WNC Report Card predicted that the trends will continue, with the population of the western counties expected to increase by 22.7 percent between 2010 and 2030. At this rate, 47,500 acres of land will be converted from forests and farms into subdivisions, shopping malls, roads, and office parks by 2030.[38]

The resulting parcelization and forest fragmentation in North Carolina mirrors what has happened in Vermont. Reduced habitat makes birds and other wildlife more vulnerable to predators. The shrinking green space is severely affecting bats, which find it more difficult to find areas for roosting and foraging. As a result, three of North Carolina's sixteen bat species are on the federal government's endangered species list.[39] Increased road traffic causes greater mortality of wildlife, which are also affected by noise and air pollution. Martin added, "We have seen more problematic encounters between black bears and people."[40] As the forests have become fragmented, bear habitat has shrunk, driving the animals closer to human settlements.

According to the *WNC Report Card*, "Forest fragmentation also can impair water and air quality."[41] As roads, sidewalks, and parking lots are built, water from storms runs off and carries chemical-laden lawn fertilizers and pesticides into rivers and streams. At the same time, heavier traffic increases emissions and air pollution, and fewer trees means that more carbon dioxide and other pollutants remain in the atmosphere, worsening climate change and global warming.

In North Carolina, as in Vermont, conservationists have recognized the effect of the conversion of forests and other rural lands, and the federal government, state government, and private timberland owners have collaborated to develop strategies to protect western North Carolina's forests. The Forest Service's forest inventory and analysis program—in existence since the 1930s—monitors forests for plant diversity, wildlife habitat, infestations of insects, and other indicators of forest health. In addition, the agency's Eastern Forest Environmental Threat Assessment Center, located in Asheville, tracks the threats posed by development, pollution, invasive species, and other environmental threats and issues reports to forest managers. The forest health monitoring program adds more data about forests by conducting aerial surveys and inventories of tree populations.[42]

Another priority in North Carolina is to encourage tree planting and sustainable forestry. One notable initiative has been the American Tree Farm System, which has been in existence nationally since the 1940s and promotes forest stewardship among some eighty-eight thousand private timberland owners who control twenty-six million acres of forestland nationally. North Carolina's Division of Forest Resources, which supervises the program in the state, works with private timberland owners to develop plans to encourage the growth and management of forests on a sustainable basis and certifies properties with the Tree Farm sign.[43] As a result of such efforts, the volume of timber in western North Carolina actually increased between 1984 and 2006 by 39 percent on private timberlands and 35 percent on public forestlands. In addition, the total volume of timber grew from 6.1 billion cubic feet to 8.4 billion cubic feet. Statistics also show that the state's forests are being managed on a sustainable basis. Timber harvesting takes 1 percent of the stock of trees per annum, whereas the growth of trees adds 3.9 percent annually to the total volume of timber. This important trend helps counteract the loss of forestland to development.[44]

Martin also emphasized the priority of conserving farmland as well as forestland. "The two go hand in hand," he commented, "and they're important in preserving North Carolina's rural culture, which is a major priority for people who have deep roots in this land." To slow the trend toward development, the Wilderness Society in North Carolina helps private forestland owners manage their lands. "We don't want them to sell off their lands to developers," Martin explained, "because these forestlands serve as buffers around the national forests."[45] The Wilderness Society also supports the expansion of wilderness on North Carolina's national forests, which have not gained new wilderness protection since 1984. As of 2012, the Forest Service in North Carolina was about to start writing a new forest plan, during which proposals for new wilderness would be considered.

Lessons Learned

The issues of development, parcelization, and fragmentation near eastern national forests have shot into prominence since 1990, and conservationists, the Forest Service, and state environmental agencies are only beginning to understand these issues and their effects. Making the situation even more challenging is that collecting accurate data is difficult, partly because data collection involves numerous governmental agencies.

On the more hopeful side, conservationists are armed with a vastly greater understanding of the ecological impact of uncontrolled and unregulated development. Ecologists emphasize that to protect biodiversity, society needs to preserve ecosystem connectivity and protect the natural links among forests, farmland, prairie, and savannas. A key element is preserving the migratory corridors for wildlife, particularly as climate change drives more species farther north and causes increased north-south migrations. The initiatives undertaken by conservationists in Vermont and North Carolina point to several promising strategies.

1. *Gather the data.* Vermont's Forest Roundtable, the Vermont Natural Resources Council, and state agencies have collected information about commercial and residential development near the Green Mountain National Forest, and their reports have clearly spelled out the ecological and economic effects on the forest. Vermont's approach to data collection can serve as a model for other states.

2. *Educate the public about the importance of protecting large ecosystems.* The boundaries that surround national forests are human artifacts, and we must continually remind ourselves that for wildlife, flowers, and trees, such boundaries do not exist. When houses and commercial buildings are constructed near national forests, they become, for better or worse, part of a forest ecosystem and affect that ecosystem. To build public support for protecting lands near eastern national forests, conservationists must explain the connection between people's properties and nearby forest ecosystems.

3. *Share information across states about effective ways to manage growth near national forests.* In Vermont, conservationists, government officials, and managers of private timber corporations and other enterprises have collaborated to develop multipronged strategies that include zoning, land purchases, conservation easements, and tax policies that discourage runaway development and encourage forest conservation near national forests. Conservationists have also issued easily accessible reports that elucidate the problems and propose mitigation strategies. Tax laws that assess the value of farms and forests for their actual or current use—rather than for the highest potential use—can encour-

age owners to keep their lands. Another federal policy that encourages private owners to protect forests is the Forest Legacy Program, which was initiated in 1990 and is administered by the Forest Service. In this program, the federal government collaborates with the states to provide incentives to grant conservation easements and create legally binding agreements not to develop property. The policy has been an outstanding success, enabling the protection of more than 2.2 million acres of forestland in forty-three states and territories.[46]

4. *Protect the culture.* The Wilderness Society's Brent Martin spoke eloquently of the critical importance of preserving local rural culture: the music, the folklore, and the values that have thrived for generations in Appalachia. The preservation of this culture goes hand in hand with forest conservation.

The challenges posed by the development of rural lands near eastern national forests may at first glance seem to be unrelated to the forests themselves. After all, such development takes place outside the boundaries of the forests, which are protected by the Organic Act, the Weeks Act, and other legislation. The development of adjacent lands, however, has a very real and potentially harmful effect on forests by exacerbating air and water pollution; fragmenting wildlife habitat; introducing alien invasive species; threatening plant diversity; and increasing economic pressures to extract timber, oil, and natural gas from national forests.

A major theme of this book has been that the eastern national forests are qualitatively different from those of the West because they are located near major population centers; experience heavy use; and bear the unmistakable marks of the past, such as old railroad grades and plugged oil wells. Through the efforts of dedicated citizens, professional foresters, ecologists, political leaders, and other conservationists, the United States has undertaken an extraordinary effort to restore and protect those forests. In total, the forests represent only a minor part of the land mass of the United States east of the 100th meridian, yet the significance of the forests far outweighs the amount of property they consume. They are precious islands of publicly owned ecosystems that serve as a model for the management of healthy forests and the protection of significant portions of America's biodiversity.

For these reasons, we must be aware of and continue to learn more about the effects of development and conversion of rural lands near eastern national forests. In Vermont and North Carolina, individuals and groups have come together to examine development trends and spell out their effects on all kinds of forests, from publicly owned national forests to the private forestlands that provide so much of the nation's timber and other wood products. The initiatives in these two states can serve as a model for similar efforts in other areas of the East, the South, and the Lake states, efforts that are essential to ensuring that the eastern national forests remain vibrant, healthy, and verdant refuges for future generations.

Conclusion

During March and April every year, West Virginians indulge in a culinary tradition that dates back generations: eating ramps. Ramps are wild leeks and are related on the family tree of plants to garlic, chives, and onions. In early spring, they begin to sprout in the forests—a sure harbinger of warmer weather—and folks harvest them. The locals say that ramps are truly tasty, but they definitely have a strong flavor and are an acquired taste, and if you eat enough of them, the odor oozes out of the pores of your skin. Nearly every town in West Virginia has its annual ramps festival, where people eat ramps cooked with potatoes or meat or in any number of other ways.[1]

Ramps, in short, are an integral part of the forest heritage of West Virginia. The celebration of this local delicacy tells us that even in our hypermodern and ultraurbanized and suburbanized world, forests and the culture that is nurtured by them remain an essential and distinctive part of the American landscape. Certainly forests supply us with wood, which continues to have innumerable uses even in our fossil-fuel-driven modern economy, yet forests are so much more. They are the places where we sojourn to re-create ourselves, to attain physical and mental and spiritual restoration by hunting, fishing, hiking, climbing, camping, canoeing, kayaking, swimming, and just plain relaxing. When we enter the forest, we enter something ancient, primordial, and it touches something ancient and primordial in ourselves.

Moreover, as scientists have learned with ever-growing depth and breadth, forests are essential to the health of our planet and of the thousands of plant and animal species that make their homes here. Forests are Earth's lungs, storing carbon dioxide and releasing the oxygen that all living things require, and they are the planet's refuges, providing food and water and shelter for thousands of species.

For these reasons, the story of the restoration of America's eastern national forests since 1900 has been nothing short of extraordinary. In the early twentieth century, the eastern forests were largely cut over and blackened by massive forest fires. Over the one hundred years since the passage of the Weeks Act, these very same forests have regenerated themselves as nature has worked its remarkable restorative powers, aided by the actions of citizens and conservation leaders, like Gifford Pinchot, who collaborated and compromised to formulate practical laws and apply science-based strategies that set the stage for this remarkable recovery. Today, Region 9 of the U.S. Forest Service, which encompasses the East and the Lake states, boasts seventeen national forests. Region 8, the Southern Region, features another thirty-five national forests. Together, the national forests of the East, the South, and the Lake states comprise some 45.1 million acres of forestland located in twenty-six states.[2] Most of these forests are a legacy of the Weeks Act, the Clarke-McNary Act, and other laws passed in the early twentieth century.

A few numbers will capture the scope of eastern forest restoration. In 1630, as European-American settlement was beginning to transform the landscape, North America was blessed with 1,037 million acres of forestland, which constituted approximately 46 percent of the total land area. In the North, forests covered approximately 72 percent of the land, but by 1907, the amount of forest had been reduced to 32 percent of the land. In the South, forests constituted 66 percent of the land in 1630, but by 1907, that number had dropped to 46 percent.[3]

Since forest conservation started in earnest in the early twentieth century, however, the forests of the East have rebounded so that they now cover 42 percent of the land area. In the South, urbanization since World War II has continued to remove forestland, and forests now constitute some 40 percent of the land. According to a recent Forest Service study, *Forest Resources of the United States, 2007*, though, the decline in the South "could have been much greater, but millions of acres have been planted under various Federal- and State-sponsored incentive programs that encourage tree planting."[4] Forest restoration has been particularly intensive in recent years, with the amount of forestland increasing by 4 percent nationwide since 1987.[5]

The United States has also placed large amounts of forestland in preserves, meaning that the lands are not used for timber harvesting, mining, or grazing. Such preserves include wilderness areas, national parks, and national monuments. As of 2007, outside Alaska, the United States had about 41.5 million acres of preserved forestland. The majority of these preserves—31.7 million acres—are in the West. The eastern national forests, however, include approximately 9.4 million acres of preserved forestlands, with about 6 million acres in the Northeast and Lake states and another 3.4 million acres in the South. In addition, nongovernmental organizations, such as The Nature Conservancy, own large amounts of forestland that is managed for biodiversity and nontimber uses. The amount of preserved forestland has tripled since 1953 and now constitutes about 10 percent of all forests. Roadless areas represent an additional 32 million acres of public lands, most of which will probably never be harvested for timber. Roadless areas make up about 31 percent of the lands in the National Forest System.[6]

Gifford Pinchot articulated two key principles that have guided the Forest Service's stewardship of the national forests: (1) management of the forests for "the greatest good, for the greatest number, for the longest time" and (2) using the forests for multiple uses. The principles have served as the polar stars for those who have navigated the direction of the national forests, and the federal government, state governments, and many private timberland owners now manage forestlands in sustainable ways. In 1900, the United States had fifty million to eighty million acres of cutover lands, and forest fires were consuming twenty million to fifty million acres a year. Today, fires burn far less acreage, approximately two million to ten million acres a year.[7] Lands that were once cut over now support second- and third-growth forests that have been economically productive for decades. Some of these forests have been harvested as many as three times since 1900.[8] Government agencies have managed the eastern national forests on a sustainable basis and have disseminated knowledge and methods of forestry that have also improved the management of private timberlands.

Because of sustainable methods of forestry, net growth of trees in the United States now exceeds net removal by approximately 70 percent. Today, U.S. forests contain 50 percent more biomass—the trunks, forks, branches, limbs, and tops of trees—per acre than they did in 1953. In the East, biomass per acre has increased by almost 100 percent. Moreover, the annual growth of trees in forests is about four times greater now than it was in 1920, which is a testament not only to sustainable forestry methods but to vastly improved control of wildfires. In addition, the production of wood products has become much more efficient. Sawmills typically generate wood products at a rate that is two to three times greater per log than it was in 1900. Much less wood is left on the floor of a forest, and with modern engineering and architectural techniques, builders use lower amounts of wood per square foot than they did a hundred years ago. In addition, the post–World War II era has seen the development of certification programs by forestry associations and environmental organizations, which regularly assess the management of private and public forests against standards of forest sustainability.[9]

The uses of the national forests have undergone a marked shift from commodity production to recreation and habitat protection, a phenomenon reflected in the case studies in this book. In 1977, national forests represented 19 percent of all timberland and supplied 17 percent of the timber in the country. By 2007, the national forests supplied only 2 percent of the country's timber, reflecting a fundamental shift in management priorities and public demands for recreation and protection of wildlife, fish, and vegetation.[10] The account of timber harvesting in the Holly Springs National Forest examined in chapter 8 reflected this shift, with the Forest Service adapting its management in response to citizens' input. Private timberland owners in the United States and timber suppliers in Canada and other countries now provide a greater percentage of the country's timber needs. In 2006, private timberland owners supplied approximately fifteen billion cubic feet of timber and pulp, whereas national forests provided less than one billion cubic feet.[11]

Wildlife management and ecosystem protection are greater priorities for the eastern national forests than they once were, and this shift is partly a legacy of the scientific knowledge about forests that has accumulated since the passage of the Weeks Act. During the debates over that law, pioneering conservationists like Philip Ayres argued that forests were essential to protecting watersheds. His insights and those of other early conservationists have been vindicated. Today, water from lakes, rivers, streams, and aquifers within national forests serves approximately sixty million Americans. In 2008, James Burchfield and Martin Nie, who are on the faculty of the College of Forestry and Conservation at the University of Montana, issued *National Forests Policy Assessment*, a comprehensive review of policies regarding national forests. They wrote, "In addition to consuming water for their own growth, forests create above-and-below-ground conditions for water storage, circulate water vapor back to the atmosphere, and regulate the amount and timing of water yield."[12]

Equally critical is the role of the eastern national forests in protecting the biodiversity of America's ecosystems. In *The Lands Nobody Wanted: The Legacy of the Eastern National Forests*, forest historian William E. Shands wrote, "Thanks to the natural resiliency of eastern forests and Forest Service stewardship, the land again supports stands of trees and diverse wildlife."[13] As chapter 11 explained, remote wildernesses on national forests have helped protect wildlife, such as the gray wolf, that do not thrive near human populations. The wolf recovery success explained in that chapter has been echoed in myriad other efforts to protect wildlife by preserving old growth and protecting riparian areas and wetlands. The Forest Service has undertaken efforts to manage and protect habitat to encourage populations of the red-cockaded woodpecker, the bald eagle, the northern spotted owl, the black-footed ferret, the grizzly bear, and other threatened and endangered species.[14]

A further benefit of the restored eastern national forests is their pivotal role in mitigating global warming. Trees sequester carbon, which is an important factor in the emerging market in carbon credits to reduce greenhouse gases. The Forest Service, which is charged with estimating the amount of carbon sequestration in forests, uses its forest inventory and analysis program to conduct inventories and provide the required data. According to the agency, forests in the East and Lake states store about 175 million metric tons of carbon dioxide (CO_2) a year, and those in the South sequester almost 250 million metric tons of CO_2. Forests in the West sequester about 320 million metric tons of CO_2.[15] Without this sequestration, the impact of greenhouse gases on the earth would be significantly more pronounced than it already is.

No account of the benefits of the eastern national forests would be complete without highlighting their importance in meeting people's recreational demands. The Forest Service monitors recreation use, and for the years 2005 through 2009, it reported that annual visits to national forests in the South exceeded 28 million. Most heavily used were the four national forests of North Carolina, which received more than 7.5 million visits.[16] Region 9, which encompasses the East and Lake states, reported more than 15 million annual visits during the same period. The Huron-Manistee National Forests in Michigan recorded more than 4 million visits, and the Green Mountain and Finger Lakes National Forests experienced more than 2.8 million visits.[17] The most common recreational activities were viewing natural features, hiking, observing wildlife, pleasure driving, fishing, hunting, picnicking, camping, studying nature, motorized water activity, visiting nature centers, touring historic sites, and bicycling.

As the eastern national forests enter their second century of existence, they face daunting challenges, some of which are continuations of old problems and some of which are emerging because of changing ecological and economic conditions. Seven of these challenges stand out. Certainly one of the most pressing is that of ongoing climate change. The Intergovernmental Panel on Climate Change and the U.S. Climate Change Science Program have predicted that during the twenty-first century, temperatures in North America will rise between 3.5 degrees and 7.2 degrees Fahrenheit.[18] This warming trend will cause longer and more intense heat waves and higher temperatures in both days and nights, with significant effects on vegetation, wildlife, and forest succession. The country can also expect to experience more extreme weather events such as floods, droughts, and windstorms as well as more intense wildfires. The Oregon Forest Resource Institute has identified three forest-management strategies to counter climate change: mitigation, adaptation, and conservation. These strategies all share the goal of increasing carbon sequestration by expanding forests, protecting ecosystems, and using products from wood fiber to store carbon.

The second challenge is growing population density and development of areas near forests. As chapter 14 explained, this phenomenon complicates the job of fighting fires, which need to be put out to save lives and property. In addition, as that chapter also made clear, development near national forests fragments habitats for wildlife and vegetation and reduces the "linkage zones," the protected areas that allow animals to migrate safely.[19] Development also leads to increasingly problematic human-wildlife confrontations, as is evidenced in the problems between wolves and humans in part of the Michigan's Upper Peninsula.

The third problem is that of increasing wildfires, a phenomenon related to climate change and economic development near national forests. In recent years, wildfires have grown more destructive because of droughts, insects, disease, invasive plants such as cheat grass, and the tighter density of forests, which contain more fuel per acre. From 1997 through 2006, wildfires burned approximately six million acres a year on forests in the United States, but from 2004 through 2006, the acreage burned exceeded eight million, reflecting an alarming upward trend.[20] In 2007, the Forest Service spent $1.8 billion to fight forest fires.[21] Development near national forests exacerbates the problem of wildfires because firefighters have a more difficult time reducing fuel through prescribed burns near residential areas. Community wildfire protection plans have been developed to improve fire protection near population centers, but the funding has not been commensurate with the dangers. In addition, because of climate change, winters are shorter and snow cover is reduced. As a result, forests have less moisture and burn hotter.

The fourth problem is the threat posed by invasive species and diseases. Climate change is allowing certain insects, such as the pine bark beetle, to live longer than it did previously. This insect in particular has caused extensive damage in forests in the West. In the East, the emerald ash borer, the hemlock wooly adelgid, the Asian long-horned beetle, and the gypsy moth are all serious threats to the forests. Biological controls show promise, but the Forest Service faces enormous pressures—and serious budget constraints—to develop new solutions to deal with the problem of alien invasive species. Invasive vegetation is just as much of a problem. Kudzu, for example, has spread rapidly throughout the South, and Asian bittersweet has caused similar problems in the northern states. Controlling invasive vegetation costs billions of dollars each year. Burchfield and Nie cited the lack of "an effective national early warning system" as a major problem in identifying alien species early enough to prevent their spread.[22]

Fifth, water quality in the eastern national forests faces significant dangers from acid rain, air pollution, drainage from mining activities, and sedimentation from human activities such as road building. As chapter 12 explained, oil and natural-gas extraction and the projected increase in hydraulic fracturing use enormous amounts of water, and neither industry nor government has yet to develop ways of ensuring that the water recycled back into forests is free of contamination from chemicals.

The sixth challenge is the growing recreational demands placed on the eastern national forests because of the proximity of those forests to large metropolitan areas. The most contentious issues surround motorized vehicles within national forests. Snowmobilers and users of off-road vehicles view their activities as legitimate uses of the national forests, whereas opponents fume at the resulting ground disturbance, noise, and pollution. The Forest Service is writing plans for motorized use into forest plans to try to improve the regulation of these vehicles, while wilderness advocates argue for more wilderness because these lands are off-limits to motorized vehicles. Maintaining trails for motorized use is also a significant cost for the Forest Service.

The seventh challenge raises a specter that could significantly impede the Forest Service and state forest agencies in their efforts to maintain the quality of our forests, and that challenge is budgetary. In *National Forests Policy Assessment*, Burchfield and Nie wrote, "The U.S. should invest in our National Forests and the agency responsible for managing them. The USFS must be funded at responsible levels in the future."[23] The authors pointed out that the Forest Service has been required to meet many more regulations, such as compliance with the National Environmental Protection Act (NEPA) standards, without receiving the commensurate staffing or budget. Without adequate personnel, the Forest Service has at times lacked the resources to develop strategies in a timely fashion to meet NEPA requirements. It seems unfair to aim criticism at the Forest Service, however, because the agency has not received the additional funding to meet these environmental requirements. In addition, the Forest Service often pays for forest restoration projects through timber-harvesting programs. Burchfield and Nie believe that the U.S. Congress should allocate funds to pay for restoration projects, which could relieve some of the controversy over timber-harvesting programs.[24] In addition, because of the growing environmental pressures on forests, the Forest Service must conduct more monitoring programs, but such programs often face the budget axe, which inhibits the agency from being as proactive as it could be. Another area that requires more funding is the supervision of recreational activities and enforcement of laws and regulations in the national forests, a need that inevitably accompanies increased recreational use of the forests. To counter development near national forests and other forestlands, increased funding of the Forest Legacy Program would expand the amount of green space that is preserved from development.[25]

As the eastern national forests face these challenges in their second century, the Forest Service, citizen-conservationists, environmental organizations, and private industry face the same questions that faced conservationists a century ago. What is the greatest good of the forests for the greatest number of people in the long run? How much resource extraction is adequate, and what role should public forestlands play in supplying the country with timber, oil, and natural gas? How much recreation and wilderness are enough? Addressing these issues effectively requires greater environmental literacy—accomplished through education and responsible media coverage—so that citizens have a better understanding of the complexities of land management. As the evolution of eastern national forests has shown, practical solutions emerge most productively in an environment that values collaboration, compromise, and respect for scientific knowledge. Those values have been key ingredients in the very real progress that part II of this book has reported, changes that resulted in a modified timber-harvesting policy in Holly Springs National Forest, the development of ecologically beneficial fire policies in Florida's national forests, wilderness protection in Minnesota and West Virginia, the recovery of the wolf population in Michigan, and the development of strategies to counteract the emerald ash borer throughout the Lake states and the Northeast. The issue of oil and natural-gas extraction in the Allegheny National Forest remains controversial, but if the past is any indication, the data-gathering process will unfold, and the resulting information and inclusion of the full spectrum of opinions will yield solutions to protect the integrity of that beautiful forest while meeting domestic energy demands. Similarly, researchers are just beginning to study the effect of economic development on forest fragmentation and parcelization. This growing body of knowledge will inevitably help shape the decisions of professional forest managers.

Gifford Pinchot was a visionary who poured his love of forests into one of the great achievements in American environmental history: the restoration of the country's eastern national forests to provide timber and other resources for an expanding nation, protect thousands of valuable species of flora and fauna, and extend healthful recreation for millions of people. The forests face challenges that are, without question, cause for great concern, yet a philosophical foundation is firmly in place to guide the United States in meeting those challenges in innovative and productive ways. Pinchot's advice—to consider the best uses of the forests for the most people in the long run—is as wise today as when he served as the first chief forester of the Forest Service. If we heed his words, the eastern national forests will thrive far into the future and continue their rightful role in helping protect the health of the planet.

NOTES

Introduction

1. Benton MacKaye, "Our White Mountain Trip: Its Organization and Methods," in *Log of Camp Mossilauke, 1904* (Wentworth, NH: Camp Moosilauke, 1904), 11.

2. James Fickle, *Mississippi Forests and Forestry* (Jackson: University Press of Mississippi, 2001), 244–46.

3. William E. Shands and Robert G. Healy, *The Lands Nobody Wanted: A Conservation Foundation Report* (Washington, DC: The Conservation Foundation, 1977), 1.

4. "Weeks Law Lands in the National Forest System," *USDA Forest Service Briefing Paper* (March 2, 2011).

5. Fickle, *Mississippi Forests*, 244.

6. Max Oelschlaeger, *The Idea of Wilderness* (New Haven: Yale University Press, 1991), 105.

7. Alexis de Tocqueville, *Democracy in America*, vol. 2 (New York: Vintage Books, 1945), 115.

Chapter 1

1. William Cronon, *Changes in the Land: Indians, Colonists, and the Ecology of New England* (New York: Hill and Wang, 2003), 35.

2. Frederick W. Kilbourne, *Chronicles of the White Mountains* (Boston: Houghton Mifflin, 1916), 379.

3. New Hampshire Forestry Commission, *Report of the Forestry Commission of New Hampshire, January Session, 1891* (Manchester: John B. Clarke, 1891), 29.

4. C. Francis Belcher, *Logging Railroads of the White Mountains* (Boston: Appalachian Mountain Club Books, 1980), 84.

5. Bill Gove, *J. E. Henry's Logging Railroads* (Littleton, NH: Bondcliff Books, 1998), 5.

6. Ibid., 27.

7. Ibid., 20–22.

8. Ibid., 14.

9. Ibid., 27–28.

10. Ibid., 8, 27–28.

11. D. W. MacCleery, *American Forests: A History of Resiliency and Recovery* (Durham, NC: Forest History Society, 2002), 58.

12. Alfred K. Chittenden, *Forest Conditions of Northern New Hampshire (Bureau of Forestry Bulletin No. 55)* (Washington, DC: US Department of Agriculture, 1905), 62, 70.

13. Ibid., 62.

14. Gove, *J. E. Henry's Logging Railroads*, 107.

15. Ernest Russell, "The Wood-Butchers," *Colliers,* May 9, 1909, 20.

16. Gove, *J. E. Henry's Logging Railroads*, vii.

17. Julius H. Ward, "White Mountain Forests in Peril," *Atlantic Monthly*, February 1893, 249–50.

18. Kilbourne, *Chronicles*, 380.

19. Philip W. Ayres, *Commercial Importance of the White Mountain Forests* (Washington, DC: U.S. Department of Agriculture, 1909), 6.

20. New Hampshire Forestry Commission, *Report*, 12.

21. Chittenden, *Forest Conditions*, 70.

22. Kilbourne, *Chronicles*, 381–82.

23. Chittenden, *Forest Conditions*, 66.

24. "10,000 People Are Idle," *North Adams (Massachusetts) Transcript*, March 3, 1896.

25. Thomas Cole, quoted in Louis Legrand Noble, *The Life and Works of Thomas Cole*, ed. Elliot S. Vesell (Hensonville, NY: Black Dome Press, 1997), 65.

26. Moses Sweetser, *Chisholm's White Mountain Guide Book* (Portland, ME: Chisholm Brothers, 1898), 16.

27. Gifford Pinchot, *Breaking New Ground* (Washington, DC: Island Press, 1998), 5, 91.

28. Francis Parkman, "The Forests of the White Mountains," *Garden and Forest* 1, no. 1 (February 29, 1888): 2.

29. *Daily Record and American*, quoted in Charles Sprague Sargent, "Destruction of Forests in New Hampshire," *Garden and Forest* 2, no. 52 (February 20, 1889): 86.

30. New Hampshire Forestry Commission, *Report*, 20.

31. Ibid., 8.

32. Ward, "White Mountain Forests in Peril," 248.

33. David Govatski, "The Weeks Act and the Creation of the White Mountain National Forest," http://whitemountainhistory.org/Weeks_Act.html, accessed April 7, 2012.

34. Iris Baird, *Looking Out for Our Forests: The Evolution of a Plan to Protect New Hampshire's Woodlands from Fire* (Lancaster, NH: Baird Backwoods Construction Publications, 2005), 2.

35. J. E. Henry, quoted in Ernest Russell, "The Wood-Butchers," *Collier's*, May 9, 1909, 19.

36. New Hampshire Forestry Commission, *Report*, 15.

Chapter 2

1. Ben Eastman, *Congressional Globe, 1851–1852*, quoted in Timothy Bawden, "The Northwoods: Back to Nature?" in *Wisconsin Land and Life*, ed. Robert C. Ostergren and Thomas R. Vale (Madison: University of Wisconsin Press, 1997), 454.

2. A. R. Reynolds, *The Daniel Shaw Lumber Company: A Case Study of the Wisconsin Lumbering Frontier* (New York: New York University Press, 1957), 3–7.

3. Agnes Larson, *History of the White Pine Industry in Minnesota* (Minneapolis: University of Minnesota Press, 1949), 5.

4. Reynolds, *The Daniel Shaw Lumber Company*, 4–13.

5. William Gerald Rector, *Log Transportation in the Lake States Lumber Industry, 1840–1918* (Glendale, CA: Arthur H. Clark, 1953), 59.

6. Donald I. Dickmann and Larry A. Leefers, *The Forests of Michigan* (Ann Arbor: University of Michigan Press, 2003), 120.

7. Timothy Bawden, "The Northwoods: Back to Nature?" in *Wisconsin Land and Life*, ed. Robert C. Ostergren and Thomas R. Vale (Madison: University of Wisconsin Press, 1997), 451.

8. Rector, *Log Transportation*, 51–52.

9. Larson, *History*, 6.

10. Rector, *Log Transportation*, 48.

11. Charles E. Brown, *Paul Bunyan Classics: Authentic Original Stories Told in the Old Time Logging Camps of the Wisconsin Pineries* (Madison: Wisconsin Folklore Society, 1945), 3.

12. Dickmann and Leefers, *The Forests of Michigan*, 119.

13. Ibid., 119–24.

14. Ibid., 121.

15. *Proceedings of the Forestry Convention, Held in Grand Rapids, Michigan, January 26 and 27, 1888, Under the Auspices of the Independent Forestry Commission* (Lansing: Department of Botany and Forestry, Agricultural College, Michigan, 1888), 4.

16. Dickmann and Leefers, *The Forests of Michigan*, 142–46.

17. Bawden, "The Northwoods," 453.

18. Rector, *Log Transportation*, 43–44.

19. Reynolds, *The Daniel Shaw Lumber Company*, 65.

20. Ralph W. Hidy, Frank Ernest Hill, and Allan Nevins, *Timber and Men: The Weyerhaeuser Story* (New York: Macmillan, 1963), 5–7.

21. Hidy, Hill, and Nevins, *Timber and Men*, 50, 62.

22. Reynolds, *The Daniel Shaw Lumber Company*, 1, 7–18.

23. Larson, *History*, 13–15.

24. Clifford Ahlgren and Isabel Ahlgren, *Lob Trees in the Wilderness* (Minneapolis: University of Minnesota Press, 1984), 101.

25. Larson, *History*, 53–57.

26. Ibid., 39.

27. Ahlgren and Ahlgren, *Lob Trees in the Wilderness*, 99.

28. Larson, *History*, 75.

29. Ibid., 85.

30. Ibid., 51.

31. I. A. Lapham, "The Forest Trees of Wisconsin," in *Transactions of the Wisconsin State Agricultural Society for the Years 1854-5-6-7* (Madison: Atwood and Rublee, Printers, 1857), 203.

32. Clifford Ahlgren and Isabel Ahlgren, "The Human Impact," in *The Great Lakes Forest: An Environmental and Social History*, ed. Susan L. Flader (Minneapolis: University of Minnesota Press, 1983), 40.

33. Dickmann and Leefers, *The Forests of Michigan*, 153.

34. Ibid., 152.

35. Arthur Hill, "The Greatest Enemy of the Forest Reserve: Fire," in *Report of the Michigan Forestry Commission for the Years 1903–4* (Lansing: Michigan Forestry Commission, 1905), quoted in Donald I. Dickmann and Larry A. Leefers, *The Forests of Michigan* (Ann Arbor: University of Michigan Press, 2003), 152.

36. Dickmann and Leefers, *The Forests of Michigan*, 149–64.

37. David M. Gates, C. H. D. Clarke, and James T. Harris, "Wildlife in a Changing Environment," in *The Great Lakes Forest: An Environmental and Social History*, ed. Susan L. Flader (Minneapolis: University of Minnesota Press, 1983), 62–71.

38. William Henry, quoted in Timothy Bawden, "The Northwoods: Back to Nature?" in *Wisconsin Land and Life*, ed. Robert C. Ostergren and Thomas R. Vale (Madison: University of Wisconsin Press, 1997), 455.

39. Bawden, "The Northwoods," 457–58.

40. Benton MacKaye, "Powell as Unsung Lawgiver," 3, MacKaye Family Papers, Dartmouth College Library, quoted in Larry Anderson, *Benton MacKaye: Conservationist, Planner, and Creator of the Appalachian Trail* (Baltimore: Johns Hopkins University Press, 2002), 82.

41. Bawden, "The Northwoods," 460.

42. Ahlgren and Ahlgren, *Lob Trees in the Wilderness*, 107–8.

43. Dickmann and Leefers, *The Forests of Michigan*, 169.

44. Filibert Roth, *Forestry Condition and Interests of Wisconsin (Bulletin 16)* (Washington, DC: Government Printing Office, 1910), 22, 25, 30.

45. Bernard E. Fernow, introduction to *Forestry Condition and Interests of Wisconsin (Bulletin 16)*, by Filibert Roth (Washington, DC: Government Printing Office, 1910), 10.

46. Roth, *Forestry Condition*, 27–28.

47. Ibid., 48–49.

48. Newell Searle, "Minnesota State Forestry Association," *Minnesota History* (Spring 1974): 16–19.

49. Ahlgren and Ahlgren, *Lob Trees in the Wilderness*, 101–3.

50. *Proceedings of the Forestry Convention*, 4.

51. Gifford Pinchot, *Breaking New Ground* (Washington, DC: Island Press, 1998), 26.

Chapter 3

1. Florence Cope Bush, *Dorie: Woman of the Mountains* (Knoxville: University of Tennessee Press, 1992), 31–32.

2. Ibid., 84.

3. Ibid., 132.

4. Ibid., 133.

5. Ibid., 134–35.

6. Horace Kephart, *Our Southern Highlanders* (New York: Macmillan, 1922), 51.

7. Diane Flaugh (program manager, Great Smoky Mountains National Culture Resources), telephone interview by Christopher Johnson, January 29, 2009.

8. Ronald L. Lewis, *Transforming the Appalachian Countryside: Railroads, Deforestation, and Social Change in West Virginia, 1880–1920* (Chapel Hill: University of North Carolina Press, 1998), 3.

9. Albert E. Cowdrey, *This Land, This South: An Environmental History* (Lexington: University Press of Kentucky, 1996), 112.

10. Bush, *Dorie*, 131.

11. Robert S. Lambert, "Logging the Great Smokies," *Tennessee Historical Quarterly* 20, no. 4 (December 1961): 354.

12. Cowdrey, *This Land, This South*, 120.

13. Lewis, *Transforming the Appalachian Countryside*, 50.

14. Daniel S. Pierce, *The Great Smokies: From Natural Habitat to National Park* (Knoxville: University of Tennessee Press, 2000), 25.

15. Ronald D. Eller, *Miners, Millhands, and Mountaineers: Industrialization of the Appalachian South, 1880–1930* (Knoxville: University of Tennessee Press, 1982), 91–92.

16. Lambert, "Logging," 352–53.

17. Eller, *Miners, Millhands, and Mountaineers*, 90–91.

18. Ibid., 104.

19. Lewis, *Transforming the Appalachian Countryside*, 81–82.

20. Dudley W. Crawford, "The Coming of the Railroad to Asheville 70 Years Ago," *Asheville Citizen*, October 29, 1950, quoted in Ronald D. Eller, *Miners, Millhands, and Mountaineers: Industrialization of the Appalachian South, 1880–1930* (Knoxville: University of Tennessee Press, 1982), 99.

21. Eller, *Miners, Millhands, and Mountaineers*, 101.

22. William G. Robbins, *American Forestry: A History of National, State, and Private Cooperation* (Lincoln: University of Nebraska Press, 1985), 21.

23. A. E. Brown, quoted in "Appalachian Forests: Effects of Inroads of Lumbermen Upon Them," *Manufacturers' Record* (January 20, 1910): 52.

24. Eller, *Miners, Millhands, and Mountaineers*, 95–98.

25. Lewis, *Transforming the Appalachian Countryside*, 143.

26. Eller, *Miners, Millhands, and Mountaineers*, 104.

27. William McLellan Ritter, "Early Days in West Virginia," *Southern Lumberman* 167 (December 15, 1945), quoted in Ronald L. Lewis, *Transforming the Appalachian Countryside: Railroads, Deforestation, and Social Change in West Virginia, 1880–1920* (Chapel Hill: University of North Carolina Press, 1998), 91.

28. Eller, *Miners, Millhands, and Mountaineers*, 104–5.

29. Clarence O. Vance, quoted in Stephen Wallace Taylor, *The New South's New Frontier: A Social History of Economic Development in Southwestern North Carolina* (Gainesville: University Press of Florida, 2001), 29.

30. Ibid., 30.

31. Stephen Wallace Taylor, *The New South's New Frontier: A Social History of Economic Development in Southwestern North Carolina* (Gainesville: University Press of Florida, 2001), 32.

32. Lewis, *Transforming the Appalachian Countryside*, 181.

33. Emory Wriston, quoted in Ronald L. Lewis, *Transforming the Appalachian Countryside: Railroads, Deforestation, and Social Change in West Virginia, 1880–1920* (Chapel Hill: University of North Carolina Press, 1998), 264.

34. Theodore Roosevelt, *Message from the President of the United States Transmitting a Report of the Secretary of Agriculture in Relation to the Forests, Rivers, and Mountains of the Southern Appalachian Region* (Washington, DC: Government Printing Office, 1902), 13.

35. Ibid., 14.

36. H. B. Ayres and W. W. Ashe, "Appendix: Forests and Forest Conditions in the Southern Appalachians," in Theodore Roosevelt, *Message from the President of*

the United States Transmitting a Report of the Secretary of Agriculture in Relation to the Forests, Rivers, and Mountains of the Southern Appalachian Region (Washington, DC: Government Printing Office, 1902), 45.

37. Roosevelt, *Message from the President*, 25.

38. Ayres and Ashe, "Appendix: Forests and Forest Conditions," 57.

39. Ibid., 56.

40. Donald E. Davis, *Where There Are Mountains: An Environmental History of the Southern Appalachians* (Athens: University of Georgia Press), 169.

41. Lewis, *Transforming the Appalachian Countryside*, 265.

42. Roosevelt, *Message from the President*, 38.

43. Cowdrey, *This Land, This South*, 129.

44. Ayres and Ashe, "Appendix: Forests and Forest Conditions," 58.

45. Lewis, *Transforming the Appalachian Countryside*, 277–78.

46. Roosevelt, *Message from the President*, 32–33.

47. Bush, *Dorie*, 194.

Chapter 4

1. Edward Everett Hale, *The Life and Letters of Edward Everett Hale*, vol. 2 (Boston: Little Brown, 1917), 393.

2. Theodore Roosevelt, "First Annual Message to the Senate and the House of Representatives, *December 3, 1901, Roosevelt State Papers as Governor and President* (New York: Scribner, 1925), 118.

3. David Lowenthal, "Introduction," in *Man and Nature*, by George Perkins Marsh (Seattle: University of Washington Press, 2003), xviii–xix.

4. George Perkins Marsh, *Man and Nature* (Seattle: University of Washington Press, 2003), 36.

5. Ibid., 79.

6. Ibid., 171.

7. Ibid., 186–87.

8. Ibid., 202.

9. Henry Clepper, *Origins of American Conservation* (New York: Ronald Press, 1966), 9–12.

10. Samuel P. Hays, *Conservation and the Gospel of Efficiency: The Progressive Conservation Movement, 1890–1920* (Cambridge: Harvard University Press, 1959), 28.

11. Harold K. Steen, "The Beginning of the National Forest System," in *American Forests: Nature, Culture, and Politics*, ed. Char Miller (Lawrence: University Press of Kansas, 1997), 60.

12. Gifford Pinchot, *Breaking New Ground* (Washington, DC: Island Press, 1998), 107–8.

13. William E. Shands and Robert G. Healy, *The Lands Nobody Wanted: A Conservation Foundation Report* (Washington, DC: The Conservation Foundation, 1977), 11.

14. Pinchot, *Breaking New Ground*, 2.

15. Charles Sprague Sargent, "Mr. Vanderbilt's Forest," *Garden and Forest* 8 (December 4, 1895): 481.

16. Pinchot, *Breaking New Ground*, 135–36.

17. Edmund Morris, *The Rise of Theodore Roosevelt* (New York: Modern Library, 1979), 388.

18. Pinchot, *Breaking New Ground*, 254–56.

19. James Wilson to Gifford Pinchot, February 1, 1905, U.S. Forest Service History, "Wilson Letter, 1905," 4.

20. Pinchot, *Breaking New Ground*, 264–65.

21. A. E. Brown, *Manufacturer's* Record 57 (January 20, 1910), quoted in Ronald D. Eller, *Miners, Millhands, and Mountaineers: Industrialization of the Appalachian South, 1880–1930* (Knoxville: University of Tennessee Press, 1982), 110–11.

22. Locke Craig, quoted in Daniel Pierce, *The Great Smokies: From Natural Habitat to National Park* (Knoxville: University of Tennessee Press, 2000), 41.

23. Theodore Roosevelt, *Message from the President of the United States Transmitting a Report of the Secretary of Agriculture in Relation to the Forests, Rivers, and Mountains of the Southern Appalachian Region* (Washington, DC: Government Printing Office, 1902), 2.

24. Julius J. Ward, "White Mountain Forests in Peril," *Atlantic Monthly*, February 1893, 251.

25. Ibid., 254.

26. Marcia Schmidt Blaine, "Public Forests: Joseph B. Walker, Philip Ayres, and the White Mountain National Forest," in *Beyond the Notches: Stories of Place in New Hampshire's North Country*, ed. Mike Dickerman and Kay Morgan (Littleton, NH: Bondcliff Books, 2011), 260.

27. Charles D. Smith, "The Mountain Lover Mourns: Origins of the Movement for a White Mountain National Forest 1880–1903," *New England Quarterly* 33, no. 1 (March 1961): 44.

28. Philip W. Ayres, "Philip Wheelock Ayres: Chronology" (unpublished manuscript from the Society for the Protection of New Hampshire Forests, Concord, undated), 4–7.

29. Smith, "Mountain Lover Mourns," 47.

30. Ibid., 51–52.

31. Ibid., 52.

32. Raleigh Barlowe, "Changing Land Use and Policies: The Lake States," in

The Great Lakes Forest: An Environmental and Social History, ed. Susan L. Flader (Minneapolis: University of Minnesota Press, in association with the Forest History Society, 1983), 161.

33. John T. Rich, *Messages of the Governors of Michigan* (Lansing: Michigan Historical Commission, 1925-27), quoted in William B. Botti and Michael D. Moore, *Michigan's State Forests: A Century of Stewardship* (East Lansing: Michigan State University Press, 2006), 4.

34. Timothy Bawden, "The Northwoods: Back to Nature?" in *Wisconsin Land and Life*, ed. Robert C. Ostergren and Thomas R. Vale (Madison: University of Wisconsin Press, 1997), 461–62.

35. Ibid., 462.

36. William Watts Folwell, *A History of Minnesota*, vol. 4 (St. Paul: Minnesota Historical Society, 1930), 387.

37. Ibid., 388.

38. Ibid., 393.

39. Tim Brady, "The Real Story of Chippewa National Forest," Minnesota Department of Natural Resources, http://www.dnr.state.mn.us/volunteer/nov dec04/chippewanf.html, 1, accessed November 7, 2010.

40. Ibid., 3.

41. Ibid., 2–3.

Chapter 5

1. Paul Schneider, *The Adirondacks: America's First Wilderness* (New York: Henry Holt, 1997), 227.

2. *Report of the Forestry Commission of New Hampshire, January Session, 1891* (Manchester: John B. Clarke, 1891), 8, 13, 20.

3. John Ise, *The United States Forest Policy* (New Haven: Yale University Press, 1923), 208–9.

4. Harold K. Steen, *The U.S. Forest Service: A History* (Durham, NC: Forest History Society; Seattle: University of Washington Press, 2004), 124.

5. Charles D. Smith, "Gentlemen, You Have My Scalp," *American Forests*, February 1962, 18.

6. Ise, *United States Forest Policy*, 209.

7. Smith, "Gentlemen," 18.

8. Philip Ayres, *Commercial Importance of the White Mountain Forests* (Washington, DC: US Department of Agriculture, 1909), 3.

9. Frederick Weyerhaueser, quoted in Philip Ayres, "Reasons for a National Forest Reservation in the White Mountains," *The Northern: Official Organ of the New Hampshire and Maine Federation of Women's Clubs* 1, no. 9 (November 1905): 292.

10. Philip Ayres, "National Forests in the Eastern Mountains: A Brief Account of the Weeks Law of 1911 and of Its Extension Pending in Congress, the McNary-Woodfull Bill" (unpublished manuscript, photocopy from the Society for the Protection of New Hampshire Forests, Concord, undated), 2.

11. Ayres, "National Forests," 3.

12. Smith, "Gentlemen," 19.

13. Joseph G. Cannon, quoted in Charles D. Smith, "Gentlemen, You Have My Scalp," *American Forests*, February 1962, 18–19.

14. Booth Mooney, *Mr. Speaker* (Chicago: Follett, 1964), 94.

15. L. White Busbey, *Uncle Joe Cannon: The Story of a Pioneer American* (New York: Holt, 1927), iii.

16. Mooney, *Mr. Speaker*, 101.

17. Busbey, *Uncle Joe Cannon*, 128–29.

18. Joseph G. Cannon, quoted in Blair Bolles, *Tyrant from Illinois: Uncle Joe Cannon's Experiment with Personal Power* (New York: Norton, 1950), 119.

19. Charles G. Washburn, *The Life of John W. Weeks* (Boston: Houghton Mifflin, 1928), 28–45.

20. John W. Weeks, quoted in Charles G. Washburn, *The Life of John W. Weeks* (Boston: Houghton Mifflin, 1928), 76.

21. Joseph G. Cannon, quoted in Charles G. Washburn, *The Life of John W. Weeks* (Boston: Houghton Mifflin, 1928), 77.

22. Ibid., 78.

23. C. R. McKim, quoted in William E. Shands, *The Lands Nobody Wanted: The Legacy of the Eastern National Forests* (Milford, PA: Pinchot Institute for Conservation, 1991), 13.

24. "Three Drowned," *Daily Courier* (Connellsville, PA), March 13, 1907, http://www.gendisasters.com/data1/pa/floods/pittsburgh-flood-mar1907.html, accessed February 11, 2009.

25. "Thousands Are Homeless," *Washington Post*, March 15, 1907, http://www.gendisasters.com/data1/pa/floods/pittsburgh-flood-mar1907.html, accessed February 11, 2009.

26. "Fire with Flood," *Washington Post*, March 16, 1907, http://www.gendisasters.com/data1/pa/floods/pittsburgh-flood-mar1907.html, accessed February 11, 2009.

27. "Thousands Are Homeless."

28. "The City a Scene of Desolation," *Wheeling Daily News*, March 16, 1907, http://wheeling.weirton.lib.wv.us/history/events/floods/1907/1907e.html, accessed February 11, 2009.

29. "An Explanation," *Wheeling Daily News*, March 16, 1907, http://wheeling

.weirton.lib.wv.us/history/events/floods/1907/1907e.html, accessed February 11, 2009.

30. Dave Saville, "Floods and Deforestation." *Mountain State Sierran: West Virginia Chapter Newsletter,* May 2004, http://westvirginia.sierraclub.org/newsletter /archives/2004/05/a_007.html, 1–2, accessed February 11, 2009.

31. H. A. Pressey and E. W. Myers, "Appendix C: Hydrology of the Southern Appalachians," in Theodore Roosevelt, *Message from the President of the United States Transmitting a Report of the Secretary of Agriculture in Relation to the Forests, Rivers, and Mountains of the Southern Appalachian Region* (Washington, DC: Government Printing Office, 1902), 129.

32. Frederick W. Kilbourne, *Chronicles of the White Mountains* (Boston: Houghton Mifflin, 1916), 392–93.

33. Theodore Roosevelt, "Opening Address by the President," *Proceedings of a Conference of Governors* (Washington, DC: Government Printing Office, 1909), 4, 6.

34. "Declaration," *Proceedings of a Conference of Governors* (Washington, DC: Government Printing Office, 1909), 193.

35. *Fifty Years of National Forest Protection and Development under the Weeks Law* (Washington, DC: Government Printing Office, undated), 3.

36. Ayres, "National Forests," 14.

37. Kilbourne, *Chronicles,* 394.

38. "Weeks Act Chronology," www.foresthistory.org/ASPNET/Policy/.

39. *Weeks Law of 1911,* Public Law 961, 62nd Congress, 1st session (1911). http://memory.loc.gov/cgi-bin/ampage?collId=amrvl&fileName=vl024//amrvlv l024.db&recNum=0&itemLink=r?ammem/AMALL:@field%28DOCID+@lit%28 jjh94-000053%29%29&linkText=0, 961, accessed September 12, 2012.

40. Steen, *U.S. Forest Service,* 126.

41. "The Battle for the Weeks Bill," *American Forestry* 16, no. 4 (March 1910): 133.

42. L. C. Glenn, "Forests as Factors in Stream Flow," *American Forestry* 16, no. 4 (March 1910): 217.

43. Willis S. Moore, quoted in Filibert Roth, "The Appalachian Forests and the Moore Report," *American Forestry* 16, no. 4 (March 1910): 213.

44. Filibert Roth, "The Appalachian Forests and the Moore Report," *American Forestry* 16, no. 4 (March 1910): 214.

45. Philip W. Ayres, "Philip Wheelock Ayres: Chronology" (unpublished manuscript; photocopy from the Society for the Protection of New Hampshire Forests, Concord, undated), 14.

46. W. B. Heyburn, quoted in "The Passage of the Appalachian Bill," *American Forests,* March 1911, 165.

47. "The 1910 Fires," *U.S. Forest Service History*, http://www.foresthistory.org/ASPNET/Policy/Fire/FamousFires/1910Fires.aspx, 1–4, accessed September 8, 2012.

48. Cannon, quoted in Smith, "Gentlemen," 19.

49. *Weeks Law of 1911*, Sections 1–13.

50. "Editorial: The Appalachian Bill," *American Forestry* 17, no. 3 (March 1911): 168.

51. Philip W. Ayres, quoted in Marcia Schmidt Blaine, "Public Forests: Joseph B. Walker, Philip Ayres, and the White Mountain National Forest," in *Beyond the Notches: Stories of Place in New Hampshire's North Country*, ed. Mike Dickerman and Kay Morgan (Littleton, NH: Bondcliff Books, 2011), 265.

Chapter 6

1. "Prompt Action Needed," *New York Times*, May 11, 1911, 10.

2. Ibid.

3. Samuel P. Hays, *Conservation and the Gospel of Efficiency: The Progressive Conservation Movement, 1890–1920* (Cambridge: Harvard University Press, 1959), 2.

4. "The Appalachian Forests: Putting the New Law into Operation," *American Forestry* 17, no. 5 (May 1911): 290–91.

5. Ibid., 292–93.

6. Ibid., 290–91.

7. Ibid., 288–90.

8. Ibid.

9. "The Appalachian Forests," *American Forestry* 17, no. 6 (June 1911): 364.

10. "Dr. Smith's Attitude on Forest Reserves," *New York Times*, June 22, 1911, 6.

11. Otis Smith, quoted in "Beneficial Effect of Forests on Stream Flow," *Municipal Engineering* 43, no. 1 (July 1912): 47.

12. David E. Conrad, *The Land We Cared For: A History of the Forest Service's Eastern Region* (Milwaukee: USDA Forest Service Region 9, 1997), 36.

13. David K. Rice, quoted in William E. Shands and Robert G. Healy, *The Lands Nobody Wanted: A Conservation Foundation Report* (Washington, DC: The Conservation Foundation, 1977), 16.

14. Thomas Cox, quoted in Shelley Smith Mastran and Nan Lowerre, *Mountaineers and Rangers: A History of Federal Forest Management in the Southern Appalachians, 1900–1981* (Washington, DC: US Department of Agriculture Forest Service, 1983), http://www.foresthistory.org/ASPNET/Publications/region/8/history/contents.aspx, 55, accessed February 2, 2011.

15. James Denman, quoted in Shelley Smith Mastran and Nan Lowerre, *Mountaineers and Rangers: A History of Federal Forest Management in the Southern*

Appalachians, 1900–1981 (Washington, DC: US Department of Agriculture Forest Service, 1983), http://www.foresthistory.org/ASPNET/Publications/region/8/history/index.aspx, 55, accessed February 2, 2011.

16. William L. Hall, "To Remake the Appalachians," *The World's Work*, July 1914, 328.

17. Donald Edward Davis, *Where There Are Mountains: An Environmental History of the Southern Appalachians* (Athens: University of Georgia Press, 2000), 172.

18. "Pisgah Forest Purchased," *American Forestry* 20, no. 6 (June 1914): 425.

19. William B. Greeley, quoted in "Reforesting Cut-Over Pinelands," *American Forestry* 18, no. 10 (October 1912): 674.

20. Inman Eldredge, quoted in Shelley Smith Mastran and Nan Lowerre, *Mountaineers and Rangers: A History of Federal Forest Management in the Southern Appalachians, 1900–1981* (Washington, DC: US Department of Agriculture Forest Service, 1983), http://www.foresthistory.org/ASPNET/Publications/region/8/history/index.aspx, 42, accessed February 2, 2011.

21. Verne Rhoades, quoted in Shelley Smith Mastran and Nan Lowerre, *Mountaineers and Rangers: A History of Federal Forest Management in the Southern Appalachians, 1900–1981* (Washington, DC: US Department of Agriculture Forest Service, 1983), http://www.foresthistory.org/ASPNET/Publications/region/8/history/index.aspx, 57, accessed February 2, 2011.

22. Shelley Smith Mastran and Nan Lowerre, *Mountaineers and Rangers: A History of Federal Forest Management in the Southern Appalachians, 1900–1981* (Washington, DC: US Department of Agriculture Forest Service, 1983), 60.

23. Philip W. Ayres, "New England's Federal Forest Reserve," *American Forestry* 21, no. 7 (July 1915): 804–6.

24. Ibid., 803.

25. "First Purchase of White Mountain Lands Under the Weeks Law," *American Forestry* 18, no. 2 (July 1912): 440.

26. "A White Mountain Purchase," *American Forestry* 20, no. 10 (October 1914): 733.

27. George E. Zink, "Who Is Kate Sleeper?" *Wonalancet Outdoor Club Newsletter*, April 1995, 3–5.

28. Douglas McVicar, "Who Is Tainter?" *Wonalancet Outdoor Club Newsletter*, May 1994, 3.

29. Conrad, *The Land We Cared For*, 46.

30. William L. Hall, "The White Mountain Forest and How It Is to Be Made Useful," *American Forestry* 19, no. 9 (September 1913): 620.

31. Ibid., 623.

32. Ibid.

33. William L. Hall, "Improving White Mountain Forests," *American Forestry* 21, no. 2 (February 1915): 117.

34. Ibid., 123.

35. Ibid., 123–26.

36. Joseph J. Jones, "Transforming the Cutover: The Establishment of National Forests in Northern Michigan," *Forest History Today*, Spring/Fall 2011, 48.

37. John A. Doelle, "The Land Riddle of the Lake States," *American Forests and Forest Life* 31, no. 381 (September 1925): 515–16.

38. Ibid., 516.

39. Conrad, *The Land We Cared For*, 59–60.

40. "Chippewa National Forest: History," USDA Forest Service, http://www.fs.usda.gov/wps/portal/fsinternet/!ut/p/c5/04_SB8K8Xll, accessed February 20, 2011.

41. Ayres, "New England's Federal Forest Reserve," 803.

42. Ibid., 806.

43. Sally K. Fairfax, Lauren Gwin, Mary Ann King, and Leigh Raymond, *Buying Nature: The Limits of Land Acquisition as a Conservation Strategy* (Cambridge: MIT Press, 2005), 89.

44. Ibid., 90.

45. William E. Shands and Robert G. Healy, *The Lands Nobody Wanted: A Conservation Foundation Report* (Washington, DC: The Conservation Foundation, 1977), 16.

46. Harold L. Steen, *The U.S. Forest Service: A History* (Durham, NC: Forest History Society; Seattle: University of Washington Press, 2004), 203.

47. Shands and Healy, *The Lands Nobody Wanted*, 16.

48. Conrad, *The Land We Cared For*, 70.

49. Ibid.

50. John A. Salmond, *The Civilian Conservation Corps, 1933–1942: A New Deal Case Study* (Durham, NC: Duke University Press, 1967), 35.

51. Leslie Alexander Lacy, T*he Soil Soldiers: The Civilian Conservation Corps in the Great Depression* (Radnor, PA: Chilton, 1976), 140–41.

Chapter 7

1. Joel Gardner, interview by Christopher Johnson, March 11, 2011, at Holly Springs National Forest District Office, Oxford, MS.

2. Ibid.

3. Ibid.

4. Caren Briscoe, interview by Christopher Johnson, March 11, 2011, at Holly Springs National Forest District Office, Oxford, MS.

5. "Weeding and Thinning Young Forest Stands," *Maine Forestry*, April 5, 2011, http://maineforestry.net/forestry%20items/thin_weed_forests.htm, accessed September 9, 2012.

6. Menominee Tribal Enterprises, *The Menominee Forest-Based Sustainable Development Tradition* (Keshena, WI: Sustainable Development Institute, 2003), 12.

7. Clemson University Department of Forest Resources, *Responses to Wildlife to Clearcutting and Associated Treatment in the Eastern United States*, Technical Paper No. 19, March 2000, http://www.clemson.edu/extfor/timber_production/fortp19 .htm, 1. Accessed September 9, 2012.

8. David E. Conrad, *The Land We Cared For: A History of the Forest Service's Eastern Region* (Milwaukee: USDA Forest Service, Region 9, 1997), 231–33.

9. Ibid., 142.

10. Ibid., 234–35.

11. Rives "Buddy" Lowery, interview by Christopher Johnson, March 11, 2011, at Holly Springs National Forest District Office, Oxford, MS.

12. Gardner, interview.

13. "Interest Growing in Cutting Back on Miss. Clear-Cuts," *Memphis Commercial Appeal*, January 2, 1999, http://nl.newsbank.com/nl-search/we/Archives?p _product=CA&p_theme=ca&p_action=search&p_maxdocs=200&s_trackval=CA &s_search_type=keyword&s_dispstring=Interest%20Growing%20in%20cutting %20back%20on%20miss.%20clear-cuts%20AND%20date%28all%29&p_field_ad vanced-0=&p_text_advanced-0=%28Interest%20Growing%20in%20cutting%20 back%20on%20miss.%20clear-cuts%29&xcal_numdocs=20&p_perpage=10&p _sort=YMD_date:D&xcal_useweights=no, accessed September 9, 2012.

14. Tom Charlier, "Chain Saws Buzzing in Miss. Forest Sound Like Massacre to Ex-Urbanites," *Memphis Commercial Appeal*, March 9, 1998, http://nlnewsbank .com/nl-search/we/Archives?p_action-doc&pOdoci, accessed March 21, 2011.

15. Gary Yeck, quoted in Tom Charlier, "Chain Saws Buzzing in Miss. Forest Sound Like Massacre to Ex-Urbanites," *Memphis Commercial Appeal*, March 9, 1998, http://nlnewsbank.com/nl-search/we/Archives?p_action-doc&pOdoci, accessed March 21, 2011.

16. Ibid.

17. Aldo Leopold, *A Sand County Almanac* (New York: Ballantine Books, 1970), 230.

18. Ann Phillippi, telephone interview by Christopher Johnson, March 28, 2011.

19. Ibid.

20. Charlier, "Chain Saws Buzzing."

21. Andy Mahler, telephone interview by Christopher Johnson, March 30, 2011.

22. Steven Higgs, *Eternal Vigilance: Nine Tales of Environmental Heroism in Indiana* (Bloomington: Indiana University Press, 1995), 11.

23. Mahler, interview.

24. Ibid.

25. Joe Glisson, telephone interview by Christopher Johnson, March 31, 2011.

26. Ibid.

27. *The South's Fourth Forest: Alternatives for the Future* (Washington, DC: USDA Forest Service, 1988), 113.

28. Brad W. Smith, Patrick D. Miles, Charles H. Perry, and Scott A. Pugh., *Forest Resources of the United States, 2007: A Technical Document Supporting the Forest Service 2010 RPA Assessment* (Washington, DC: US Department of Agriculture Forest Service, 2009), 179.

29. Ibid.

30. Ray Vaughan, telephone interview by Christopher Johnson, March 25, 2011.

31. "Interest Growing in Cutting Back on Miss. Clear-Cuts."

32. James E. Fickle, *Mississippi Forests and Forestry* (Jackson: University Press of Mississippi, 2001), 241.

33. "Interest Growing in Cutting Back on Miss. Clear-Cuts."

34. Fickle, *Mississippi Forests*, 241.

35. Ibid., 242–43.

36. Ibid., 242.

37. Greg Brown and Chuck Harris, "Professional Foresters and the Land Ethic, Revisited," *Journal of Forestry* 96, no. 1 (January 1998): 7.

38. Ibid., 7–11.

39. "Understanding the RPA," *Journal of Forestry* 89, no. 6 (June 1991): 17.

40. *2000 RPA Assessment of Forest and Range Lands* (Washington, DC: US Department of Agriculture Forest Service, 2000), 1.

41. Ibid., 1.

42. "Interest Growing in Cutting Back on Miss. Clear-Cuts."

43. Vaughan, interview.

44. Glisson, interview.

45. Mahler, interview.

46. United States Department of Agriculture Office of Communications, "Agriculture Secretary Vislack Presents National Vision for America's Forests," August 14, 2009, http://www.usda.gov/wps/portal/usda/usdahome?content idonly+true&c, accessed March 25, 2011.

47. Vaughan, interview.

48. Ibid.

49. Ibid.

50. *National Forests in Mississippi: Revised Land and Resource Management Plan, Proposed Action Description*, http://www.fs.usda.gov/Internet/FSE_DOCUMENTS/stelprdb5220872.pdf, accessed September 9, 2012.

Chapter 8

1. Steve Parrish, interview by Christopher Johnson, March 9, 2011, at Apalachicola National Forest District Office, Crawfordsville, FL.

2. Deirdre Dether, *Prescribed Fire Lessons Learned: Escape Prescribed Fire Reviews and Near Miss Incidents* (Tucson, AZ: Wildlife Fire Lessons Learned Center, 2005), http://www.wildfirelessons.net/documents/rx_fire_ll_escapes_review.pdf, 2, accessed September 10, 2012.

3. David Carle, *Burning Questions: America's Fight with Nature's Fire* (Westport, CT: Praeger, 2002), 38.

4. Cynthia Fowler and Evelyn Konipik, "The History of Fire in the Southern United States," *Human Ecology Review* 14, no. 2 (2007): 171.

5. Herman H. Chapman, "Forest Fires and Forestry in the Southern States," *American Forestry* 18, no. 8 (August 1912): 512.

6. Ibid.

7. Ibid., 516.

8. Herbert L. Stoddard, *The Bobwhite Quail: Its Habits, Preservation, and Increase* (New York: Scribner, 1946), 407.

9. Ibid., 407–8.

10. Carle, *Burning Questions*, 42.

11. Ibid., 45.

12. Stephen J. Pyne, *Fire in America: A Cultural History of Wildland and Rural Fire* (Princeton, NJ: Princeton University Press, 1982), 115.

13. E. L. Demmon, "The Silvicultural Aspects of the Forest-Fire Problem in the Longleaf Pine Region," *Journal of Forestry* 33 (March 1935): 330.

14. Shirley Allen, quoted in Henry E. Hardtner, "A Tale of a Root—A Root of a Tale, Root Hog or Die," *Journal of Forestry* 33 (March 1935): 360.

15. Ed Komarek, quoted in Henry E. Hardtner, "A Tale of a Root—A Root of a Tale, Root Hog or Die," *Journal of Forestry* 33 (March 1935): 360.

16. David Carle, *Burning Questions: America's Fight with Nature's Fire* (Westport, CT: Praeger, 2002), 50.

17. Kevin Pierce, "Charting Lightning's Future," *Florida Environment Radio*, http://www.floridaenvironment.com/programs/fe00703.htm, accessed September 10, 2012.

18. Jerome A. Jackson, "Biopolitics, Management of Federal Lands, and the Conservation of the Red-Cockaded Woodpecker," *American Birds* 40, no. 5 (Winter 1986): 1162.

19. Richard B. Dusenbury, *A History of the Osceola National Forest* (unpublished manuscript in files of Osceola National Forest, dated February 28, 1979), 164–65.

20. Fowler and Konipik, "History of Fire," 173–74.

21. Carle, *Burning Questions*, 53.

22. Joseph P. Ferguson, "Prescribed Fire on the Apalachicola Ranger District: The Shift from Dormant Season to Growing Season and Effects on Wildfire Suppression," in *Proceedings from Twentieth Tall Timbers Fire Ecology Conference. Fire in Ecosystem Management: Shifting the Paradigm from Suppression to Prescription*, ed. Teresa L. Pruden and Leonard A. Brennan (Tallahassee: Tall Timbers Research Station, 1998), 120.

23. "Welcome to Tall Timbers Research Station and Land Conservancy" (Tall Timbers, FL: Stewards of Wildlife and Wildlands), http://www.talltimbers.org/welcome.html, accessed April 25, 2012.

24. David A. Cleaves, Jorge Martin, and Terry K. Haines, *Influences on Prescribed Burning Activity and Costs in the National Forest System* (Asheville, NC: Southern Research Station, USDA Forest Service, 2000), 6–7.

25. Ibid., 13.

26. Ferguson, "Prescribed Fire," 121–22.

27. Mike Herrin, interview by Christopher Johnson, March 7, 2011, at Ocala National Forest District Office, Silver Springs, FL.

28. "Florida Forest Trees: Sand Pine," http://www.sfrc.ufl.edu/4h/Sand_pine/sandpine.htm, 1, accessed September 10, 2012.

29. Jackson, "Biopolitics," 1162–63.

30. Ibid., 1163–64.

31. Ibid., 1165.

32. Ibid., 1164–65.

33. Daowei Zhang and Sayeed R. Mehmood, "Safe Harbor for the Red-Cockaded Woodpecker." *Journal of Forestry* 100, no. 5 (July/August 2002), 24.

34. Parrish, interview.

35. Ibid.

36. Ibid.

37. Chuck Hess, interview by Christopher Johnson, March 9, 2011, at Apalachicola National Forest District Office, Crawfordsville, FL.

38. Ibid.

39. David Dorman, interview by Christopher Johnson, March 8, 2011, at Osceola National Forest District Office, Olustee, FL.

40. Jackson, "Biopolitics," 1163.

41. Parrish, interview.

42. Dorman, interview.

43. Robin J. Innis, "Gopherus Polyphemus," in *Fire Effects Information System* (Fort Collins, CO: Rocky Mountain Research Station, Fire Sciences Laboratory, 2009), http://www.fs.fed.us/database/feis, accessed June 2, 2011.

44. Mike Drayton, interview by Christopher Johnson, March 7, 2011, at Ocala National Forest District Office, Silver Springs, FL.

45. Ibid.

46. John B. Loomis, Lucas S. Bair, and Armando González-Cabán, "Prescribed Fire and Public Support," *Journal of Forestry* 99, no. 11 (November 2001): 21–22.

47. Drayton, interview.

48. Ibid.

49. *Premo MK III Operations and Service Manual* (Delta, British Columbia: SEI Industries, 2012), http://www.sei-ind.com/sites/default/files/pdf/2012_Premo _MK_III_Manual.pdf, 21, accessed September 10, 2012.

Chapter 9

1. David E. Conrad, *The Land We Cared For: A History of the Forest Service's Eastern Region* (Milwaukee: US Forest Service Region 9, 1997), 231.

2. Toddi A. Steelman, "The Monongahela Controversy and Decision," in *Forests and Forestry in the Americas: An Encyclopedia*, http://wiki.safnet.org/index .php/The_Monongahela_Controversy_and_Decision, 2–3, accessed September 10, 2012.

3. Ed Cliff, "Visitors from the Monongahela," http://www.foresthistory.org /blogs/Monongahela.pdf, accessed September 10, 2012.

4. Steelman, "Monongahela Controversy," 2.

5. Cliff, "Visitors."

6. Arthur Carhart, "Memorandum for Mr. Leopold, District 3," December 10, 1919, Carhart Collection, Denver Public Library, quoted in Tom Wolf, *Arthur Carhart: Wilderness Prophet* (Boulder: University Press of Colorado, 2008), 66.

7. Doug Scott, *The Enduring Wilderness* (Golden, CO: Fulcrum Publishing, 2004), 35–40.

8. Ibid., 54.

9. *Wilderness Act*, Public Law 88-577, 88th Congress, 2nd session (September 3, 1964), 1.

10. Ibid., 4.

11. William E. Shands and Robert G. Healy, *The Lands Nobody Wanted: A Con-*

servation Foundation Report (Washington, DC: The Conservation Foundation, 1977), 46.

12. Conrad, *The Land We Cared For,* 243.

13. Helen McGinnis, quoted in David P. Elkinton, *Fighting to Protect the Highlands: The First Forty Years of the West Virginia Highlands Conservancy* (Blacksburg, VA: Pocahontas Press, 2007), 19.

14. David P. Elkinton, *Fighting to Protect the Highlands: The First Forty Years of the West Virginia Highlands Conservancy* (Blacksburg, VA: Pocahontas Press, 2007, 24–25.

15. Steve Milauskas, "Wilderness Areas and Timber," *Timber and Timber Harvesting in West Virginia* (Ripley: West Virginia Forestry Association, August 2002), 1.

16. Thomas King, *The Highlands Voice,* October 1969, quoted in David P. Elkinton, *Fighting to Protect the Highlands: The First Forty Years of the West Virginia Highlands Conservancy* (Blacksburg, VA: Pocahontas Press, 2007), 27.

17. Elkinton, *Fighting,* 32–39.

18. Ibid., 42.

19. Ibid., 44.

20. Ron Hardway, *The Highlands Voice,* January 1975, quoted in David P. Elkinton, *Fighting to Protect the Highlands: The First Forty Years of the West Virginia Highlands Conservancy* (Blacksburg, VA: Pocahontas Press, 2007), 45.

21. Conrad, *The Land We Cared For,* 253.

22. Scott, *The Enduring Wilderness,* 81.

23. Conrad, *The Land We Cared For,* 246.

24. Elkinton, *Fighting,* 51–58.

25. Mary Wimmer, telephone interview by Christopher Johnson, November 29, 2011.

26. Dennis Hanson, "Some Plain Dealing Pays Off," *Sierra,* January/February 1987, 21.

27. Mike Costello, telephone interview by Christopher Johnson, November 1, 2011.

28. Elkinton, *Fighting,* 62–63.

29. Ibid., 66.

30. Wimmer, interview.

31. Costello, interview.

32. Beth Little, telephone interview by Christopher Johnson, November 11, 2011.

33. Costello, interview.

34. Ibid.

35. Wimmer, interview.

36. Ibid.

37. "Congress Passes Historic Wild Monongahela Act," West Virginia Wilderness Coalition, March 25, 2009, http://www.wvwild.org/pressroom_final _passage.htm, accessed November 7, 2011.

38. Wimmer, interview.

39. William Cronon, "The Trouble with Wilderness; or, Getting Back to the Wrong Nature," in *Uncommon Ground: Rethinking the Human Place in Nature*, ed. William Cronon (New York: Norton, 1996), 85.

40. Wimmer, interview.

Chapter 10

1. Chel Anderson, interview by Christopher Johnson, July 15, 2011, at Boundary Waters Canoe Area Wilderness.

2. Ibid.

3. Ibid.

4. Ibid.

5. "History of the Quetico Superior," Quetico Superior Foundation, http://www.queticosuperior.org/abouttheregion/history.html, 2, accessed May 31, 2010.

6. Ted Hall, "'Ober,' Boundary Waters' Friend, Creator of a Magic Island," *Minneapolis Star Tribune*, September 12, 1983, http://eober.org/Oberholtzer/Biography.shtml, 2, accessed August 6, 2011.

7. Tom Wolf, *Arthur Carhart: Wilderness Prophet* (Boulder: University Press of Colorado, 2008), 117.

8. Arthur H. Carhart, "Recreation in the Forests," *American Forestry* 26, no. 317 (May 1920): 269.

9. Ibid.

10. Mark Harvey, "Sound Politics: Wilderness, Recreation, and Motors in the Boundary Waters, 1945–1964," *Minnesota History*, Fall 2002, 130.

11. Ernest Oberholtzer, "Oral History Interviews with Ernest C. Oberholtzer," Minnesota Historical Society, quoted in Joe Paddock, *Keeper of the Wild: The Life of Ernest Oberholtzer* (St. Paul: Minnesota Historical Society Press, 2001), 163.

12. Ibid., 164–66.

13. Ibid., 166–73.

14. Ibid., 177–86.

15. William Nolan, quoted in Joe Paddock, *Keeper of the Wild: The Life of Ernest Oberholtzer* (St. Paul: Minnesota Historical Society Press, 2001), 199.

16. Ibid., 208.

17. Bruce Kerfoot, interview by Christopher Johnson, July 12, 2011, in Grand Marais, MN.

18. Clifford Ahlgren and Isabel Ahlgren, *Lob Trees in the Wilderness* (Minneapolis: University of Minnesota Press, 1984), 152.

19. Ibid., 157.

20. Ibid., 161.

21. Kevin Proescholdt, Rip Rapson, and Miron L. Heinselman, *Troubled Waters: The Fight for the Boundary Waters Canoe Area Wilderness* (St. Cloud, MN: North Star Press of St. Cloud, 1995), 20.

22. Ibid., 20–22.

23. Ibid., 31.

24. Ibid., 59.

25. Ibid., 70.

26. Ibid., 73–75.

27. Ibid., 158–63.

28. Ibid., 167.

29. Ibid., 218–19.

30. Ibid., 266–70.

31. Stephen Wilbers, *Boundary Waters Chronology*, http://www.wilbers.com /ChronologyWildernessManagement.htm, accessed August 18, 2011.

32. Anderson, interview.

33. *Sulfide Mining in Northeastern Minnesota* (Minneapolis: Friends of the Boundary Waters Wilderness, 2011).

34. Minnesota Department of Natural Resources and US Army Corps of Engineers, *NorthMet Project Draft Environmental Impact Statement* (St. Paul: Author, October 2009), http://files.dnr.state.mn.us/input/environmentalreview /polymet/draft_eis/summary_document.pdf, S-1, accessed September 10, 2011.

35. Ibid., S-1.

36. Ibid., S-12–S-14.

37. Bharat Mathur, "Letter from District 5 of the United States Environmental Protection Agency to Colonel Jon L. Christensen of the St. Paul District of the U.S. Army Corps of Engineers," February 18, 2010, http://www.friendscvsf.org /NorthMet%20Project%2020090387.pdf, 3, accessed September 10, 2012.

38. Ibid., 2–4.

39. "PolyMet Revises Proposed Project to Ease Concerns, Reduce Costs," *Business North.com*, September 10, 2011, http://ww.businessnorth.com/briefing .asp?RID-3873, accessed September 10, 2011.

40. "Agencies Delay Completion of PolyMet Environment Document," *Business

North.com, May 24, 2012, http:www.businessnorth.com/briefing.asp?RID-4667, accessed September 11, 2012.

41. Paul Dancic, interview by Christopher Johnson, July 12, 2011, in Ely, MN.

42. *Sulfide Mining in Northeastern Minnesota.*

43. Bill Hansen, telephone interview by Christopher Johnson, June 13, 2011.

44. Anderson, interview.

Chapter 11

1. "Wolf Facts," California Wolf Center, http://www.californiawolfcenter.org/learn/wolf-facts, accessed October 10, 2011.

2. Rolf O. Peterson and Paolo Ciucci, "The Wolf as a Carnivore," in *Wolves: Behavior, Ecology, and Conservation*, ed. L. David Mech and Luigi Boitani (Chicago: University of Chicago Press, 2003), 113.

3. Ibid.

4. Barry Lopez, *Of Wolves and Men* (New York: Scribner, 1978), 12–13.

5. Dean Beyer Jr., Rolf O. Peterson, John A. Vucetich, and James H. Hammill, "Wolf Population Changes in Michigan," in *Recovery of Gray Wolves in the Great Lakes Region of the United States: An Endangered Species Success Story*, ed. Adrian P. Wydeven, Timothy R. Van Deelen, and Edward J. Heske (New York: Springer, 2010), 73.

6. Brian J. Roell, Dean E. Beyer Jr., Patrick E. Lederle, Donald H. Lonsway, and Kristie L. Sitar, *Michigan Wolf Management 2009 Report* (Lansing: Michigan Department of Natural Resources and Environment, 2010), 12.

7. Keith Schneider, "Wolves Making a Comeback in Northern Michigan," Michigan Land Use Institute, October 31, 2005, http://www.mlui.org/print.asp?fileid=16933, 1, accessed June 10, 2010.

8. USDA Forest Service, Eastern Region, "Threatened, Endangered and Sensitive Species," http://www.fs.fed.us/r9/wildlife/tes/faq.shtml, 1, accessed May 4, 2012.

9. US Fish and Wildlife Service, "Kirtlands Warbler Wildlife Management Area: Northern Lower Peninsula, MI," http://www.fws.gov/refuges/profiles/index.cfm?id=31513, accessed September 11, 2012.

10. Center for Biological Diversity, "Dwarf Cinquefoil," *Endangered Species Act Works*, http://www.biologicaldiversity.org/campaigns/esa_works/profile_pages/DwarfCinquefoil.html, accessed September 11, 2012.

11. E. Billings, *Canadian Naturalist and Geologist*, 1856, quoted in Stanley Paul Young, *The Wolves of North America. Part I: Their History, Life Habits, Economic Status, and Control* (Washington, DC: Wildlife Institute, 1944), 122.

12. Beyer et al., "Wolf Population Changes," 65–69.

13. Curt Meine, *Aldo Leopold: His Life and Work* (Madison: University of Wisconsin Press, 1988), 3.

14. Aldo Leopold, "The Varmint Question," in *The River of God and Other Essays by Aldo Leopold,* ed. Susan L. Flader and J. Baird Callicott (Madison: University of Wisconsin Press, 1991), 47.

15. Aldo Leopold, "Conservationist in Mexico," in *The River of God and Other Essays by Aldo Leopold,* ed. Susan L. Flader and J. Baird Callicott (Madison: University of Wisconsin Press, 1991), 241–42.

16. Aldo Leopold, *A Sand County Almanac* (New York: Ballantine Books, 1970), 138–39.

17. Olaus Murie, quoted in Curt Meine, *Aldo Leopold: His Life and Work* (Madison: University of Wisconsin Press, 1988), 286.

18. Rolf O. Peterson, *The Wolves of Isle Royale: A Broken Balance* (Ann Arbor: University of Michigan Press, 2007), 22.

19. Durward L. Allen and L. David Mech, "Wolves versus Moose on Isle Royale," *National Geographic,* February 1963, 203.

20. Ibid., 212.

21. Ibid., 219.

22. Peterson, *The Wolves of Isle Royale,* 141–42.

23. Ibid., 142.

24. US Fish and Wildlife Service, "Gray Wolf Recovery in Minnesota, Wisconsin, and Michigan," July 2009, http://www.fws.gov/midwest/wolf/aboutwolves /r3wolfrec.htm, 1, accessed September 11, 2012.

25. Steven J. Fritts, "Wolf Depredation on Livestock in Minnesota," http:// www.npwrc.usgs.gov/resource/mammals/minnwolf/index.htm, 3, accessed September 11, 2012.

26. Beyer et al., "Wolf Population Changes," 70.

27. USDA Forest Service, Eastern Region.

28. Beyer et al., "Wolf Population Changes," 76.

29. Tom Weise, telephone interview by Christopher Johnson, September 30, 2011.

30. Beyer et al., "Wolf Population Changes," 73–74.

31. Ibid., 75.

32. Weise, interview.

33. David J. Mladenoff, Murray K. Clayton, Sarah D. Pratt, Theodore A. Stickley, and Adrian P. Wydeven, "Change in Occupied Wolf Habitat in the Northern Great Lakes Region," *Recovery of Gray Wolves in the Great Lakes Region of the United States: An Endangered Species Success Story,* ed. Adrian P. Wydeven, Timothy R. Van Deelen, and Edward J. Heske (New York: Springer, 2010), 135.

34. Michigan Gray Wolf Recovery Team, *Michigan Gray Wolf Recovery and Management Plan* (Lansing: Michigan Department of Natural Resources, 1997), 28.

35. Ibid., 29.

36. Ibid., 29–30.

37. USDA Forest Service, *Ottawa National Forest Fiscal Year 2008 Monitoring and Evaluation Report* (Ironwood, MI: US Department of Agriculture, 2009), 29.

38. USDA Forest Service, *Hiawatha National Forest Executive Summary of the Final Environmental Impact Statement* (Escanaba, MI: U.S. Department of Agriculture, 2006), ES-24–ES-25.

39. USDA Forest Service, *Hiawatha National Forest 2006 Monitoring and Evaluation Report* (Escanaba, MI: US Department of Agriculture, 2006), 21.

40. USDA Forest Service, *Forest Service Manual 2600: Wildlife, Fish, and Sensitive Plant Habitat Management* (Washington, DC: U.S. Department of Agriculture, 2005), 4.

41. USDA Forest Service, *Hiawatha National Forest Executive Summary of the Final Environmental Impact Statement*, ES-24–ES-25.

42. USDA Forest Service, *Ottawa National Forest 2008 Monitoring and Evaluation Report*, 26–32.

43. Pat Hallfrisch, telephone interview by Christopher Johnson, September 26, 2011.

44. Weise, interview.

45. Roell et al., *Michigan Wolf Management 2009 Report*, 6.

46. Ibid., 6–7.

47. Ibid., 8.

48. Beyer et al., "Wolf Population Changes," 76–77.

49. Roell et al., *Michigan Wolf Management 2009 Report*, 2.

50. USDA Forest Service, *Ottawa National Forest 2008 Monitoring and Evaluation Report*, 30.

51. Thomas P. Rooney and Dean P. Anderson, "Are Wolf-Mediated Trophic Cascades Boosting Biodiversity in the Great Lakes Region?" in *Recovery of Gray Wolves in the Great Lakes Region of the United States: An Endangered Species Success Story*, ed. Adrian P. Wydeven, Timothy R. Van Deelen, and Edward J. Heske (New York: Springer, 2010), 208–12.

52. Kevin Schanning, "Human Dimensions: Public Opinion Research Concerning Wolves in the Great Lakes States of Michigan, Minnesota, and Wisconsin," in *Recovery of Gray Wolves in the Great Lakes Region of the United States: An Endangered Species Success Story*, ed. Adrian P. Wydeven, Timothy R. Van Deelen, and Edward J. Heske (New York: Springer, 2010), 254.

53. Weise, interview.

54. Roell et al., *Michigan Wolf Management*, 10.

55. Jess Edberg, telephone interview by Christopher Johnson, October 7, 2011.

56. Ibid.

Chapter 12

1. George Zimmermann, quoted in Jon Hurdle, "Pennsylvania Lawsuit Says Drilling Polluted Water," *Reuters Business and Financial News*, November 9, 2009, http://www.reuters.com/assets/print?aid-USTRE5A80PP20091109, 1, accessed February 13, 2012.

2. Ibid.

3. Pennsylvania Independent Oil and Gas Association, "Allegheny National Forest Private Mineral Estates," http://www.pioga.org/publication_files/winter 2012-mayer-allegheny-national-forest.pdf, 2, September 12, 2012.

4. Pennsylvania Oil and Gas Association, "Pennsylvania's Traditional Oil and Natural Gas Industry," http://www.pioga.org/pa-oil-gas/traditional/, accessed September 12, 2012.

5. Charles G. Groat and Thomas W. Grimshaw, *Fact-Based Regulation for Environmental Protection in Shale Gas Development* (Austin: Energy Institute at the University of Texas at Austin, 2012), 4.

6. Ibid., 1.

7. Rich Miller, Asjylyn Loder, and Jim Polson, "Americans Gaining Energy Independence with U.S. as Top Producer," *Bloomberg News*, February 6, 2012, http://www.bloomberg.com/news/2012-02-07/americans-gaining-energy-inde pendence-with-u-s-as-top-producer.html, accessed September 12, 2012.

8. "How Exporting LNG Could Bring Serious Wealth to the U.S.," *Oil and Gas Investments Bulletin*, April 2, 2012, http://oilandgas-investments.com/2012 /natural-gas/exporting-lng-liquid-natural-gas-wealth, accessed May 14, 2012.

9. "U.S. Energy-Related CO_2 Emissions in Early 2012 Lowest Since 1002," U.S. Energy Information Administration, August 1, 2012, http://www.eia.gov/today inenergy/detail.cfm?id=7350#tabs)co2emission, accessed September 17, 2012.

10. John McFerrin, "Hydraulic Fracturing Coming to the Monongahela National Forest?" *West Virginia Highlands Voice*, February 10, 2012, http://wvhigh lands/org/wv_voice/?p=4522, 1, accessed May 9, 2012.

11. Southern Environmental Law Center, "Hydraulic Fracturing: "'Fracking' Looms as a New Threat to Clean Water in the Southeast," April 27, 2012, http:// www.southernenvironment.org/cases/hydraulic_fracturing, accessed May 9, 2012.

12. Southern Environmental Law Center, "BLM Putting Alabama Forest at Risk from Proposed Gas Fracking," April 10, 2012, http://www.southernenviron ment.org/cases/hydraulic_fracturing, accessed September 12, 2012.

13. Pennsylvania Independent Oil and Gas Association, "Allegheny National Forest Private Mineral Estates," 2.

14. Cathy Pedler, "Energy Is [NOT] a Public Use of the Forest," manuscript given to Christopher Johnson on March 10, 2012, 1.

15. Don Hopey, "Many Tapped Oil and Gas Sites in Allegheny National Forest Are Fragile Areas, Say Officials," *Pittsburgh Post-Gazette*, March 17, 2012.

16. Pedler, "Energy," 1.

17. Leanne Marten, "Declaration of Leanne Marten," Minard Run Oil Company et al., Plaintiffs, v. U.S. Department of Agriculture et al., Defendants, United States District Court, Western District of Pennsylvania, Filed June 25, 2009, 3.

18. USDA Forest Service, *Allegheny National Forest Final Environmental Impact Statement to Accompany the Land and Resource Management Plan* (Warren, PA: Allegheny National Forest, March 2007), 11.

19. Marten, "Declaration," 3.

20. Veronica A. Lopez, e-mail to William Belitskus, April 13, 2011.

21. USDA Forest Service, *Allegheny National Forest Final Environmental Impact Statement*, 34.

22. Ibid., 27.

23. Ibid., 25.

24. Karen Atwood, quoted in James Johnston, "Blight on the Land," *Forest Magazine*, Fall 2007, http:www.fseee.org/component/content/article/200236, 2, accessed March 23, 2012.

25. Jan Burkness, quoted in James Johnston, "Blight on the Land," *Forest Magazine*, Fall 2007, http:www.fseee.org/component/content/article/200236, 3, accessed March 23, 2012.

26. Forest Service Employees for Environmental Ethics, "Stop Hydrofracking on Public Lands," http://www.fseee.org/index.php/stay-informed/projects/1002790, 2, accessed September 12, 2012.

27. Groat and Grimshaw, *Fact-Based Regulation*, 24.

28. "Marcellus Shale: Appalachian Basin Natural Gas Play," *Geology.com*, http://geology.com/articles/marcellus-shale.shtml, 2–3, accessed March 22, 2012.

29. Ibid., 1.

30. Anthony Ingraffea, "Hydrofracking and the Marcellus Shale" (lecture, Northampton Community College, Bethlehem, PA, March 17, 2012).

31. "Marcellus Shale: Appalachian Basin Natural Gas Play," *Geology.com*, 3.

32. Theo Colborn, Carol Kwiatkowski, Kim Schultz, and Mary Bachran, "Natural Gas Operations from a Public Health Perspective," *Human and Ecological Risk Assessment: An International Journal* 17, no. 5 (2011): 1040–41.

33. Ibid.

34. Forest Service Employees for Environmental Ethics, "Stop Hydrofracking on Public Lands," 1.

35. Colborn et al., "Natural Gas Operations," 1042.

36. Ibid., 1040.

37. "Editorial: The Halliburton Loophole," *New York Times*, November 2, 2009.

38. Colborn, "Natural Gas Operations," 1039.

39. Ibid., 1045.

40. "Hydraulic Fracturing 101," *Earthworks*, http://www.earthworksaction .org/issues/detail/hydraulic_fracturing_101, accessed February 13, 2012.

41. Colborn, "Natural Gas Operations," 1045.

42. Chevron Corporation, "Responsible Gas Development: Protecting People and the Environment," http://www.chevron.com/deliveringenergy/naturalgas /shalegas/responsibleshalegasdevelopment/, 1, accessed September 12, 2012.

43. Ibid., 2.

44. Pennsylvania Independent Oil and Gas Association, "Protecting the Environment and Pennsylvania's Water Resources," www.pioga.org/publication_files /pioga-environmental-fact-sheet.pdf, 2, accessed March 22, 2012.

45. Groat and Grimshaw, *Fact-Based Regulation*, 18.

46. Ingraffea, lecture.

47. Dominic C. DiGiulio, Richard T. Wilkin, Carlyle Miller, and Gregory Oberley, *Investigation of Ground Water Contamination Near Pavillion, Wyoming: Draft Report* (Ada, OK: U.S. Environmental Protection Agency: Office of Research and Development, National Risk Management Research Laboratory, 2011), xi.

48. "Methane," U.S. Environmental Protection Agency, http://www.epa.gov /methane/, accessed March 25, 2012.

49. Stephen G. Osborn, Avner Vengosh, Nathaniel R. Warner, and Robert B. Jackson, "Methane Contamination of Drinking Water Accompanying Gas-Well Drilling and Hydraulic Fracturing," *PNAS Early Edition*, April 14, 2011, http:// www.ncbi.nlm.nih.gov/pmc/articles/PMC3100993/, 1–2, accessed September 12, 2012.

50. Ibid., 4.

51. Ibid.

52. "Hydraulic Fracturing 101," 9.

53. Ibid., 4–10.

54. Groat and Grimshaw, *Fact-Based Regulation*, 27.

55. Ibid., 28.

56. Don Hopey, "Drilling Stalled in Allegheny National Forest," *Pittsburgh*

Post-Gazette, March 16, 2008, http://old.post-gazette.com/pg/09075/955874-85
.stm, 1, accessed March 23, 2012.

57. Ibid.

58. "Forest Service to Follow Law on Allegheny National Forest," Allegheny Defense Project, http://www.alleghenydefense.org/hchronicles/?m=200905, 1, accessed September 12, 2011.

59. Pedler, "Energy," 1.

60. Bradley R. Jones, Minard Run Oil Company v. United States Forest Service, *3rd U.S. Circuit, September 20, 2011, WL 4389220*, http://publiclandlawreview
.files.wordpress.com/2011/07/cs2-jones_final.pdf, 2, accessed August 30, 2012.

61. Ibid., 3.

62. Ibid., 5–7.

63. Forest Service Employees for Environmental Ethics, *Brief in Support of Summary of Motion for Summary Judgement in Case 1:09-cv-00125-SJM, Document 112-1,* filed in the United States District Court for the Western District of Pennsylvania, March 6, 2012, 9.

64. William Belitskus, interview by Christopher Johnson, March 7, 2012, Allegheny National Forest, PA.

65. USDA Forest Service, *Revised Environmental Assessment: Trans Superior Resources, Inc. Private Minerals Exploration in the Matchwood Tower Road Area* (Ontonagon, MI: Ottawa National Forest, 2007), 1.

66. Ibid.

67. Jacob Perryman, "Judge Tosses Ban on ANF Drilling," *Times-Observer* [Warren, PA], September 8, 2012, http://www.timesobserver.com/page.content/detail
/id/559499/Judge-tos, accessed September 11, 2012.

68. Earl Hagström, "Hydraulic Fracturing Litigation on the Rise," *Sedgwick Law Firm Publications and Presentations*, September 2011, http://www.sdma.com
/hydraulic-fracturing-litigation-is-on-the-rise-09-19-2011/, 2, accessed September 12, 2012.

69. Forest Service Employees for Environmental Ethics, "Stop Hydrofracking on Public Lands," 2.

70. *Weeks Law of 1911*, Public Law 961, 62nd Congress, 1st session (1911). http://memory.loc.gov/cgi-bin/ampage?collId=amrvl&fileName=vl024//amrvlv
l024.db&recNum=0&itemLink=r?ammem/AMALL:@field%28DOCID+@lit%28
jjh94-000053%29%29&linkText=0, 961, Section 1.

71. USDA Forest Service, *Allegheny National Forest Final Environmental Impact Statement*, 11.

72. James Wilson, letter to Gifford Pinchot ("The Forester"), February 1, 1905.

Chapter 13

1. Therese Poland, telephone interview by Christopher Johnson, July 8, 2011.

2. Therese Poland, "Twenty Million Ash Trees Later: Current Status of Emerald Ash Borer in Michigan," *Newsletter of the Michigan Entomological Society* 52, nos. 1 and 2 (April 2007), http://pbadupws.nrc.gov/docs/ML1126/ML112630307.pdf, 11, accessed September 12, 2012.

3. "Emerald Ash Borer," Ottawa National Forest, http://www.fs.usda.gov/detail/ottawa/home/?cid=STELPRDB5111749, accessed September 12, 2012.

4. "Many Insect Pests and Diseases Threaten Michigan's Forests," Michigan Department of Natural Resources and Environment, September 22, 2005, http://www.michigan.gov/dnr/0,4570,7-153-10366_46403_59160-128400—,00.html, 1, accessed September 12, 2012.

5. Harold A. Mooney, "Invasive Alien Species: The Nature of the Problem," in *Invasive Alien Species: A New Synthesis,* ed. Harold A. Mooney, Richard Mack, Jeffrey A. McNeely, Laurie E. Neville, Peter Johan Schei, and Jeffrey K. Waage (Washington, DC: Island Press, 2005), 1.

6. Ibid., 2.

7. David Pimentel, Rodolfo Zuniga, and Doug Morrison, "Update on the Environmental and Ecological Costs Associated with Alien-Invasive Species in the United States," *Ecological Economics* 52, no. 3 (February 2005): 282.

8. "Invasive Species," National Wildlife Federation, http://www.nwf.org/Wildlife/Wildlife-Conservation/Threats-to-Wildlife/Invasive-Species.aspx, accessed September 12, 2012.

9. Mooney, "Invasive Alien Species," 6–9.

10. "Invasive Species."

11. Pimentel, Zuniga, and Morrison, "Update," 281.

12. USDA Forest Service, "Hemlock Woolly Adelgid," *Pest Alert*, August 2005, 1.

13. *National Strategy and Implementation Plan for Invasive Species Management* (Washington, DC: US Department of Agriculture, 2004), i.

14. Therese M. Poland and Deborah G. McCullough, "Emerald Ash Borer: Invasion of the Urban Forest and the Threat to North America's Ash Resource," *Journal of Forestry* 104, no. 3 (April/May 2006): 120.

15. Ibid., 118–21.

16. Ibid., 118.

17. Poland, interview.

18. Poland and McCullough, "Emerald Ash Borer," 119.

19. Victor Mastro and Richard Reardon, compilers, *Emerald Ash Borer Research and Technology Development Meeting, Romulus, Michigan, October 5–6, 2004* (Morgantown, WV: Forest Health Technology Enterprise Team, 2004), iii.

20. Ibid.

21. "Forest Health Technology Enterprise Team," http://www.fs.fed.us/forest health, accessed June 30, 2011.

22. Andrew J. Storer, Elizabeth E. Graham, Michael D. Hyslop, and Robert L. Heyd, "Michigan Emerald Ash Borer Detection Survey," in *Emerald Ash Borer Research and Technology Development Meeting, October 5–6, Romulus, Michigan*, comp. Victor Masto and Richard Reardon (Morgantown, WV: Forest Health Technology Team, 2004), http://www.fs.fed.us/foresthealth/technology/pdfs /2004EAB.pdf, 7, accessed June 15, 2011.

23. Poland, interview.

24. Ibid.

25. Rodrigo J. Mercader, "Comparing Potential Management Options to Slow the Spread of EAB Populations in Localized Sites," in *Emerald Ash Borer Research and Technology Development Meeting, October 20–21, 2009, Pittsburgh, Pennsylvania*, comp. David Lance (Morgantown, WV: Forest Health Technology Team, 2010), http://www.fs.fed.us/foresthealth/technology/pdfs/2009EAB.pdf, 40, accessed June 15, 2011.

26. Robin A. J. Taylor, Leah S. Bauer, Deborah L. Miller, and Robert A. Haack, "Emerald Ash Borer Flight Potential," in *Emerald Ash Borer Research and Technology Development Meeting, October 5–6, Romulus, Michigan*, comp. Victor Masto and Richard Reardon (Morgantown, WV: Forest Health Technology Team, 2004), http://www.fs.fed.us/foresthealth/technology/pdfs/2004EAB.pdf, 15, accessed June 15, 2011.

27. Poland, interview.

28. Michael D. Hyslop and Andrew J. Storer, "Assessment of an Emerald Ash Borer Infestation in Houghton County, Michigan: Development of an Apparently Isolated Population," in *Emerald Ash Borer Research and Technology Development Meeting, October 20–21, 2009, Pittsburgh, Pennsylvania*, comp. David Lance et al. (Morgantown, WV: Forest Health Technology Team, 2010), http://www.fs.fed.us /foresthealth/technology/pdfs/2009EAB.pdf, 29, accessed June 15, 2011.

29. *National Strategy and Implementation Plan for Invasive Species Management*, ii–iii.

30. Poland and McCullough, "Emerald Ash Borer," 122.

31. Poland, "Twenty Million Ash Trees Later," 11.

32. Poland and McCullough, "Emerald Ash Borer," 121–22.

33. Poland, "Twenty Million Ash Trees Later," 11.

34. Robert A. Haack and Toby R. Petrice, "Evaluation of Various Insecticides Applied to the Bark to Control Emerald Ash Borer," in *Emerald Ash Borer Research and Technology Development Meeting, October 5–6, Romulus, Michigan*, comp. Victor

Masto and Richard Reardon (Morgantown, WV: Forest Health Technology Team, 2004), http://www.fs.fed.us/foresthealth/technology/pdfs/2004EAB.pdf, 5–36, accessed June 15, 2011.

35. Houping Liu, Leah S. Bauer, and Deborah L. Miller, "2004 Update on Studies of BotaniGard® for Control of Emerald Ash Borer Adults and Larvae," in *Emerald Ash Borer Research and Technology Development Meeting, October 5–6, Romulus, Michigan* comp. Victor Masto and Richard Reardon (Morgantown, WV: Forest Health Technology Team, 2004), http://www.fs.fed.us/foresthealth/technology/pdfs/2004EAB.pdf, 41–42, accessed June 15, 2011.

36. Mercader, "Comparing Potential Management Options," 41–42.

37. Poland, interview.

38. Ibid.

39. David Cappaert, Deborah McCullough, and Therese Poland, "The Upside of the Emerald Ash Borer Catastrophe: A Feast for Woodpeckers," in *Emerald Ash Borer Research and Technology Development Meeting, October 5–6, Romulus, Michigan*, comp. Victor Masto and Richard Reardon (Morgantown, WV: Forest Health Technology Team, 2004), http://www.fs.fed.us/foresthealth/technology/pdfs/2004EAB.pdf, 69, accessed June 15, 2011.

40. Steven A. Katovich, "The Upper Peninsula Slam Projects—An Integrated Strategy to Slow Ash Mortality in Emerald Ash Borer Outlier Sites," in *Emerald Ash Borer Research and Technology Development Meeting, October 20–21, 2009, Pittsburgh, Pennsylvania*, comp. David Lance (Morgantown, WV: Forest Health Technology Team, 2010), http://www.fs.fed.us/foresthealth/technology/pdfs/2009EAB.pdf, 42, accessed June 15, 2011.

41. Steven A. Katovich, telephone interview by Christopher Johnson, September 13, 2012.

42. Poland and McCullough, "Emerald Ash Borer," 122.

43. *National Strategy and Implementation Plan for Invasive Species Management*, 10.

44. "Connecticut Purple Loosestrife Program," University of Connecticut Cooperative Extension System, http://www.purpleloosestrife.uconn.edu, accessed July 1, 2011.

Chapter 14

1. Susan Stein et al., *National Forests on the Edge: Development Pressures on America's National Forests and Grasslands* (Portland, OR: U.S. Department of Agriculture, Forest Service, Pacific Northwest Research Station, 2007), 2.

2. Ibid., 7.

3. Ibid., 9–10.

4. Ibid., 14.

5. Ibid., 3.

6. Sustainable Forests Partnership, "What Is Forest Fragmentation?" July 16, 2009, http://sfp.cas.psu.edu/fragmentation/what.htm, accessed September 13, 2012.

7. Stein, *National Forests*, 18–19.

8. Jamey Fidel, telephone interview by Christopher Johnson, January 24, 2012.

9. J. M. Hagan, L. C. Irland, and A. A. Whitman, *Changing Timberland Ownership in the Northern Forest and Implication for Biodiversity (Report #MCCS-0FCP-2005-1)* (Brunswick, ME: Manomet Center for Conservation Sciences, 2005), iii.

10. Fidel, interview.

11. "Vermont's Largest Conservation Project: The Champion Lands," Vermont Land Trust, www.vlt.org/land-weve-conserved/champion, accessed January 24, 2012.

12. "Vermont's Housing Market: Trends and Perspectives" (handout by Phil Dodd for Forest Roundtable, October 18, 2006), cited in Jamey Fidel, *Roundtable on Parcelization and Forest Fragmentation* (Montpelier: Vermont Natural Resources Council, 2007), 5.

13. "Vermont Housing Data," http://www.housingdata.org/profile/profile MainResult.php?submitted-stateProfile.

14. Jamey Fidel, *Roundtable on Parcelization and Forest Fragmentation* (Montpelier: Vermont Natural Resources Council, 2007), 8.

15. Ibid., 4.

16. Katie Johnston, "Elusive Ski Season Cut Short," *Boston Globe*, March 23, 2012.

17. Ibid., 9.

18. Ibid., 10–11.

19. USDA Forest Service, "Ecosystem Services," http://www.fs.fed.us/ecosystemservices/, accessed September 13, 2012.

20. Fidel, *Roundtable*, 14.

21. Ibid., 12.

22. Fidel, interview.

23. Ibid.

24. George Leoniak, *Critical Paths: Enhancing Road Permeability for Wildlife in Vermont* (Montpelier: Vermont Natural Resources Council, 2009), 6.

25. Ibid., 7.

26. Ibid., 4–7.

27. Fidel, interview.

28. North East State Foresters Association, *The Economic Importance of Ver-*

mont's Forests, December 2004, cited in Jamey Fidel, *Roundtable on Parcelization and Forest Fragmentation* (Montpelier: Vermont Natural Resources Council, 2007), 16.

29. Vermont Forest Products Council Blueprint for Action, http://vtrural.org /webfm_send/147, accessed September 13, 2012, 8.

30. David Brynn, telephone interview by Christopher Johnson, February 14, 2012.

31. Wendell Berry, "Private Property and the Common Wealth," in *Another Turn of the Crank* (Berkeley, CA: Counterpoint Press, 1995), 57.

32. Vermont Family Forests, *Vermont Family Forests 2010 Annual Report* (Bristol: Vermont Family Forests, 2010), 1.

33. Brynn, interview.

34. Ibid.

35. Fidel, *Roundtable*, 15.

36. USDA Forest Service, "Land Use Change," *WNC Report Card on Forest Sustainability* (Asheville, NC: Southern Research Station Headquarters, 2010), http:// www.wncforestreportcard.org/browse/biodiversity/land-use, 1–2, accessed May 18, 2012.

37. Brent Martin, telephone interview by Christopher Johnson, January 26, 2012.

38. USDA Forest Service, "Land Use Change," 2.

39. Edwin J. Jones, Mark A. Megalos, and J. Chris Turner, *Working with Wildlife #21: Bats* (Raleigh: North Carolina Cooperative Extension Service of North Carolina State University), http://www.ces.ncsu.edu/forestry/pdf/www/www21.pdf, 1, accessed September 13, 2012.

40. Martin, interview.

41. USDA Forest Service, "Forest Fragmentation," *WNC Report Card on Forest Sustainaility* (Asheville, NC: Southern Research Station Headquarters, 2010), http://www.wncforestreportcard.org/browse/biodiversity/forest-fragmentation, 2, accessed May 18, 2012.

42. USDA Forest Service, "Planning and Monitoring," *WNC Report Card on Forest Sustainability* (Asheville, NC: Southern Research Station Headquarters, 2010), http://www.wncforestreportcard.org/browse/policy/planning-monitoring, 1, accessed May 18, 2012.

43. Ibid., 1–2.

44. USDA Forest Service, "Volume of Timber," *WNC Report Card on Forest Sustainability* (Asheville, NC: Southern Research Station Headquarters, 2010), http:// www.wncforestreportcard.org/browse/production/timber-volume, 1, accessed May 18, 2012.

45. Martin, interview.

46. USDA Forest Service, "Forest Legacy Program: Protecting Private Forest Lands from Conversion to Non-Forest Uses," http://www.fs.fed.us/spf/coop/pro grams/loa/flp.shtml, accessed September 13, 2012.

Conclusion

1. Carol Ann Gillespie, *Mountain Mists: Appalachian Folkways of West Virginia* (Parsons, WV: McClain Printing, 2009), 15–18.

2. USDA Forest Service, "Table 2: Regional Areas Summary," www.fs.fed.us /land/staff/lar/LAR2011/LAR_Table_02.pdf, accessed June 12, 2012.

3. W. Brad Smith, Patrick D. Miles, Charles H. Perry, and Scott A. Pugh, *Forest Resources of the United States, 2007: A Technical Document Supporting the Forest Service 2010 RPA Assessment* (Washington, DC: US Department of Agriculture, 2009), 13.

4. Ibid., 14.

5. Ibid.

6. Ibid., 14–16.

7. D. W. MacCleery, *American Forests: A History of Resiliency and Recovery* (Durham, NC: Forest History Society, 2002), 58.

8. Smith et al., *Forest Resources*, 30.

9. Ibid., 30–33.

10. Ibid., 103.

11. Ibid., 102–3.

12. James Burchfield and Martin Nie, *National Forests Policy Assessment: Report to Montana Senator Jon Tester* (Missoula: University of Montana College of Forestry and Conservation, 2008), 11.

13. William E. Shands, *The Lands Nobody Wanted: The Legacy of the Eastern National Forests* (Milford, PA: Pinchot Institute for Conservation, 1991), 2.

14. USDA Forest Service, *Managing Multiple Uses on National Forests, 1905– 1995*, http://www.foresthistory.org/ASPNET/Publications/multiple_use/chap6 .htm, 82–85, accessed September 13, 2012.

15. Smith et al., *Forest Resources*, 46–47.

16. USDA Forest Service, *Visitor Use Report: Southern Region (R8)* (Washington, DC: US Department of Agriculture, 2012), 10.

17. USDA Forest Service, *Visitor Use Report: Eastern Region (R9)* (Washington, DC: US Department of Agriculture, 2012), 10.

18. Burchfield and Nie, *National Forests Policy Assessment*, 8.

19. Ibid., 9.

20. Smith et al., *Forest Resources*, 51.
21. Burchfield and Nie, *National Forests Policy Assessment*, 10.
22. Ibid., 12.
23. Ibid., 13.
24. Ibid., 14.
25. Ibid.

Adams, Sherman. *The Weeks Act: A 75th Anniversary Appraisal.* New York: Newco-
   men Society of the United States, 1986.
Ahlgren, Clifford, and Isabel Ahlgren. *Lob Trees in the Wilderness.* Minneapolis:
   University of Minnesota Press, 1984.
Allen, Durward L., and L. David Mech. "Wolves versus Moose on Isle Royale."
   National Geographic, February 1963, 200–219.
Allin, Craig W. *The Politics of Wilderness Preservation.* Westport, CT: Greenwood
   Press, 1982.
American Forestry. "The Appalachian Forests." *American Forestry* 17, no. 7 (July
   1911): 380–83.
——. "Editorial: The Appalachian Bill." *American Forestry* 17, no. 3 (March 1911):
   168–71.
——. "First Purchase of White Mountain Lands Under the Weeks Law." *American
   Forestry* 18, no. 2 (July 1912): 440.
——. "The Passage of the Appalachian Bill." *American Forestry* 17, no. 3 (March
   1911): 164–67.
——. "Pisgah Forest Purchased." *American Forestry* 20, no. 6 (June 1914): 425–29.
——. "Reforesting Cut-Over Pinelands." *American Forestry* 18, no. 10 (October
   1912): 674–75.
——. "A White Mountain Purchase." *American Forestry* 20, no. 10 (October 1914):
   733.
"Appalachian Forests: Effects of Inroads of Lumbermen Upon Them." *Manufactur-
   ers' Record,* January 20, 1910, 52.
Ayres, H. B., and W. W. Ashe. "Forests and Forest Conditions in the Southern
   Appalachians." In Theodore Roosevelt, *Message from the President of the United
   States Transmitting a Report of the Secretary of Agriculture in Relation to the Forests,
   Rivers, and Mountains of the Southern Appalachian Region,* 45–59. Washington,
   DC: Government Printing Office, 1902.
Ayres, Philip W. *Commercial Importance of the White Mountain Forests.* Washington,
   DC: US Department of Agriculture, 1909.
——. "National Forests in the Eastern Mountains: A Brief Account of the Weeks
   Law of 1911 and of Its Extension Pending in Congress, the McNary-Woodruff
   Bill." Unpublished manuscript; photocopy from the Society for the Protection
   of New Hampshire Forests, Concord, undated.

———. "New England's Federal Forest Reserve." *American Forestry* 21, no. 7 (July 1915): 803–12.

———. *Philip Wheelock Ayres: Chronology*. Unpublished manuscript; photocopy from the Society for the Protection of New Hampshire Forests, Concord, undated.

———. "Reasons for a National Forest Reservation in the White Mountains." *The Northern: Official Organ of the New Hampshire and Maine Federation of Women's Clubs* 1, no. 9 (November 1905): 286–92.

Baird, Iris. *Looking Out for Our Forests: The Evolution of a Plan to Protect New Hampshire's Woodlands from Fire*. Lancaster, NH: Baird Backwoods Construction Publications, 2005.

Bawden, Timothy. "The Northwoods: Back to Nature?" In *Wisconsin Land and Life*, edited by Robert C. Ostergren and Thomas R. Vale, 450–69. Madison: University of Wisconsin Press, 1997.

Belcher, C. Francis. *Logging Railroads of the White Mountains*. Boston: Appalachian Mountain Club Books, 1980.

"Beneficial Effects of Forests on Stream Flow." *Municipal Engineering* 43, no. 1 (July 1912), 47.

Beyer, Dean, Jr., Rolf O. Peterson, John A. Vucetich, and James H. Hammill. "Wolf Population Changes in Michigan." In *Recovery of Gray Wolves in the Great Lakes Region of the United States: An Endangered Species Success Story*, edited by Adrian P. Wydeven, Timothy R. Van Deelen, and Edward J. Heske, 65–85. New York: Springer, 2010.

Blaine, Marcia Schmidt. "The Public Forests." In *Beyond the Notches: Stories of Place in New Hampshire's North Country*, edited by John R. Harris, Kay Morgan, and Mike Dickerman, 258–266. Littleton, NH: Bondcliff Books, 2011.

Bolles, Blair. *Tyrant from Illinois: Uncle Joe Cannon's Experiment with Personal Power*. New York: Norton, 1950.

Botti, William B., and Michael D. Moore. *Michigan's State Forests: A Century of Stewardship*. East Lansing: Michigan State University Press, 2006.

Brown, Greg, and Chuck Harris. "Professional Foresters and the Land Ethic, Revisited." *Journal of Forestry* 96, no. 1 (January 1998): 4–12.

Brown-Nuñez, Christine, and Jonathan G. Taylor. *Americans' Attitudes Toward Wolves and Wolf Reintroduction: An Annotated Bibliography*. Washington, DC: US Department of the Interior, 2002.

Burchfield, James, and Martin Nie. *National Forests Policy Assessment: Report to Montana Senator Jon Tester*. Missoula: University of Montana College of Forestry and Conservation, 2008.

Bush, Florence Cope. *Dorie: Woman of the Mountains*. Knoxville: University of Tennessee Press, 1992.

Cappaert, David, Deborah McCullough, and Therese Poland. "Emerald Ash Borer Life Cycle: A Reassessment." In *Emerald Ash Borer Research and Technology Development Meeting, October 5–6, Romulus, Michigan*, compiled by Victor Masto and Richard Reardon, 19–20. Morgantown, WV: Forest Health Technology Team, 2004. http://www.fs.fed.us/foresthealth/technology/pdfs/2004EAB.pdf. Accessed June 15, 2011.

———. "The Upside of the Emerald Ash Borer Catastrophe: A Feast for Woodpeckers." In *Emerald Ash Borer Research and Technology Development Meeting, October 5–6, Romulus, Michigan*, compiled by Victor Masto and Richard Reardon, 69–70. Morgantown, WV: Forest Health Technology Team, 2004. http://www.fs.fed.us/foresthealth/technology/pdfs/2004EAB.pdf. Accessed June 15, 2011.

Carhart, Arthur H. "Recreation in the Forests." *American Forestry* 26, no. 317 (May 1920): 268–72.

———. "Sanctuary!" *American Forests* 31, no. 375 (March 1925): 147–50.

Carle, David. *Burning Questions: America's Fight with Nature's Fire*. Westport, CT: Praeger, 2002.

Chapman, Herman H. "Forest Fires and Forestry in the Southern States." *American Forestry* 18, no. 8 (August 1912): 510–17.

Charlier, Tom. "Chain Saws Buzzing in Miss. Forest Sound Like Massacre to Ex-Urbanites." *Memphis Commercial Appeal*, March 9, 1998. http://nlnewsbank.com/nl-search/we/Archives?p_action-doc&pOdoci. Accessed March 21, 2011.

Chittenden, Alfred K. *Forest Conditions of Northern New Hampshire (Bureau of Forestry Bulletin No. 55)*. Washington, DC: US Department of Agriculture, 1905.

Clarkson, Roy B. *Tumult on the Mountains: Lumbering in West Virginia, 1770–1920*. Parsons, WV: McClain Printing, 1964.

Cleaves, David A., Jorge Martin, and Terry K. Haines. *Influences on Prescribed Burning Activity and Costs in the National Forest System*. Asheville, NC: Southern Research Station, USDA Forest Service, 2000.

Clemson University Department of Forest Resources. *Responses to Wildlife to Clearcutting and Associated Tratment in the Eastern United States: Technical Paper No. 19*. March 2000. http://www.clemson.edu/extfor/timber_production/fort p19.htm. Accessed September 9, 2012.

Clepper, Henry. *Origins of American Conservation*. New York: Ronald Press, 1966.

Colborn, Theo, Carol Kwiatkowski, Kim Schultz, and Mary Bachran. "Natural Gas Operations from a Public Health Perspective." *Human and Ecological Risk Assessment: An International Journal* 17, no. 5 (2011): 1039–56.

Conrad, David E. *The Land We Cared For: A History of the Forest Service's Eastern Region*. Milwaukee: US Forest Service Region 9, 1997.

Cowdrey, Albert E. *This Land, This South: An Environmental History*. Lexington: University Press of Kentucky, 1996.

Cronon, William. *Changes in the Land: Indians, Colonists, and the Ecology of New England*. New York: Hill and Wang, 2003.

———. "The Trouble with Wilderness; or, Getting Back to the Wrong Nature." In *Uncommon Ground: Rethinking the Human Place in Nature*, edited by William Cronon, 69–90. New York: Norton, 1996.

Davis, Donald Edward. *Homeplace Geography: Essays for Appalachia*. Macon, GA: Mercer University Press, 2006.

———. *Where There Are Mountains: An Environmental History of the Southern Appalachians*. Athens: University of Georgia Press, 2000.

Demmon, E. L. "The Silvicultural Aspects of the Forest-Fire Problem in the Longleaf Pine Region." *Journal of Forestry* 33 (March 1935): 323–31.

Dickmann, Donald I., and Larry A. Leefers. *The Forests of Michigan*. Ann Arbor: University of Michigan Press, 2003.

DiGiulio, Dominic C., Richard T. Wilkin, Carlyle Miller, and Gregory Oberley. *Investigation of Ground Water Contamination Near Pavillion, Wyoming: Draft Report*. Ada, OK: US Environmental Protection Agency, Office of Research and Development, National Risk Management Research Laboratory, 2011.

Doelle, John A. "The Land Riddle of the Lake States." *American Forests and Forest Life* 31, no. 381 (September 1925): 515–18+.

Dusenbury, Richard B. *A History of the Osceola National Forest*. From files of Osceola National Forest. February 28, 1979.

Elkinton, David P. *Fighting to Protect the Highlands: The First Forty Years of the West Virginia Highlands Conservancy*. Blacksburg, VA: Pocahontas Press, 2007.

Eller, Ronald D. *Miners, Millhands, and Mountaineers: Industrialization of the Appalachian South, 1880–1930*. Knoxville: University of Tennessee Press, 1982.

Fairfax, Sally K., Lauren Gwin, Mary Ann King, Leigh Raymond, and Laura A. Watt. *Buying Nature: The Limits of Land Acquisition as a Conservation Strategy*. Cambridge: MIT Press, 2005.

Ferguson, Joseph P. "Prescribed Fire on the Apalachicola Ranger District: The Shift from Dormant Season to Growing Season and Effects on Wildfire Suppression." In *Proceedings from 20th Tall Timbers Fire Ecology Conference. Fire in Ecosystem Management: Shifting the Paradigm from Suppression to Prescription*, edited by Teresa L. Pruden and Leonard A. Brennan, 120–26. Tallahassee: Tall Timbers Research Station, 1998.

Fernow, Bernard E. "Introduction." In *Forestry Condition and Interests of Wisconsin (Bulletin 16)*, by Filibert Roth, 7–20. Washington, DC: Government Printing Office, 1898.

Fickle, James E. *Mississippi Forests and Forestry*. Jackson: University Press of Mississippi, 2001.

Fidel, Jamey. *Roundtable on Parcelization and Forest Fragmentation*. Montpelier: Vermont Natural Resources Council, 2007.

Flader, Susan L., ed. *The Great Lakes Forest: An Environmental and Social History*. Minneapolis: University of Minnesota Press in association with the Forest History Society, 1983.

Flint, Tim. "Michigan's Emerald Ash Borer Response Project." In *Emerald Ash Borer Research and Technology Development Meeting, October 5–6, Romulus, Michigan*, compiled by Victor Masto and Richard Reardon, 6. Morgantown, WV: Forest Health Technology Team, 2004. http://www.fs.fed.us/foresthealth/tech nology/pdfs/2004EAB.pdf. Accessed June 15, 2011.

Forest History Society. *Weeks Act Centennial Issue*. Special issue of *Forest History Today*, Spring/Fall 2011.

Forest Service Employees for Environmental Ethics. Brief in Support of Summary of Motion for Summary Judgement in Case 1:09-cv-00125-SJM, Document 112-1. Filed in the United States District Court for the Western District of Pennsylvania, March 6, 2012.

———. "Stop Hydrofracking on Public Lands." http://www.fseee.org/index .php/stay-informed/projects/1002790. Accessed September 12, 2012.

Fowler, Cynthia, and Evelyn Konipik. "The History of Fire in the Southern United States." *Human Ecology Review* 14, no. 2 (2007): 165–76.

Fries, Robert F. *Empire in Pine: The Story of Lumbering in Wisconsin, 1830–1900*. Madison: State Historical Association of Wisconsin, 1951.

Frome, Michael. *Strangers in High Places: The Story of the Great Smoky Mountains*. Knoxville: University of Tennessee Press, 1980.

Glenn, L. C. "Forests as Factors in Stream Flow." *American Forestry* 16, no. 4 (March 1910): 217–24.

Gove, Bill. *J. E. Henry's Logging Railroads*. Littleton, NH: Bondcliff Books, 2012.

Groat, Charles G., and Thomas W. Grimshaw. *Fact-Based Regulation for Environmental Protection in Shale Gas Development*. Austin: Energy Institute at the University of Texas at Austin, 2012.

Hagström, Earl. "Hydraulic Fracturing Litigation on the Rise." *Sedgwick Law Firm Publications and Presentations*, September 2011. http://www.sdma.com /hydraulic-fracturing-litigation-is-on-the-rise-09-19-2011/. Accessed September 12, 2012.

Hall, William L. "The Appalachian Work." *American Forestry* 18, no. 3 (March 1912): 192.

———. "Improving White Mountain Forests." *American Forestry* 21, no. 2 (February 1915): 117–26.

———. "To Remake the Appalachians." *The World's Work*, July 1914, 321–38.

———. "The White Mountain Forest and How It Is to Be Made Useful." *American Forestry* 19, no. 9 (September 1913): 620–25.

Hanson, Dennis. "Some Plain Dealing Pays Off." *Sierra*, January/February 1987, 20–22.

Hardtner, Henry E. "A Tale of a Root—A Root of a Tale, Root Hog or Die." *Journal of Forestry* 33 (March 1935): 351–60.

Harvey, Mark. "Sound Politics: Wilderness, Recreation, and Motors in the Boundary Waters, 1945–1964." *Minnesota History*, Fall 2002, 130–45.

Hays, Samuel P. *Conservation and the Gospel of Efficiency: The Progressive Conservation Movement, 1890–1920*. Cambridge: Harvard University Press, 1959.

Higgs, Steven. *Eternal Vigilance: Nine Tales of Environmental Heroism in Indiana*. Bloomington: Indiana University Press, 1995.

Holtrop, Joel D. "Allegheny National Forest 2007 Revised Land and Resource Management Plan Appeal Decision." Washington, DC: USDA Forest Service, February 15, 2008.

Hopey, Don. "Drilling Stalled in Allegheny National Forest." *Pittsburgh Post-Gazette*, March 16, 2008. http://old.post-gazette.com/pg/09075/955874-85.stm. Accessed March 23, 2012.

———. "Many Tapped Oil and Gas Sites in Allegheny National Forest Are Fragile Areas, Say Officials," *Pittsburgh Post-Gazette*, March 17, 2012.

Hurdle, Jon. "Pennsylvania Lawsuit Says Drilling Polluted Water." Reuters News Agency, November 9, 2009. http://www.reuters.com/assets/print?aid-USTRE5A80PP20091109. Accessed February 13, 2012.

"Hydraulic Fracturing: 'Fracking' Looms as a New Threat to Clean Water in the Southeast." Southern Environmental Law Center, April 27, 2012. http://www.southernenvironment.org/cases/hydraulic_fracturing. Accessed May 9, 2012.

"Hydraulic Fracturing of Oil and Gas Wells Drilled in Shale." *Geology.com* http://geology.com/articles/hydraulic-fracturing/. Accessed February 13, 2012.

Ingraffea, Anthony. "Hydrofracking and the Marcellus Shale." Lecture. Northampton Community College, Bethlehem, PA, March 17, 2012.

Ise, John. *The United States Forest Policy*. New Haven: Yale University Press, 1923.

Jackson, Jerome A. "Biopolitics, Management of Federal Lands, and the Conservation of the Red-Cockaded Woodpecker." *American Birds* 40, no. 5 (Winter 1986): 1162–68.

Johnson, John E. *The Boa Constrictor of the White Mountains, or the Worst "Trust" in the World* (pamphlet, July 4, 1900; reprinted in *New England Homestead*, December 8, 1900.

Johnston, James. "Blight on the Land." *Forest Magazine*, Fall 2007. http:www.fseee
.org/component/content/article/200236. Accessed March 23, 2012

Jones, Bradley R. Minard Run Oil Company v. United States Forest Service, 3rd
U.S. Circuit, September 20, 2011, WL 4389220. http://publiclandlawreview
.files.wordpress.com/2011/07/cs2-jones_final.pdf. Accessed March 22, 2012.

Jones, Joseph J. "Transforming the Cutover: The Establishment of National Forests in Northern Michigan." *Forest History Today*, Spring/Fall 2011, 48–55.

Kephart, Horace. *Our Southern Highlanders*. New York: Macmillan, 1922.

Kilbourne, Frederick W. *Chronicles of the White Mountains*. Boston: Houghton Mifflin, 1916.

Kimball, Jay. "Congress Releases Report on Toxic Chemicals Used in Fracking." http://8020vision.com/2011/04/17/congress-releases-report-on-toxic
-chemicals-used-in-fracking. Accessed February 20, 2012.

Kneipp, L. F. "Uncle Sam Buys Some Forests: How the Weeks Law of Twenty-Five
Years Ago Is Building Up a Great System of National Forests in the East." *American Forests*, October 1936, 443+.

Lacy, Leslie Alexander. *The Soil Soldiers: The Civilian Conservation Corps in the Great Depression*. Radnor, PA: Chilton, 1976.

Lambert, Robert S. "Logging the Great Smokies." *Tennessee Historical Quarterly* 20, no. 4 (December 1961): 350–63.

Lance, David, James Buck, Denise Binion, Richard Reardon, and Victor Mastro, comps. *Emerald Ash Borer Research and Technology Development Meeting, Port Huron, Michigan, October 20–21, 2009*. Morgantown, WV: Forest Health Technology Enterprise Team, 2010.

Lapham, I. A. "The Forest Trees of Wisconsin," *Transactions of the Wisconsin State Agricultural Society for the Years 1854-5-6-7*, 194–204. Madison: Atwood and Rublee, Printers [1857].

Larson, Agnes M. *History of the White Pine Industry in Minnesota*. Minneapolis: University of Minnesota Press, 1949.

Leoniak, George, Tina Scharf, Jamey Fidel, George Gay, Forrest Hammond, and Jens Hilke. *Critical Paths: Enhancing Road Permeability for Wildlife in Vermont*. Montpelier: Vermont Natural Resources Council, 2009.

Leopold, Aldo. "Conservationist in Mexico." In *The River of the Mother of God and Other Essays by Aldo Leopold*, edited by Susan L. Flader and J. Baird Callicott, 239–44. Madison: University of Wisconsin Press, 1991.

———. *A Sand County Almanac*. New York: Ballantine, 1970.

———. "The Varmint Question." In *The River of God and Other Essays by Aldo Leopold*, edited by Susan L. Flader and J. Baird Callicott, 47–48. Madison: University of Wisconsin Press, 1991.

Lewis, Jamie. "How Turkeys Changed Forest History." *Peeling Back the Bark: Exploring the Collections, Acquisitions, and Treasures of the Forest History Society.* http://www.fhsarchives.wordpress.com/2009/11/25/how-turkeys-changed -forest-history. Accessed November 22, 2011.

Lewis, Ronald L. "Railroads, Deforestation, and the Transformation of Agriculture in the West Virginia Back Counties, 1880–1920." In *Appalachia in the Making: The Mountain South in the Nineteenth Century*, edited by Mary Beth Pudup, Dwight B. Billings, and Altina L. Waller, 297–320. Chapel Hill: University of North Carolina Press, 1995.

——. *Transforming the Appalachian Countryside: Railroad, Deforestation, and Social Change in West Virginia, 1880–1920.* Chapel Hill: University of North Carolina Press, 1998.

Loomis, John B., Lucas S. Bair, and Armando González-Cabán. "Prescribed Fire and Public Support." *Journal of Forestry* 99, no. 11 (November 2001): 18–22.

MacCleery, D. W. *American Forests: A History of Resiliency and Recovery.* Durham, NC: Forest History Society, 2002.

MacKaye, Benton. "Our White Mountain Trip: Its Organization and Methods." In *Log of Camp Moosilauke, 1904*, 4–11. Wentworth, NH: Camp Moosilauke, 1904.

"Many Insect Pests and Diseases Threaten Michigan's Forests." Michigan Department of Natural Resources and Environment, September 22, 2005. http://www.michigan.gov/dnr/0,4570,7-153-10366_46403_59160-128400—,00.html. Accessed September 12, 2012.

Marsh, George Perkins. *Man and Nature.* Seattle: University of Washington Press, 2003.

Marshall, Robert. "The Problem of the Wilderness." *Scientific Monthly* 30, no. 2 (February 1930): 141–48.

Marten, Leanne. "Declaration of Leanne Marten." Minard Run Oil Company, et al., Plaintiffs, vs. US Department of Agriculture, et al., Defendants. United States District Court, Western District of Pennsylvania. Filed June 25, 2009.

Mastran, Shelley Smith, and Nan Lowerre. *Mountaineers and Rangers: A History of Federal Forest Management in the Southern Appalachians, 1900–1981.* Washington, DC: US Department of Agriculture Forest Service, 1983.

Mastro, Victor, and Richard Reardon, compilers. *Emerald Ash Borer Research and Technology Development Meeting, Port Huron, Michigan, September 30–October 1, 2003.* Morgantown, WV: Forest Health Technology Enterprise Team, 2004.

——. *Emerald Ash Borer Research and Technology Development Meeting, Romulus,*

Michigan, October 5–6, 2004. Morgantown, WV: Forest Health Technology Enterprise Team, 2004.

McFerrin, John. "Hydraulic Fracturing Coming to the Monongahela National Forest?" *West Virginia Highlands Voice*, February 10, 2012. http://wvhighlands/org/wv_voice/?p=4522. Accessed May 9, 2012.

Mech, L. David. Foreword to *The Wolves of Isle Royale: A Broken Balance*, by Rolf O. Peterson. Ann Arbor: University of Michigan Press, 2007.

———. "Long-Term Research on Wolves in the Superior National Forest." In *Recovery of Gray Wolves in the Great Lakes Region of the United States: An Endangered Species Success Story*, edited by Adrian P. Wydeven, Timothy R. Van Deelen, and Edward J. Heske, 15–34. New York: Springer, 2010.

Mech, L. David, and Luigi Boitani, eds. *Wolves: Behavior, Ecology, and Conservation.* Chicago: University of Chicago Press, 2003.

Meine, Curt. *Aldo Leopold: His Life and Work.* Madison: University of Wisconsin Press, 1988.

———. "Early Wolf Research and Conservation in the Great Lakes Region." In *Recovery of Gray Wolves in the Great Lakes Region of the United States: An Endangered Species Success Story*, edited by Adrian P. Wydeven, Timothy R. Van Deelen, and Edward J. Heske, 1–14. New York: Springer, 2010.

Mercader, Rodrigo J. "Comparing Potential Management Options to Slow the Spread of EAB Populations in Localized Sites." In *Emerald Ash Borer Research and Technology Development Meeting, October 20–21, 2009, Pittsburgh, Pennsylvania*, compiled by David Lance, James Buck, Denise Binion, Richard Reardon, and Victor Mastro, 40–41. Morgantown, WV: Forest Health Technology Team, 2010. http://www.fs.fed.us/foresthealth/technology/pdfs/2009EAB.pdf. Accessed April 15, 2011.

Michigan Gray Wolf Recovery Team. *Michigan Gray Wolf Recovery and Management Plan.* Lansing: Michigan Department of Natural Resources, 1997.

Milauskas, Steve. "Fact Sheet: Wilderness Areas and Timber." *Timber and Timber Harvesting in West Virginia.* Ripley: West Virginia Forestry Association, August 2002.

Miller, Char, ed. *American Forests: Nature, Culture, and Politics.* Lawrence: University of Kansas Press, 1997.

Minnesota Department of Natural Resources and U.S. Army Corps of Engineers. *NorthMet Project Draft Environmental Impact Statement.* St. Paul: Author, October 2009. http://files.dnr.state.mn.us/input/environmentalreview/polymet/draft_eis/summary_document.pdf. Accessed September 10, 2011.

Mladenoff, David J., Murray K. Clayton, Sarah D. Pratt, Theodore A. Sickley, and Adrian P. Wydeven. "Change in Occupied Wolf Habitat in the Northern Great

Lakes Region." In *Recovery of Gray Wolves in the Great Lakes Region of the United States: An Endangered Species Success Story*, edited by Adrian P. Wydeven, Timothy R. Van Deelen, and Edward J. Heske, 119–38. New York: Springer, 2010.

Mooney, Harold A. "Invasive Alien Species: The Nature of the Problem." In *Invasive Alien Species: A New Synthesis*, edited by Harold A. Mooney, Richard Mack, Jeffrey A. McNelly, Laurie E. Neville, and Peter John Schei, 1–15. Washington, DC: Island Press, 2005.

Morris, Edmund. *The Rise of Theodore Roosevelt*. New York: Modern Library, 1979.

——. *Theodore Rex*. New York: Random House, 2001.

Moses, George H. "Pullman, New Hampshire: A Lumber Camp." *Granite Monthly*, May 1895.

National Forests in Mississippi: Revised Land and Resource Management Plan Proposed Action Description. http://www.fs.usda.gov/Internet/FSE_DOCUMENTS/stel prdb5220872.pdf. Accessed September 9, 2012.

National Strategy and Implementation Plan for Invasive Species Management. Washington, DC: US Department of Agriculture, 2004.

National Wildlife Refuge. "Building Corridors and Critical Paths for Vermont Wildlife." http://www.nwf.org/Wildlife/What-We-Do/Wildlife-Conservation /Wild. Accessed January 6, 2012.

Nelder, Chris. "The Questionable Economics of Shale Gas." *Smart Planet*, December 14, 2011. http://sz0161.wc.mail.comcast.net/zimbra/mail?app=mail#7. Accessed February 20, 2012.

New Hampshire Forestry Commission. *Report of the Forestry Commission of New Hampshire, January Session, 1891*. Manchester: John B. Clarke, 1891.

New York Times. "Dr. Smith's Attitude on Forest Reserves," June 22, 1911, 6.

——. "Editorial: The Halliburton Loophole," November 2, 2009.

——. "Nation May Buy Biltmore Lands," September 17, 1912, 5.

——. "'Not a Rubber Stamp,'" June 23, 1911, 10.

——. "Prompt Action Needed," May 11, 1911, 10.

——. "Strategic Forests," February 13, 1911, 10.

Oelschlaeger, Max. *The Idea of Wilderness*. New Haven: Yale University Press, 1991.

Olson, Sigurd F. *Of Time and Place*. Minneapolis: University of Minnesota Press, 1982.

Osborn, Stephen G., Avner Vengosh, Nathaniel R. Warner, and Robert B. Jackson. "Methane Contamination of Drinking Water Accompanying Gas-Well Drilling and Hydraulic Fracturing." *PNAS Early Edition*, April 14, 2011. www.pnas.org /cgi.doi/10.1073/pnas.11100682108. Accessed March 21, 2012.

Ostergren, Robert C., and Thomas R. Vale, eds. *Wisconsin Land and Life*. Madison: University of Wisconsin Press, 1997.

Paddock, Joe. *Keeper of the Wild: The Life of Ernest Oberholtzer*. St. Paul: Minnesota Historical Society Press, 2001.

Parkman, Francis. "The Forests of the White Mountains." *Garden and Forest* 1, no. 1 (February 29, 1888): 2.

Paxton, Percy J. *The National Forests and Purchase Units of Region Eight*. Unpublished manuscript, Forest History Society, 1950.

Pedler, Cathy. "Atlas Energy Lives Dangerously." Allegheny Defense Project, November 24, 2010. http://org2.democracyinaction.org/o/6155/p/salsa/web /common/public. Accessed March 8, 2012.

——. "Energy Is [NOT] a Public Use of the Forest." Manuscript given to Christopher Johnson on March 10, 2012.

Pennsylvania Independent Oil and Gas Association. "Allegheny National Forest Private Mineral Estates." http://www.pioga.org/publication_files/winter2012 -mayer-allegheny-national-forest.pdf. Accessed September 12, 2012.

——. "Pennsylvania's Traditional Oil and Natural Gas Industry." http://www .pioga.org/pa-oil-gas/traditional/. Accessed September 12, 2012.

——. "Protecting the Environment and Pennsylvania's Water Resources." www .pioga.org/publication_files/pioga-environmental-fact-sheet.pdf. Accessed March 22, 2012.

Peterson, Rolf O. *The Wolves of Isle Royale: A Broken Balance*. Ann Arbor: University of Michigan Press, 2007.

Peterson, Rolf O., and Paolo Ciucci. "The Wolf as a Carnivore." In *Wolves: Behavior, Ecology, and Conservation*, edited by L. David Mech and Luigi Boitani, 104–30. Chicago: University of Chicago Press, 2003.

Pierce, Daniel S. *The Great Smokies: From Natural Habitat to National Park*. Knoxville: University of Tennessee Press, 2000.

Pimentel, David, Rodolfo Zuniga, and Doug Morrison. "Update on the Environmental and Ecological Costs Associated with Alien-Invasive Species in the United States." *Ecological Economics* 52, no. 3 (February 2005): 273–88.

Pinchot, Gifford. *Breaking New Ground*. Washington, DC: Island Press, 1998.

Poland, Therese M. "Twenty Million Ash Trees Later: Current Status of Emerald Ash Borer in Michigan." *Newsletter of the Michigan Entomological Society* 52, nos. 1 and 2 (April 2007): 10–14. http://pbadupws.nrc.gov/docs/ML1126/ML 112630307.pdf. Accessed September 12, 2011.

Poland, Therese M., and Deborah G. McCullough. "Emerald Ash Borer: Invasion of the Urban Forest and the Threat to North America's 'Ask' Resource." *Journal of Forestry* 104, no. 3 (April/May 2006): 118–24.

Proceedings of the Forestry Convention, Held in Grand Rapids, Michigan, January 26 and 27, 1888, Under the Auspices of the Independent Forestry Commission. Lansing: Department of Botany and Forestry, Agricultural College, Michigan, 1888.

Proescholdt, Kevin, Rip Rapson, and Miron L. Heinselman. *Troubled Waters: The Fight for the Boundary Waters Canoe Area Wilderness*. St. Cloud, MN: North Star Press of St. Cloud, 1995.

Pyne, Stephen J. *Fire in America: A Cultural History of Wildland and Rural Fire*. Princeton, NJ: Princeton University Press, 1982.

Rector, William Gerald. *Log Transportation in the Lake States Lumber Industry, 1840–1918*. Glendale, CA: Arthur H. Clark, 1953.

Report of the Forestry Commission of New Hampshire, January Session, 1891. Manchester: John B. Clarke, 1891.

Report of the Forestry Commission of the State of Wisconsin. Madison: Democrat Printing Company, 1889.

Reynolds, A. R. *The Daniel Shaw Lumber Company: A Case Study of the Wisconsin Lumbering Frontier*. New York: New York University Press, 1957.

Robbins, William G. *American Forestry: A History of National, State, and Private Cooperation*. Lincoln: University of Nebraska Press, 1985.

Roell, Brian J., Dean E. Beyer Jr., Patrick E. Lederle, Donald H. Lonsway, and Kristie L. Sitar. *Michigan Wolf Management 2009 Report*. Lansing: Michigan Department of Natural Resources and Environment, 2010.

Rohe, Randall. "Lumbering: Wisconsin's Northern Urban Frontier. In *Wisconsin Land and Life*, edited by Robert C. Ostergren and Thomas R. Vale, 221–40. Madison: University of Wisconsin Press, 1997.

Rooney, Thomas P., and Dean P. Anderson. "Are Wolf-Mediated Trophic Cascades Boosting Biodiversity in the Great Lakes Region?" In *Recovery of Gray Wolves in the Great Lakes Region of the United States: An Endangered Species Success Story*, edited by Adrian P. Wydeven, Timothy R. Van Deelen, and Edward J. Heske, 205–15. New York: Springer, 2010.

Roosevelt, Theodore. *Message from the President of the United States Transmitting a Report of the Secretary of Agriculture in Relation to the Forests, Rivers, and Mountains of the Southern Appalachian Region*. Washington, DC: Government Printing Office, 1902. http://www.foresthistory.org/ASPNET/Publications/region/8/southern. Accessed October 26, 2010.

———. "Opening Address by the President." *Proceedings of a Conference of Governors*. Washington, DC: Government Printing Office, 1909.

Roth, Filibert. "The Appalachian Forests and the Moore Report." *American Forestry* 16, no. 4 (March 1910): 209–17.

——.*Forestry Condition and Interests of Wisconsin (Bulletin 16).* Washington, DC: Government Printing Office, 1898.

——.*Michigan Forest Reserve Manual for the Information and Use of Forest Officers.* Lansing, MI: Wynkoop Hallenbeck Crawford Company, State Printer, 1904.

Russell, Ernest. "The Wood-Butchers." *Collier's*, May 9, 1909, 19–20.

Salmond, John A. *The Civilian Conservation Corps, 1933–1942: A New Deal Case Study.* Durham, NC: Duke University Press, 1967.

Sargent, Charles Sprague. "Destruction of Forests in New Hampshire." *Garden and Forest* 2, no. 52 (February 20, 1889): 86.

——. "Mr. Vanderbilt's Forest." *Garden and Forest* 8 (December 4, 1895): 481–82.

Scott, Doug. *The Enduring Wilderness.* Golden, CO: Fulcrum Publishing, 2004.

Searle, Newell. "Minnesota State Forestry Association." *Minnesota History*, Spring 1974, 16–29.

Secretary of Energy Advisory Board. *Shale Gas Production Subcommittee Second Ninety Day Report.* Washington, DC: US Department of Energy, November 18, 2011.

Shands, William E. *The Lands Nobody Wanted: The Legacy of the Eastern National Forests.* Milford, PA: Pinchot Institute for Conservation, 1991.

Shands, William E., and Robert G. Healy. *The Lands Nobody Wanted: A Conservation Foundation Report.* Washington, DC: The Conservation Foundation, 1977.

Smith, Charles D. "Gentlemen, You Have My Scalp." *American Forests*, February 1962, 16–19.

——. "The Mountain Lover Mourns: Origins of the Movement for a White Mountain National Forest 1880–1903." *New England Quarterly* 33, no. 1 (March 1961): 37–56.

Smith, W. Brad, Patrick D. Miles, Charles H. Perry, and Scott A. Pugh. *Forest Resources of the United States, 2007: A Technical Document Supporting the Forest Service 2010 RPA Assessment.* Washington, DC: US Department of Agriculture, 2009.

Society for the Protection of New Hampshire Forests. *Saving the White Mountains: The Weeks Act Then and Now.* Special issue of *Forest Notes*, Summer 2011.

Southern Environmental Law Center. "BLM Putting Alabama Forest at Risk from Proposed Gas Fracking," April 10, 2012. http://www.southernenvironment.org/cases/hydraulic_fracturing. Accessed September 12, 2012.

——. "Hydraulic Fracturing: 'Fracking' Looms as a New Threat to Clean Water in the Southeast," April 27, 2012. http://www.southernenvironment.org/cases/hydraulic_fracturing. Accessed May 9, 2012.

Steen, Harold K. "The Beginning of the National Forest System." In *American For-*

ests: Nature, Culture, and Politics, edited by Char Miller, 49–68. Lawrence: University Press of Kansas, 1997.

———, ed. *The Origins of the National Forests: A Centennial Symposium.* Durham, NC: Forest History Society, 1992.

——. *The U.S. Forest Service: A History.* Seattle: University of Washington Press, 1976.

Stein, Susan, et al. *National Forests on the Edge: Development Pressures on America's National Forests and Grasslands.* Portland, OR: U.S. Department of Agriculture, Forest Service, Pacific Northwest Research Station, 2007.

Stoddard, Herbert L. *The Bobwhite Quail: Its Habits, Preservtion and Increase.* New York: Scribner, 1946.

Stuart, Gordon W., and Livia Crowley. "The Streamflow Study that Created the White Mountain National Forest." *Historical New Hampshire* 66, no.1: 50–63.

Swain, Willis L. "The Influence of Forests on Climate and on Floods." *American Forestry* 16, no. 4 (March 1911): 224–40.

2000 RPA Assessment of Forest and Range Lands. Washington, DC: U.S. Department of Agriculture Forest Service, 2000.

Taylor, Robin A. J., Leah S. Bauer, Deborah L. Miller, and Robert A. Haack. "Emerald Ash Borer Flight Potential." In *Emerald Ash Borer Research and Technology Development Meeting, October 5–6, Romulus, Michigan,* compiled by Victor Mastro and Richard Reardon, 15–16. Morgantown, WV: Forest Health Technology Team, 2004. http://www.fs.fed.us/foresthealth/technology/pdfs/2004EAB.pdf. Accessed June 15, 2011.

Taylor, Stephen Wallace. *The New South's New Frontier: A Social History of Economic Development in Southwestern North Carolina.* Gainesville: University Press of Florida, 2001.

Treves, Adrian, Kerry A. Martin, Jane E. Wiedenhoeft, and Adrian P. Wydeven. "Dispersal of Gray Wolves in the Great Lakes Region." In *Recovery of Gray Wolves in the Great Lakes Region of the United States: An Endangered Species Success Story,* edited by Adrian P. Wydeven, Timothy R. Van Deelen, and Edward J. Heske, 191–204. New York: Springer, 2010.

United States Department of Agriculture Office of Communications. "Agriculture Secretary Vissack Presents National Vision for America's Forests." August 14, 2009. http://www.usda.gov/wps/portal/usda/usdahome?contentid only+true&c. Accessed March 25, 2011.

USDA Forest Service. *Allegheny National Forest Final Environmental Impact Statement to Accompany the Land and Resource Management Plan.* Warren, PA: Allegheny National Forest, March 2007.

———. "Ecosystem Services." http://www.fs.fed.us/ecosystemservices/. Accessed September 13, 2012.

———. "Forest Fragmentation." *WNC Report Card on Forest Sustainability*. Asheville, NC: Southern Research Station Headquarters, 2010. http://www.wncforest reportcard.org/browse/biodiversity/forest-fragmentation. Accessed May 18, 2012.

———. "Forest Health Technology Enterprise Team." http://www.fs.fed.us/forest health. Accessed June 30, 2011.

———. "Forest Legacy Program: Protecting Private Forest Lands from Conversion to Non-Forest Uses." http://www.fs.fed.us/spf/coop/programs/loa/flp.shtml. Accessed September 13, 2012.

———. *Forest Service Manual 2600: Wildlife, Fish, and Sensitive Plant Habitat Management*. Washington, DC: US Department of Agriculture, 2005.

———. *Hiawatha National Forest Executive Summary of the Final Environmental Impact Statement*. Escanaba, MI: Hiawatha National Forest, 2006.

———. *Hiawatha National Forest 2006 Monitoring and Evaluation Report*. Escanaba, MI: U.S. Department of Agriculture, 2006.

———. "Land Use Change." *WNC Report Card on Forest Sustainability*. Asheville, NC: Southern Research Station Headquarters, 2010. http://www.wncforestreport card.org/browse/biodiversity/land-use. Accessed May 18, 2012.

———. *Managing Multiple Uses on National Forests, 1905–1995*. http://www.forest history.org/ASPNET/Publications/multiple_use/chap6.htm. Accessed September 13, 2012.

———. *Ottawa National Forest Fiscal Year 2008 Monitoring and Evaluation Report*. Ironwood, MI: US Department of Agriculture, 2009.

———. "Planning and Monitoring." *WNC Report Card on Forest Sustainability*. Asheville, NC: Southern Research Station Headquarters, 2010. http://www.wnc forestreportcard.org/browse/policy/planning-monitoring. Accessed May 18, 2012.

———. *Revised Environmental Assessment: Trans Superior Resources, Inc. Private Minerals Exploration in the Matchwood Tower Road Area*. Ontonagon, MI: Ottawa National Forest, 2007.

———. *The South's Fourth Forest: Alternatives for the Future*. Washington, DC: USDA Forest Service, 1988.

———. *Visitor Use Report: Eastern Region (R9)*. Washington, DC: US Department of Agriculture, 2012.

———. *Visitor Use Report: Southern Region (R8)*. Washington, DC: US Department of Agriculture, 2012.

———. "Volume of Timber." *WNC Report Card on Forest Sustainability*. Asheville, NC:

Southern Research Station Headquarters, 2010. http://www.wncforestreport card.org/browse/production/timber-volume. Accessed May 18, 2012.

USDA Forest Service, Eastern Region. *Threatened, Endangered and Sensitive Species*. http://www.fs.fed.us/r9/wildlife/tes/faq.shtml. Accessed May 4, 2012.

US Environment Protection Agency. *Evaluation of Impacts to Underground Sources of Drinking Water by Hydraulic Fracturing of Coalbed Methane Reservoirs; National Study Final Report*, June 2004.

US Fish and Wildlife Service. *Recovery Plan for the Eastern Timber Wolf*. Twin Cities, MN: Region 3, U.S. Fish and Wildlife Service, 1992.

Van Deelen, Timothy R. "Growth Characteristics of a Recovering Wolf Population in the Great Lakes Region." In *Recovery of Gray Wolves in the Great Lakes Region of the United States: An Endangered Species Success Story*, edited by Adrian P. Wydeven, Timothy R. Van Deelen, and Edward J. Heske, 139–53. New York: Springer, 2010.

Vermont Family Forests. *Vermont Family Forests 2010 Annual Report*. Bristol: Vermont Family Forests, 2010.

Vermont Natural Resources Council and Vermont Family Forests. *Informing Land Use Planning and Forestland Conservation through Subdivision and Parcelization Trend Information*. 2010. http://svr3.acornhost.com/~vnrcorg/report. Accessed January 17, 2012.

Vucetich, John A., and Rolf O. Peterson. "Wolf and Moose Dynamics on Isle Royale." In *Recovery of Gray Wolves in the Great Lakes Region of the United States: An Endangered Species Success Story*, edited by Adrian P. Wydeven, Timothy R. Van Deelen, and Edward J. Heske, 35–48. New York: Springer, 2010.

Walker, Laurence. *The Southern Forest: A Chronicle*. Austin: University of Texas Press, 1991.

Ward, Julius H. "White Mountain Forests in Peril." *Atlantic Monthly*, February 1893, 247–55.

Washburn, Charles G. *The Life of John W. Weeks*. Boston: Houghton Mifflin, 1928.

"Weeks Act Chronology." www.foresthistory.org/ASPNET/Policy/. Accessed January 23, 2011.

"Weeks Law Lands in the National Forest System." *USDA Forest Service Briefing Paper*, March 2, 2011.

Weeks Law of 1911. Public Law 961, 62nd Congress, 1st session (1911). http://memory.loc.gov/cgi-bin/ampage?collId=amrvl&fileName=vl024//amrvlvl 024.db&recNum=0&itemLink=r?ammem/AMALL:@field%28DOCID+@lit %28jjh94-000053%29%29&linkText=0, 961. Accessed September 12, 2012.

Wilderness Act. Public Law 88-577, 88th Congress, Second Session, September 3, 1964.

Wilson, James. Letter to Gifford Pinchot ("The Forester)," February 1, 1905. [Historians generally acknowledge that Pinchot wrote this letter for Secretary of Agriculture James Wilson to sign.]

Wolf, Tom. *Arthur Carhart: Wilderness Prophet*. Boulder: University Press of Colorado, 2008.

Young, Stanley Paul. *The Wolves of North America. Part I: Their History, Life Habits, Economic Status, and Control*. Washington, DC: Wildlife Institute, 1944.

Zhang, Daowei, and Sayeed R. Mehmood. "Safe Harbor for the Red-Cockaded Woodpecker." *Journal of Forestry* 100, no. 5 (July/August 2002): 24–29.

DAVID GOVATSKI is a forester and environmental consultant who worked for the U.S. Forest Service for more than thirty years. He has conducted forest inventories, developed and supervised forest management plans, and written assessments of the environmental impact of forest plans and policies. From his base in New Hampshire, he conducts field inventories and prepares management plans for endangered plant and animal species. He is also a professional trip leader and leads canoeing, birding, and hiking expeditions throughout the United States and Canada. David is past president of the board of Weeks State Park Association in Lancaster, New Hampshire, and served as Secretary of the Weeks Act Centennial Coordinating Committee. He is a member of the Forest History Society.

CHRISTOPHER JOHNSON is a writer with extensive experience in education and publishing. He has written numerous articles on nature and the environment for *American Forests, Appalachia, Chicago Wilderness, E: The Environmental Magazine, Snowy Egret*, and other magazines and journals. In 2006, the University of New Hampshire Press published his book *This Grand and Magnificent Place: The Wilderness Heritage of the White Mountains*. He is a member of the Forest History Society.

INDEX

Abourezk, James, 221
Accessibility, 297. *See also* Roads
Adirondacks Forest Preserve, 93
ADP. *See* Allegheny Defense Project
AFSEEE. *See* Association of Forest Service Employees for Environmental Ethics
Agrarian traditions, 57, 60, 69–70
Agricultural Appropriation Act of 1905, 81
Agricultural Committee, 103–104, 107
Agriculture, 45–46, 68, 89, 243–245
Air pollution, 260, 264–265
Airplanes, 215
Aitkin, William A., 39–40
Allan, Durward, 237
Allegheny Defense Project (ADP), 256, 267–269
Allegheny Forest Alliance, 268
Allegheny National Forest
　disagreements over future of, 266–270
　hydraulic fracturing technology and, 250–251
　lessons learned in, 270–272
　mineral rights and, 208
　oil and gas drilling in, 253–256
　potential effects of hydraulic fracturing in, 256–260
Allegheny National Forest Plan, 267
Allegheny Snowmobile Loop, 256
Allen, Shirley, 172
Ambler, Chase P., 84
American Association for the Advancement of Science, 76
American chestnut, 55, 56, 164, 277
American Fisheries Society, 76
American Forest Congress, 81
American Forestry Association, 99, 114, 120
American Ornithologists Union, 76
American Tree Farm System, 309–310
Anderson, Chel, 207–209, 222, 226
Anderson, Harold, 191
Anderson, John, 30
Anderson, Wendell, 220
Andrews, Christopher C., 89–90, 134, 210
Androscoggin Purchase Unit, 130

Apalachicola National Forest, 167–169, 178, 180–181
Appalachian Mountain Club, 66, 71–72
Appalachian Mountains. *See also Specific states*
　acceleration of logging in, 62–66
　acquiring forestlands in, 124–128
　conservation movements and, 83–85
　Dorie Woodruff and, 51–53
　early legislative efforts in, 96–97, 99–100
　impacts of logging in, 66–70
　timber rush in, 54–62
　Weeks Act and, 119, 122
Appalachian National Park Association (ANPA), 66, 84–85, 95
Appalachian Trail, 1, 232
Arbor Day, 48
Arnold Arboretum, 26
Arthur, Alexander A., 60
Artists, 24–25
Ash trees, 278–279. *See also* Emerald ash borer
Ashe, William W., 67
Asheville, North Carolina, 61, 84
Asian long-horned beetle, 277, 322
Association of Forest Service Employees for Environmental Ethics (AFSEEE), 160–161
Atlas Energy, Inc., 249–250
Attitudes, public, 6, 183–184, 246–247
Atwood, Karen, 256
Ayres, Horace B., 67
Ayres, Philip W., 96–99, 109, 112, 115, 128–129, 134–135, 319

Babbitt, Sweet Home v., 180
Bachran, Mary, 260–261
Backus, Edward, 212
Backus proposal, 212–214
Bair, Lucas S., 183
Baldwin, A.T., 16
Ballast water, 275
Ballinger, Richard A., 118–119
Ballinger-Pinchot affair, 118–119
Banbury, Scott, 154

INDEX